In Situ Hybridization Protocols

METHODS IN MOLECULAR BIOLOGY™

John M. Walker, SERIES EDITOR

METHODS IN MOLECULAR BIOLOGY™

In Situ Hybridization Protocols

2ND EDITION

Edited by

Ian A. Darby

RMIT University
Bundoora, Victoria, Australia

Humana Press ✳ Totowa, New Jersey

© 2000 Humana Press Inc.
999 Riverview Drive, Suite 208
Totowa, New Jersey 07512

The content and opinions expressed in this book are the sole work of the authors and editors, who have warranted due diligence in the creation and issuance of their work. The publisher, editors, and authors are not responsible for errors or omissions or for any consequences arising from the information or opinions presented in this book and make no warranty, express or implied, with respect to its contents.

This publication is printed on acid-free paper. ∞
ANSI Z39.48-1984 (American Standards Institute) Permanence of Paper for Printed Library Materials.

Cover design by Patricia F. Cleary.
Cover illustration: Dark-field photomicrograph of a section of fibrotic kidney showing *in situ* hybridization for α1 procollagen I mRNA expression.

For additional copies, pricing for bulk purchases, and/or information about other Humana titles, contact Humana at the above address or at any of the following numbers: Tel: 973-256-1699; Fax: 973-256-8341; E-mail: humana@humanapr.com, or visit our Website at www.humanapress.com

Photocopy Authorization Policy:
Authorization to photocopy items for internal or personal use, or the internal or personal use of specific clients, is granted by Humana Press Inc., provided that the base fee of US $10.00 per copy, plus US $00.25 per page, is paid directly to the Copyright Clearance Center at 222 Rosewood Drive, Danvers, MA 01923. For those organizations that have been granted a photocopy license from the CCC, a separate system of payment has been arranged and is acceptable to Humana Press Inc. The fee code for users of the Transactional Reporting Service is: [0-89603-686-3/99 $10.00 + $00.25].

Printed in the United States of America. 10 9 8 7 6 5 4 3 2 1

Library of Congress Cataloging-in-Publication Data

In situ hybridization protocols / edited by Ian A. Darby — 2nd ed.
           p.   cm. — (Methods in molecular biology ;   v.   123)
       Includes index.
       ISBN 0-89603-686-3 (alk. paper)
   1. In situ hybridization—Laboratory manuals.   I. Darby, Ian A.   II. Series:
Methods in molecular biology (Totowa, N.J.) ;   v. 123.
QH452.8.I54   1994b
572.8'4—dc21   98–48665
                                              CIP

Preface

The technique of *in situ* hybridization, in its various forms, has been used routinely in many laboratories for a number of years. However, *in situ* hybridization has retained a reputation as one of the more difficult molecular biological techniques.

This may be caused, in part, by the hybrid nature of the technique, which often requires a mixture of molecular biological and histological skills. The two techniques are usually taught and acquired in different streams of biological science. The step-by-step and detailed protocols provided in this book, by researchers active in the field, should make it possible for both the molecular biologist with little experience of histology and the histologist with little experience of molecular biology to use the technique successfully in their laboratories.

In Situ Hybridization Protocols aims to provide useful protocols for two groups of researchers who use the technique: those who use *in situ* hybridization to localize genes to particular chromosomes and those researchers who use DNA or RNA probes to localize expression of mRNA in tissue. This new edition also encompasses a number of techniques not covered in the previous edition, such as *in situ* hybridization of whole-mount embryo specimens, *in situ* hybridization at the electron microscopic level, and *in situ* detection of DNA fragmentation in apoptosis. The final two chapters of the book provide examples of application of *in situ* hybridization techniques to research and clinical problems.

In Situ Hybridization Protocols provides ample information for novices planning to set up the *in situ* hybridization technique and use it in their laboratories for the first time, as well as giving updates of recent developments, in step-by-step protocols for those laboratories where *in situ* hybridization techniques are already used. In particular, the use of nonisotopic methods for both forms of *in situ* hybridization has advanced rapidly since the well-received previous edition.

In my own laboratory we have used *in situ* hybridization on tissue sections for a number of years; when we look back on results gained even a few years ago, there has clearly been continual improvement in the technique. I trust that those who use this new edition will find it a valuable aid in setting up the technique or improving the sensitivity and scope of applications of *in situ* hybridization in their own laboratories.

Lastly, I would like to acknowledge the contribution made by Andy Choo, who edited the first edition of the book, and thus provided a basis for the second edition, my colleagues Teresa Bisucci and Tim Hewitson for their participation in planning and writing our chapters, and all of the authors who have contributed their protocols.

Ian A. Darby

Contents

PART 3 APPLICATIONS

Contributors

ANABIL A. ACOSTA • *Centro de Estudios en Ginecologia y Reproduccion, Buenos Aires, Argentina*

NICOLETTA ARCHIDIACONO • *Istituto di Genetica, Universitá di Bari, Bari, Italy*

DANIEL G. BEDO • *Rural Division, Department of Primary Industries and Energy, Canberra, ACT, Australia*

TERESA BISUCCI • *Wound Healing and Microvascular Biology Group, Department of Human Biology and Movement Science, RMIT University, Bundoora, Victoria, Australia*

LESIL BRIHN • *Center for Pediatric Research, Division of Medical Genetics, Eastern Virginia Medical School, Norfolk, VA*

GWEN V. CHILDS • *Department of Anatomy and Neurosciences, University of Texas Medical Branch, Galveston, TX*

MARK COLEMAN • *Repatriation General Hospital, Rheumatology Research Unit, Adelaide, South Australia, Australia*

IAN A. DARBY • *Wound Healing and Microvascular Biology Group, Department of Human Biology and Movement Science, RMIT University, Bundoora, Victoria, Australia*

GUSTAVO F. DONCEL • *Jones Institute for Women's Health, Eastern Virginia Medical School, Norfolk, VA*

JONATHAN D. GLASS • *Department of Neurology, School of Medicine, Emory University, Atlanta, GA*

DARREN GRIFFIN • *Department of Biology and Biochemistry, Brunel University, Uxbridge, Middlesex, UK*

MURRAY HARGRAVE • *Centre for Molecular and Cellular Biology, The University of Queensland, Brisbane, Queensland, Australia*

HENRY H. Q. HENG • *Center for Molecular Medicine and Genetics, School of Medicine, Wayne State University, Detroit, MI*

TIM D. HEWITSON • *Department of Nephrology, Royal Melbourne Hospital, Victoria, Australia*

RONALD J. HILL • *CSIRO Molecular Science, North Ryde, New South Wales, Australia*

STANTON F. HOEGERMAN • *Department of Biology, The College of William and Mary, Williamsburg, VA*

ANTON H. N. HOPMAN • *Department of Molecular Cell Biology and Genetics, University Maastricht, Maastricht, The Netherlands*

ALLISON R. JILBERT • *Hepatitis Virus Research Laboratory, Infectious Diseases Laboratories, Institute of Medical and Veterinary Science, Adelaide, South Australia, Australia*

WILLIAM G. KEARNS • *Center for Pediatric Research, Division of Medical Genetics, Norfolk, VA, and The Johns Hopkins University School of Medicine, Baltimore, MD*

PAUL KOMMINOTH • *Department of Pathology, University Hospital Zürich, Zürich, Switzerland*

PETER KOOPMAN • *Centre for Molecular and Cellular Biology, The University of Queensland, Brisbane, Queensland, Australia*

DOMINIQUE LE GUELLEC • *Institut de Biologie et Chimie des Protéines (IBCP), Lyon, France*

CATHERINE LE MOINE • *CNRS UMR 5541 - Laboratoire d'Histologie-Embryologie, Université Victor Ségalen Bordeaux II, Bordeaux, France*

PINO MACCARONE • *Department of Genetics and Human Variation, La Trobe University, Bundoora, Victoria, Australia*

JENNIFER A. MARSHALL-GRAVES • *Department of Genetics and Human Variation, La Trobe University, Bundoora, Victoria, Australia*

JON MARTIN • *Department of Genetics, The University of Melbourne, Parkville, Victoria, Australia*

ROSALIA MARZELLA • *Instituto di Genetica, Universitá di Bari, Bari, Italy.*

GEOFFREY I. MCFADDEN • *Department of Botany, University of Melbourne, Parkville, Victoria, Australia*

PETER B. MOENS • *Department of Biology, York University, Downsview, Ontario, Canada*

MARGARET R. MOTT • *CSIRO Molecular Science, North Ryde, New South Wales, Australia*

GERARD J. NUOVO • *MGN Medical Research Laboratories, Setauket, NY*

SERGIO OEHNINGER • *Jones Institute for Women's Health, Eastern Virginia Medical School, Norfolk, VA*

MYUNG-GEOL PANG • *Biomedical Center, Korea Advanced Institute of Science and Technology, Taejon, Korea*

ANGELA PARKER • *Repatriation General Hospital, Rheumatology Research Unit, South Australia, Australia*

MARGHERITA PENNACCHIA • *Instituto di Genetica, Universitá di Bari, Bari, Italy*

A. MARIE PHILLIPS • *Department of Genetics, The University of Melbourne, Parkville, Victoria, Australia*

MARIANO ROCCHI • *Instituto di Genetica, Università di Bari, Bari, Italy*

ROBERT D. C. SAUNDERS • *Department of Anatomy and Physiology, Old Medical School, University of Dundee, Dundee, Scotland*

MALCOLM D. SMITH • *Repatriation General Hospital, Rheumatology Research Unit, Daw Park, South Australia, Australia*

COSMA SPALLUTO • *Instituto di Genetica, Università di Bari, Bari, Italy.*

ERNST J. M. SPEEL • *Department of Pathology, University Hospital Zürich, Zürich, Switzerland*

BARBARA SPYROPOULOS • *Department of Biology, York University, Downsview, Ontario, Canada*

MICHAEL STACEY • *Center for Pediatric Research, Division of Medical Genetics, Eastern Virginia Medical School, Norfolk, VA*

ROLAND TODER • *Department of Genetics and Human Variation, La Trobe University, Bundoora, Victoria, Australia*

LUIGI VIGGIANO • *Instituto di Genetica, Università di Bari, Bari, Italy*

ROSS F. WALLER • *Department of Botany, University of Melbourne, Parkville, Victoria, Australia*

GRAHAM C. WEBB • *Department of Obstetrics and Gynaecology, The Queen Elizabeth Hospital, Woodville, South Australia; The Department of Animal Science, The University of Adelaide, Waite Campus, Glen Osmond, South Australia, Australia*

STEVEN L. WESSELINGH • *Alfred Infectious Disease Unit, Alfred Hospital, Prahran, Victoria, Australia*

RIYANI WIKANINGRUM • *Repatriation General Hospital, Rheumatology Research Unit, Daw Park, South Australia, Australia*

STEPHEN A. WILCOX • *Department of Genetics and Human Variation, La Trobe University, Bundoora, Victoria, Australia*

Color Plates

Color Plates 1–5 appear as an insert following p. 176.

PART 1

CHROMOSOMAL TECHNIQUES

1

Preparation of Human Partial Chromosome Paints from Somatic Cell Hybrids

Nicoletta Archidiacono, Rosalia Marzella, Cosma Spalluto, Margherita Pennacchia, Luigi Viggiano, and Mariano Rocchi

1. Introduction

Whole chromosome painting libraries (WCPLs) have provided a very powerful tool to cytogeneticists. The technique allows the painting of specific chromosomes in metaphase spreads and in interphase nuclei *(1–4)*. The usefulness of WCPLs is particularly evident in identifying the chromosomal origin of *de novo* unbalanced translocations and marker chromosomes, or, more generally, in characterizing those cytogenetic cases in which the conventional approach based on banding techniques failed to elucidate the chromosomal rearrangement under study *(5,6)*. Such cases are frequently experienced in cancer cytogenetics *(7,8)*. WCPLs are usually derived from flow-sorted chromosomes.

Biotinylated genomic DNA from hybrid cell lines retaining specific human chromosomes has been alternatively used as starting material to obtain WCPL *(9,10)*. Human chromosomes, however, represent only a minor component of a human–rodent somatic cell hybrid, so that the sensitivity of this technique is usually unsatisfactory *(10)*. These problems can be circumvented by selective polymerase chain reaction (PCR) amplification of the human sequences using accurately designed dual-Alu primers *(11,12)*. The amplified PCR products are then labeled and used in fluorescent *in situ* hybridization (FISH) experiments (reverse-FISH).

Human–rodent somatic cell hybrids frequently retain fragments of human chromosomes as a consequence of rearrangements that occurred in vitro. This is particularly true for human–hamster somatic cell hybrids, because hamster chromosomes are prone to rearrangements. We have recently characterized, by reverse-FISH, hundreds of human-hamster hybrids, and have identified a large

From: *Methods in Molecular Biology*, Vol. 123: In Situ *Hybridization Protocols*
Edited by: I. A. Darby © Humana Press Inc., Totowa, NJ

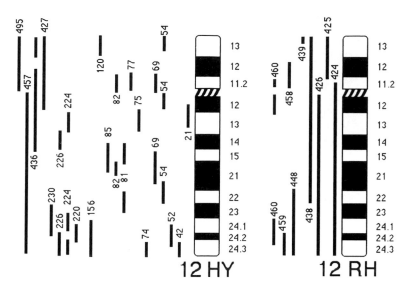

Fig. 1. Summary of our collection of fragments specific for chromosome 12. Each fragment is represented by a solid bar specifying its length and subchromosomal location on a G-band ideogram of chromosome 12. The code number for each hybrid is also shown. The hybrids are grouped in two categories: (1) 12HY: classical (nonradiation) hybrids. The rearrangement occurred in vitro by chance. Whole chromosomes or additional fragments from different chromosomes are present in these hybrids. (2) 12RH: radiation hybrids derived from the monochromosomal hybrid Y.E210TC (generated in our laboratory), retaining chromosome 12 as the only human contribution.

number of hybrids containing chromosome fragments. These hybrids can be used as starting material for the production of partial chromosome paints (PCPs), that is, paints that recognize a specific chromosomal region. They can be used as a powerful tool for cytogenetic investigations *(13,14)*. Most of these hybrids, however, contain additional fragments and/or entire human chromosomes, which prevents their efficient use. It would be therefore advantageous to obtain hybrids retaining a single fragment as the only human contribution. To generate these type of PCPs, we have started to produce radiation hybrids (RHs) from monochromosomal hybrids *(15)*. The diagram shown in **Fig. 1** summarizes as an example the PCPs from normal hybrids and RHs, specific for chromosome 12. To date RHs specific for chromosomes 2, 4, 5, 7, 10, 12, 17, 19, and X have been obtained. Thanks to support from AIRC and Telethon, DNA from these hybrids is available to the scientific community free of charge. FISH images of these PCPs can be viewed at the Internet site http://bioserver.uniba.it/fish/Cyto-

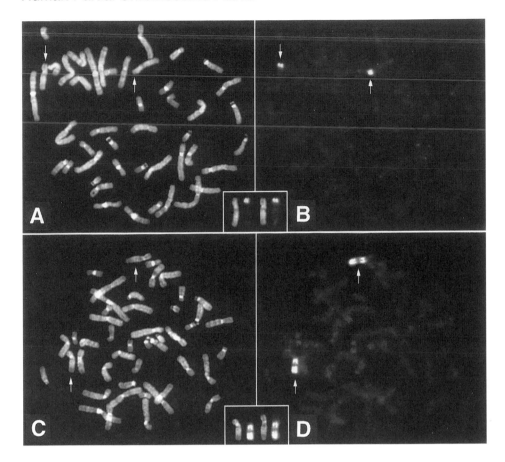

Fig. 2. FISH images of hybrids no. 425 (**A–B**), specific for 12p11.2-pter and no. 424 (**C–D**) specific for 12q. The DAPI banding (**A** and **C**) and Cy3 signals (**B** and **D**) are reported separately, in black and white, as captured by the CCD camera. DAPI-banded chromosome 12 and the corresponding Cy3 signals are shown side by side in the two boxes.

genetics/welcome.html). **Figure 2** shows, as examples, PCPs no. 425 specific for 12p11.2→12pter, and no. 424 specific for 12q.

Alu and long intervening elements (LINE) sequences are not evenly interspersed in the human genome; their primary distribution correlates with G-negative and G-positive bands respectively *(16)*, so that Alu-PCR products generate a banding pattern corresponding to R-banding *(17)*. This fact has to be kept in mind in evaluating the FISH signals of PCP obtained with the present approach.

2. Materials

2.1. PCR

1. Alu primers *(11)*:
   5' GGATT ACAGG YRTGA GCCA 3' (Y = C/T; R = A/G)
   5' RCCAY TGCAC TCCAG CCTG 3'
   stored frozen as 100 pmol/µL.
2. Genomic DNA from hybrid (100 ng) in distilled water, stored frozen.
3. 0.5-mL Test tubes suitable for the thermal cycler.
4. A set of micropipets (P20 and P200, Gilson, Villiers-le-Bel, France) and sterile tips.
5. Reagents for PCR (stored frozen):
   a. 10× dNTPs mix 2 mM each, pH 7.0;
   b. 10× Reaction buffer (usually comes with the *Taq* polymerase);
   c. *Taq* polymerase (5 U/µL) (AmpliTaq Gold, Perkin-Elmer);
   d. Sterile distilled water (autoclaved), stored at room temperature; and
   e. Light mineral oil (not necessary if the thermal cycler is equipped with a heated lid).
6. Programmable thermal cycler.
7. Agarose; ethidium bromide; gel electrophoresis apparatus, and UV transilluminator.
8. Molecular weight marker (λ-*Hin*dIII + *Phi*X–*Hae*III, 50 ng/µL each).
9. TBE buffer (900 mM Tris-base, 900 mM boric acid, 1 mM EDTA).

2.2 Nick-Translation

1. 10× Buffer: 0.5 M Tris-HCl, pH 7.8–8.0, 50 mM MgCl$_2$, 0.5 mg/mL bovine serum albumin (BSA).
2. dNTPs mix: 0.5 mM dATP, 0.5 mM dCTP, and 0.5 mM dGTP.
3a. Biotin-11-dUTP mix: 0.5 mM dTTP and 0.5 mM bio-11-dUTP (or bio-16-dUTP);
   b. Alternatively: digoxigenin (DIG)-11-dUTP mix (Boehringer Mannheim) (*see* **Note 1**);
   c. Alternatively (for direct labeling): dUTP-Cy3 (Amersham) (*see* **Note 1**).
4. Enzymes: DNA polymerase I (5 U/µL) and DNase I (2 U/µL).
5. Other chemicals: 0.1 M β-mercaptoethanol and 0.5 M EDTA.

2.3. In Situ *Hybridization*

1. Cot-I DNA, human (BRL or Boehringer Mannheim).
2. Salmon sperm DNA, 1 µg/µL.
3. 3 M Na acetate.
4. 70%, 90%, and 100% ethanol.
5. Savant concentration centrifuge.
6. Formamide and deionized formamide.
7. 50% Dextran sulfate, autoclaved.
8. 20× SSC (1× SSC = 150 mM sodium chloride, 15 mM sodium citrate, pH 7.0).

9. Vortex.
10. Coplin jar.
11. 24 × 24 mm and 24 × 50 mm coverslips.
12. Rubber cement.
13. Washing solution A: 50% formamide/2× SSC.
14. Washing solution B: 0.1× SSC.
15. Blocking solution: 3% BSA, 4× SSC, 0.1% Tween-20.
16. Detection buffer: 1% BSA, 1× SSC, 0.1% Tween-20.
17. 1 mg/mL of Avidin-Cy3 (Amersham) (or 1 mg/mL fluorescein isothiocyanate [FITC]-conjugated anti-DIG antibodies [Boheringer Mannheim]).
18. Solution C: 4× SSC, 0.1% Tween-20.
19. DAPI (4',6-diamidino-2-phenylindole, Sigma).
20. Propidium iodide (PI) (Sigma).
21. Antifade-mounting medium. 10 mL: 0.233 g of DABCO (1,4-diazabicyclo-[2.2.2]octane, Sigma), 800 μL of H_2O, 200 μL of 1 M Tris-HCl, 9 mL of glycerol.

3. Methods

3.1. Generation of PCR Products from Somatic Cell Hybrids

1. In a 0.5-mL tube (suitable for a thermocycling machine) mix in this order:
   a. 1 μL of DNA from hybrid;
   b. 37.2 μL of H_2O;
   c. 5 μL of PCR buffer 10×;
   d. 5 μL of 10× dNTPs mix;
   e. 0.8 μL of *Taq* polymerase (5 U/μL) (AmpliTaq Gold, Perkin-Elmer); and
   f. 0.5 μL for each primer (1 μ*M* final concentration).
   g. Adjust total volume to 50 μL.
   h. Overlay with a drop of light mineral oil (this step is not necessary if the thermal cycler is equipped with a heated lid).
2. In a second tube: add everything as described above, omitting the DNA. This serves as a negative control. (If more than one sample is amplified, a master mix can be prepared.)
3. Run on the thermocycling machine as follows:
   a. 10 min at 95°C (*see* **Note 2**);
   b. 30 cycles: 1 min at 94°C, 1 min at 65°C, 4 min at 72°C; and
   c. 10 min at 72°C.
4. Check the amplification products by loading a 5-μL aliquot on 1% agarose gel. Store at 4°C (*see* **Fig. 3**).

3.2. Probe Labeling by nick-translation

1. Add to a microfuge tube, on ice:
   a. 2 μg of amplified products;
   b. 10 μL of 10× nick-translation buffer;
   c. 10 μL of dNTPs mix;
   d. 5 μL of biotin mix (or 5 μL DIG mix; or 1 μL of dUTP-Cy3 mix) (*see* **Note 1**);

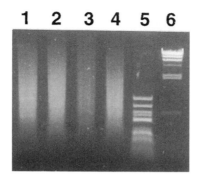

Fig. 3. Samples of Alu-PCR amplification products from four different hybrids. The PCR reaction was done in a volume of 50 µL. 5 µL were then run on a 1% agarose gel, at 100 V. The marker in lane 6 is λ-HindIII, the one on lane 5 is pCMVβHaeIII *(18)*. The first five bands of the latter are: 970, 750, 595, 544, and 447 bp. Amplified fragments range approximately from 3–4 kb to 300 bp. Discrete bands are sometimes detected in amplified DNA from hybrids which retained a small amount of human material. The run is also informative for a quantitative evaluation of PCR products, calculated approximately by visual comparison with the known amount of the marker. Quantitative evaluation can be performed better by inspecting the gel after a few minutes of running, when both the marker and the samples appear as compact bands.

    e. 10 µL of 0.1 *M* β-mercaptoethanol;
    f. Dilute (immediately before use) 1 µL of DNase I in 1mL of distilled water; add 20 µL to the nick-translation mix. (The DNase I should be calibrated to give fragments of 100–500 bp; *see* **step 4**);
    g. 1.5 µL of DNA polymerase I; and
    h. Sterile distilled water to 100 µL.
  2. Incubate at 15°C for 2 h.
  3. Place at 4°C until checked on gel.
  4. Take a 5-µL aliquot of each sample; add 4 µL of H_2O and 1 µL of 10× gel loading buffer. Run on 1% agarose gel to check fragment size (*see* **Note 3**).
  5. Stop the reaction by adding 4 µL 0.5 *M* EDTA. The labeled probe can be stored frozen for several months.

3.3. Probe Denaturation

  1. Precipitate 20 µL of labeled DNA (400 ng) with 5 µg of human Cot-I DNA (*see* **Note 4**), 3 µg of salmon sperm DNA, 0.8 µL of 3 *M* Na acetate, and 3 vol of cold (–20°C) ethanol. Leave at –80°C for 15 min. Centrifuge for 15 s (12,000*g*) at 4°C. Dry the pellet on a Savant centrifuge for a few minutes.
  2. Prepare hybridization mix (10 µL per slide). To a test tube add: 5 µL of deionized formamide, 2 µL of 50% dextran sulfate, 2 µL of distilled water, and 1 µL of 20× SSC. If more slides have to be hybridized, a master mix can be prepared.

3. Resuspend pellet in 10 µL of hybridization mix, by vortexing.
4. Denature DNA mix at 80°C for 8 min. Transfer to 37°C for 20 min, then place on ice until used.

3.4. Slide Denaturation

1a. Prepare 50 mL of denaturing solution (70% deionized formamide/2× SSC). Pour into a Coplin jar. Place in a water bath at 70°C. Check the temperature inside the jar.
2a. Prewarm slides at 60°C in dry heat oven.
3a. Immerse the slides in the denaturation solution for exactly 2 min, two slides at a time.

Alternative (faster) slide denaturation:

1b. Prepare 200 µL/slide of denaturing solution (70% deionized formamide/2× SSC).
2b. Prewarm slides at 60°C in dry oven.
3b. Put 200 µL of denaturation solution on each slide, cover with a 24 × 50 mm coverslip, incubate for exactly 2 min at 80°C in a dry oven or on an appropriate thermoblock plate.
4. Dehydrate slides in 70%, 90%, and 100% ethanol, 3 min in each solution (70% ethanol at −20°C).
5. Dry slides after dehydration.

3.5. Hybridization

1. Apply 10 µL of hybridization mix to denatured slides, avoiding air bubbles.
2. Cover with 24 × 24 mm clean coverslip; seal with rubber cement.
3. Incubate in a moist chamber overnight at 37°C.

3.6. Posthybridization Washing and Detection

Do not allow slides to dry out at any stage. All washing is performed in a Coplin jar.

1. Remove coverslips and wash 3× for 5 min in prewarmed solution A in a Coplin jar in a shaking water bath at 42°C. This first step can be omitted without significant difference.
2. Wash 3× for 5 min in prewarmed solution B in a water bath at 60°C.
3. Apply 200 µL of blocking solution per slide; cover with 24 × 50 mm coverslip; transfer the slides in a moist chamber; incubate for 30 min at 37°C.
4. Dilute stock solution of avidin-Cy3 (1 mg/mL) 1:300 in detection buffer. If DIG labeling has been performed: dilute stock solution of FITC-conjugated anti-DIG (Boheringer Mannheim), according to the manufacturer's instruction. Let coverslips slide off, then apply 200 µL of detection solution per slide. Cover with 24 × 50 mm coverslips. Transfer the slides in a dark moist chamber. Incubate at 37°C for 30 min. Important: the detection **steps 3**, **4**, and **5** are not necessary if labeling has been performed with dUTP-Cy3.

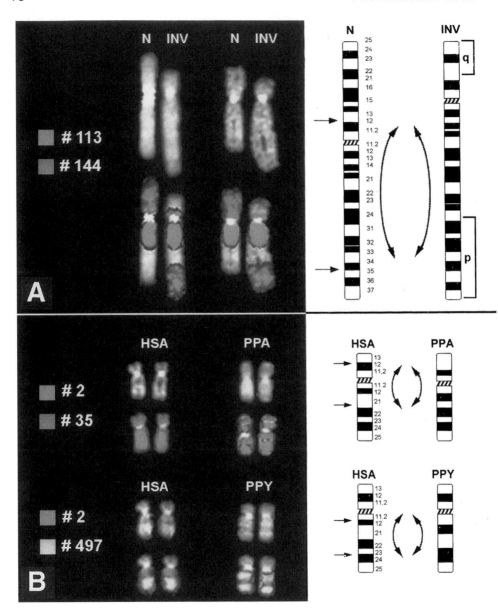

Fig. 4. (See Color Plate 1 following p. 176.) **(A)** Results of FISH experiments in a case of pericentric inversion of chromosome 2. PCP no. 113 (specific for the region 2q11–2q23, red signal) and PCP no. 114 (specific for the short arm of chromosome 2, green signal) have been used. Two pairs of homologs from distinct metaphases have been shown. Normal chromosome 2 (*N*) is on the left. In the upper row the DAPI banded chromosomes are shown without signal, to better illustrate their morphology.

continued next page

5. Remove the coverslips; rinse the slides 3× for 5 min each in prewarmed washing solution C in a water bath at 42°C.
6. Counterstain with DAPI (200 ng/mL in 2× SSC) (*see* **Note 5**) or with propidium iodide (200 ng/mL in 2× SSC), or both.
7. Rinse for 2 min in 2× SSC 0.05% Tween-20 at room temperature.
8. Apply two drops of antifade-mounting medium and cover with 24 × 50 mm coverslip. Slides can be stored for weeks or months in the dark at 4°C.

3.7. Fluorescence Microscopy

Signals from painted chromosomes are visible using an epifluorescence microscope equipped with specific filters for the fluorochromes utilized. A 100 W or preferably 50 W mercury high-pressure lamp is suitable. If a normal photographic system is used, dual- or three-band pass filters are appropriate. Multiple-bandpass filters, however, reduce the amount of light reaching the camera. If a black and white cooled charged-couple device (CCD) camera is used, filters should be distinct for each fluorochome and aligned, to guarantee an exact merging of images. Recent microscopes (such as Leica) mount perfectly aligned filters. If an older microscope is used, the image shifting problem can be circumvented using a triple-bandpass filter, a triple-band dichroic mirror, and distinct excitation filters (Chroma Technology). Cooled CCD cameras are currently the most sensitive devices. Signals from painting probes, however, are usually strong enough to be recorded using a conventional photographic camera. (*See* **Note 6** and **Figs. 2** and **3** for examples of the applications of the present technique.)

4. Notes

1. Probe labeling can be performed using biotin-16-dUTP (Boehringer Mannheim), Cy3-conjugated 11-dUTP, or digoxigenin-11-dUTP (Boehringer Mannheim). The DIG-11-dUTP is available premixed in the appropriate ratio with dNTP. The 5 µL

The breakpoints were identified as 2p13 and 2q34. **(B)** These two examples show FISH experiments performed to characterize the evolution of the phylogenetic chromosome XVII. In the upper part, FISH experiments using PCPs nos. 2 and 35, specific for the short and long arm of human chromosome, 17 respectively, show that a pericentric inversion has differentiated the human (HSA, *Homo sapiens*) and pygmy chimpanzee (PPA, *Pan paniscus*). In the lower part, PCPs no. 2 (red signal) and no. 497 (the latter specific for the 17q22-qter region, yellow signal) have been used to characterize the paracentric inversion that differentiated HSA and orangutan (PPY, Pongo pygmaeus). The original color of PCP no. 497 is green. This PCP, however, is part of PCP no. 2. The yellow color is generated by the merging of red and green signals. These analyzes suggest that the breakpoints of the pericentric inversion that differentiated HSA and PPA are located at 17p11.2 and 17q21, while those of the paracentric inversion found in PPY are located at 17q11.2 and 17q23.

indicated in the protocol refers to this mix. In the case of labeling with Cy3-dUTP: add 1 μL of Cy3-dUTP plus 2 μL of dATP/dCTP/dGTP mix, 0.5 mM each.

For cohybridization experiments we usually label the first probe with dUTP-Cy3, and the second probe with biotin-16-dUTP. The hybridization of the second probe is detected using FITC-conjugated avidin.

2. The AmpliTaq Gold (Perkin-Elmer) is a specially engineered enzyme that activates through high-temperature exposure (10 min at 95°C), thus allowing a very simple and efficient "hot start." If a normal *Taq* polymerase is used then **step 3a** should be modified accordingly (usually 4 min at 94°C).

3. Load a suitable molecular weight marker (such as *PhiX–Hae*III) on the gel. Inspect the gel on a UV transilluminator: fragments should be between 100 and 500 bp. If fragments are larger add more DNase I to the samples, incubate for an additional 30 min, and check again as described.

4. Cot-I DNA is added to the hybridization mixture to suppress hybridization of repetitive sequences. Too high a level of background signals owing to inadequate suppression may be reduced by increasing the amount of Cot-I DNA in the hybridization reaction. Nonspecific background is probably due to high molecular weight fragments or to inadequate probe purification.

5. Note that DAPI fluorescence intensity is usually inversely correlated with the denaturation of the chromosomes.

6. Applications: **Figure 4** (*see* Color Plate 1 following p. 176) shows examples of the use of PCPs in the characterization of cytogenetic rearrangements. **Figure 4A** illustrates a pericentric inversion of chromosome 2 encountered in a case of prenatal diagnosis. PCP no. 144, specific for the short arm of chromosome 2, is crucial, as it clearly indicates the nature of the rearrangement and its extension. WCPLs would be useless in these cases. **Figure 4B** shows the use of PCPs in the study of karyotype evolution of primates. It illustrates a pericentric inversion and a paracentric inversion that occurred in the evolution of phylogenetic chromosome XVII in pygmy chimpanzee and orangutan, respectively.

Acknowledgments

This work has been supported by AIRC and Telethon.

References

1. Pinkel, D., Landegent, J., Collins, C., Fuscoe, J., Seagraves, R., Lucas, J., and Gray, J. W. (1988) Fluorescence *in situ* hybridization with human chromosome-specific libraries: detection of trisomy 21 and translocation of chromosome 4. *Proc. Natl. Acad. Sci. USA* **85**, 9138–9142.

2. Lichter, P., Cremer, T., Borden, J., Manuelidis, L., and Ward, D. C. (1988) Delineation of individual human chromosomes in metaphase and interphase cells by *in situ* suppression hybridization using recombinant DNA libraries. *Hum. Genet.* **80**, 224–234.

3. Collins, C., Lin Kuo, W., Segraves, R., Fuscoe, J., Pinkel, D., and Gray, J. W. (1991) Construction and characterization of plasmid libraries enriched in sequences from single human chromosomes. *Genomics* **11**, 997–1006.

4. Vooijs, M., Yu, L.-C., Tkachuk, D., Pinkel, D., Johnson, D., and Gray, J. W. (1993) Libraries for each human chromosome, constructed from sorted-enriched chromosomes by using linker-adaptor PCR. *Am. J. Hum. Genet.* **52**, 586–597.

5. Jauch, A., Daumer, C., Lichter, P., Murken, J., Schroeder-Kurth, T., and Cremer, T. (1990) Chromosomal *in situ* suppression hybridization of human gonosomes and autosomes and its use in clinical cytogenetics. *Hum. Genet.* **85**, 145–150.

6. Weier, H. U., Lucas, J. N., Poggensee, M., Segraves, R., Pinkel, D., and Gray, J. W. (1991) Two-color hybridization with high complexity chromosome-specific probes and a degenerate alpha satellite probe DNA allows unambiguous discrimination between symmetrical and asymmetrical translocations. *Chromosoma* **100**, 371–376.

7. Cremer, T., Lichter, P., Borden, J., Ward, D. C., and Manuelidis, L. (1988) Detection of chromosome aberrations in metaphase and interphase tumor cells by *in situ* hybridization using chromosome-specific library probes. *Hum. Genet.* **80**, 235–246.

8. Gray, J. W. and Pinkel, D. (1992) Molecular cytogenetics in human cancer diagnosis. *Cancer* **69**, 1536–1542.

9. Kievits, T., Devilee, P., Wiegant, J., Wapenaar, M. C., Cornelisse, C. J., van Ommen, G. J. B., and Pearson, P. L. (1990) Direct nonradioactive *in situ* hybridization of somatic cell hybrids DNA to human lymphocyte chromosomes. *Cytometry* **11**, 105–109.

10. Boyle, A. L., Lichter, P., and Ward, D. C. (1990) Rapid analysis of mouse-hamster hybrid cell lines by *in situ* hybridization. *Genomics* **7**, 127–130.

11. Liu, P., Siciliano, J., Seong, D., Craig, J., Zhao, Y., de Jong, P. J., and Siciliano, M. J. (1993) Dual Alu PCR primers and conditions for isolation of human chromosome painting probes from hybrid cells. *Cancer Genet. Cytogenet.* **65**, 93–99.

12. Lengauer, C., Green, E. D., and Cremer, T. (1992) Fluorescence *in situ* hybridization of yac clones after Alu-PCR amplification. *Genomics* **13**, 826–828.

13. Antonacci, R., Marzella, R., Finelli, P., Lonoce, A., Forabosco, A., Archidiacono, N., and Rocchi, M. (1995) A panel of subchromosomal painting libraries representing over 300 regions of the human genome. *Cytogenet. Cell Genet.* **68**, 25–32.

14. Muller, S., Koehler, U., Wienberg, J., Marzella, R., Finelli, P., Antonacci, R., Rocchi, M., and Archidiacono, N. (1996) Comparative chromosome mapping of primate chromosomes with Alu-PCR generated probes from human/rodent somatic cell hybrids. *Chromosome Res.* **4**, 38–42.

15. Marzella, R., Viggiano, L., Ricco, A., Tanzariello, A., Fratello, A., Archidiacono, N., and Rocchi, M. (1997) A panel of radiation hybrids and YAC clones specific for chromosome 5. *Cytogenet. Cell Genet.* **77**, 232–237.

16. Korenberg, J. R. and Rykowski, M. C. (1988) Human genome organization Alu lines, and the molecular structure of metaphase chromosome bands. *Cell* **53**, 391–400.

17. Baldini, A. and Ward, D. C. (1991) *In situ* hybridization of human chromosomes with Alu-PCR products: a simultaneous karyotype for gene mapping studies. *Genomics* **9**, 770–774.

18. Schwarz,H. (1996) An inexpensive, home made DNA size standard. *TIG* **12**, 397.

2

DNA–Protein *In Situ* Covisualization
for Chromosome Analysis

Henry H. Q. Heng, Barbara Spyropoulos, and Peter B. Moens

1. Introduction

Although fluorescent *in situ* hybridization (FISH) technology has been used extensively for gene mapping and genome analysis *(1–8)*, methods that visualize the interaction of DNA and protein are required to elucidate the functional aspects of the chromosome. By displaying the physical sites of molecular reactions in the highly organized compartments of the nucleus, this methodology provides more specific information on chromatin domain, chromosome, and nuclear position than data generated by other non-*in situ* DNA–protein interaction assays such as gel retardation analysis.

We have developed a DNA–protein *in situ* codetection method to study mouse meiotic prophase chromosomes that visualizes DNA segments along the chromosomal core *(9)*. With specific DNA sequences and meiotic chromosome cores visualized simultaneously by differentially colored fluorescent tags, the interaction between the chromatin loops and protein cores is easily detected and recorded. This system reveals the characteristic patterns of chromatin loop formation of native and foreign sequences on mouse chromosomes, suggesting the possible existence of "anchor sequences" that function in chromatin loop formation *(9)*. This method has also demonstrated that the size of the chromatin loop changes relative to its position along the chromosome core, evidence that high-order structure is dictated by both the sequence of the DNA and its chromosomal position *(10)*.

Various combinations of protein markers on the chromosome and DNA probes have been used to investigate the meiotic chromosomal pairing process; the role of core protein in homologous chromatin domain interactions; as well as the relationship between chromosomal pairing, synapsis, and seg-

From: *Methods in Molecular Biology*, Vol. 123: In Situ *Hybridization Protocols*
Edited by: I. A. Darby © Humana Press Inc., Totowa, NJ

regation *(11–15)*. DNA–protein codetection has also been used to study the centromeric region of mitotic chromosomes and *de novo* formation of centromeres *(16)*, as well as chromosome behavior in mammalian oocytes *(17)*. Such strategies emphasize *in situ* codetection of DNA–protein interaction and represent another utilization of combined FISH-immunocytochemical visualization. Originally, the combination of the two technologies was used to select certain types of target cells using membrane antigens and chromosomal DNA probes *(18)*.

In this chapter, we provide a detailed protocol of DNA–protein *in situ* covisualization using mouse meiotic prophase chromosome as an example. However, the procedure can be readily adapted to study meiotic or mitotic chromosomes of other species.

2. Materials

2.1. Surface Spreading Testicular Cells

1. Multiwell slides (no. 99910090 [6-mm wells] or no. 99910095 [9-mm wells], Shandon, Pittsburg, PA, USA [1-800-245-6212]) for multiple treatments of the same material or plain glass slides for single treatments (*see* **Note 1**).
2. Dissection: dissecting tray; scissors; fine forceps (2×). A 50-mL beaker; dental wax; single-edge razor; microcentrifuge tubes (2×); centrifuge.
3. Five 10-mL Coplin jars. A 10-µL pipetor and tips; wide-bore plastic transfer pipets.
4. Phenol red indicator for pH monitoring: 0.5% phenol red in distilled H_2O. Filter and store at room temperature indefinitely. (Alkaline pH = purple-red color, acidic pH = yellow.)
5. 0.05 *M* Borate buffer stock solution: 1.91% (w/v) sodium borate in distilled H_2O, adjust pH to 9.2 with 0.5 *N* NaOH. Prepare a working solution of 0.01 *M* by diluting stock 1:4 with distilled H_2O.
6. Minimum essential medium (MEM) with Hank's salts, without L-glutamine: purchased ready-to-use from supplier (e.g., Gibco-BRL) or made from powder (10× concentrate). Adjust pH to 7.3 with 0.05 *M* borate buffer.
7. Paraformaldehyde solution: per 100 mL of distilled H_2O: 10 µL of 0.5 *N* NaOH, 30 µL of phenol red indicator and 1 g paraformaldehyde (BDH; JB EM). Bring to 55°C on a hot plate in a fume hood. Shut off the heat and place the Erlenmeyer flask in a container of cold water on the hot plate and continue stirring until all powder has dissolved. Temperature should go no higher than 60°C. If pH acidifies, add 0.01 *M* borate buffer dropwise. Filter, cool to room temperature, and adjust final pH to 8.2 with borate buffer using paper pH indicators. **CAUTION:** Paraformaldehyde is harmful if ingested and can be absorbed through skin. The fine powder is easily dispersed through the air.
8. 60 mg/mL Sodium dodecyl sulfate (SDS) stock solution with 30 µL of phenol red indicator per 100 mL of solution. Adjust pH to 8.2 with borate buffer. Store at room temperature. Depending on the degree of the chromatin dispersion desired,

use from 0% to 0.06% SDS in the first paraformaldehyde fixation (*see* **Note 2**). **CAUTION:** SDS is harmful if ingested or inhaled and irritates eyes and skin. The fine powder is easily dispersed through the air.
9. 0.4% Photo-Flo 200 (Kodak) in distilled H_2O. Add 30 μL of phenol red indicator per 100 mL of solution. Using paper pH indicator, adjust pH to 8.0 with borate buffer.
10. Spreading (hypotonic) solution: 0.5% NaCl in distilled H_2O. Adjust pH to 8.0 with borate buffer.

2.2. Immunostaining

1. Three to five Coplin jars with small stirrer bars for each. Magnetic stirring plates.
2. Antibody dilution buffer (ADB): 10% goat serum, 3% bovine serum albumin (BSA), 0.05% Triton X-100 in phosphate-buffered saline (PBS).
3. Wash buffer: 10% ADB in PBS with 1% Kodak Photo-Flo 200.
4. Triton X-100. Add 1% v/v to second of three washes.
5. Small humid chamber such as a Plexiglas box with a support for holding the slides.

2.3. FISH Detection

2.3.1. Probe Labeling

1. Biotin labeling nick-translation kit (Gibco-BRL).
2. Nick column (Pharmacia).
3. Salmon sperm DNA (100–500-bp fragments obtained by sonicating).
4. 3 *M* NaOAc.
5. 70% and 100% ethanol.

2.3.2. Hybridization

1. Hot plate (37–70°C); water baths at 37°C, 43°C, 46°C, 70°C, and 75°C; 37°C incubator.
2. 25-mL Plastic slide mailers (Surgipath).
3. Plastic slides chamber (slide holder) (CanLab).
4. RNase A (Boehringer Mannheim).
5. Ethanol: 75%, 90%, and 100%.
6. Denaturation solution: 70% deionized formamide (IBI) in 2× SSC (saline sodium citrate) (20× SSC stock solution: 3 *M* NaCl, 300 m*M* Na citrate).
7. Hybridization solution I (for use with genomic probes): 50% deionized formamide (IBI) and 10% dextran sulfate in 2× SSC.
8. Hybridization solution II (for use with repetitive DNA probes): 65% formamide and 10% dextran sulfate in 2× SSC.
9. Mouse Cot-I DNA (BRL).
10. Yeast DNA (*S. cerevisiae*, W303-1A) sonicated into 100–500-bp fragments.
11. Wash solution A (for non-repetitive DNA probe): 50% formamide in 2× SSC.
12. Wash solution B (for repetitive DNA clones): 65% formamide in 2× SSC.

13. Wash solution C: 0.1 *M* phosphate buffer, pH 8.0, with 0.1% Nonidet P-40 (Boehringer Mannheim).
14. 2× SSC (before and after 4',6-diamidino-2-phenylindole [DAPI] staining wash).

2.3.3. Detection and Amplification

1. Blocking solution: 3% BSA (Sigma Fraction V) in 4× SSC with 0.1% Tween-20.
2. Detection solution: 1% BSA and 0.1% Tween-20 in 4× SSC. Store at 4°C.
3. Avidin-FITC (fluorescein isothiocyanate, Vector): 500 µg/mL (stock solution). FITC detection working solution: 10 mL of avidin-FITC stock solution to 990 µL of detection mixture. Store in the dark at 4°C. Good for up to 6 mo.
4. Biotinylated goat anti-avidin antibody (Vector): 500 µg/mL (stock solution). Aliquots (50 µL each) can be kept at –20°C. Working solution: dilute stock solution with detection solution. The final concentration of the solution with detection mixture is 5 µg/mL.
5. Anti-digoxigenin–rhodamine Fab fragments. 200 µg/mL (stock solution). Aliquots (50 µL each) are diluted 1:10 with detection solution before use.

2.3.4. Counterstaining and Antifade

1. DAPI (Sigma): 0.2 mg/mL of stock solution in H_2O. Store in the dark at 4°C.
2. Antifade solution. ProLong Antifade, no. P-7481, Molecular Probes, Eugene, OR, USA (503-465-8300).

3. Methods

3.1. Surface Spreading Testicular Material

3.1.1. Slide Preparation

Wash the slides with a commercial glass cleaner such as Windex® immediately prior to use. Rinse in hot water, then distilled water. Rub dry with a lint-free tissue and label the slides. To reuse the slides in future experiments, wash in detergent, sonicate in a bleach/detergent solution, rinse in water and store dry. The more the slide is reused, the better the material adheres to the glass surface.

3.1.2. Preparation of Tissue

1. Remove the testes of a relatively young male (about 25 d old for rats, mice, or hamsters where there will be few spermatozoa). For animals with few meiotic nuclei, *see* **Note 2**.
2. Remove ALL fat.
3. Using a transfer pipet, run MEM over the testis. Blot off the excess MEM.
4. Hold the testis with forceps and cut open with a razor the side with the fewest blood vessels.
5. Extrude the seminiferous tubules into a drop of MEM on dental wax. Do not allow the outer casing to touch the MEM.
6. Pick up the tubules with clean tweezers and run about 3–5 mL of fresh MEM over the bundle as described above. Drain on a tissue then place the tubules in a fresh drop of MEM on the dental wax.

7. Cut the tubules several times with a new, grease-free razor blade.
8. Squeeze the tubules with clean, grease-free forceps to release the spermatocytes from the tubules.
9. Transfer the cell suspension to a 1.5-mL microfuge tube.
10. Fill the tube with MEM and draw the suspension up and down through a wide-bore plastic transfer pipet to separate the cells. Let stand 1 min until all the tubules have settled.
11. Transfer the supernatant to a clean 1.5-mL microfuge tube. Top up with fresh MEM and centrifuge 5 min at 160*g*.
12. Pour off the supernatant and gently resuspend cells in the residual MEM by tapping the side of the tube.

3.1.3. Surface Spreading

1. Fill a small Petri dish with 0.5% hypotonic NaCl solution until the surface of the liquid is convex.
2. Gently tap the cell suspension to mix and draw up 5 μL with a pipetor.
3. Wipe the pipet tip clean with a Kimwipe (Kimberley-Clark) and carefully expel the cell suspension such that a drop hangs from the pipet tip.
4. Touch the lower edge of the drop to the convex surface. Cells will spread out (*see* **Note 3**.)
5. Allow to stabilize for 10 s, then lower a slide onto the surface to pick up the cells.
6. Let sit 10 s. Roll the slide off the NaCl bath by lifting first one long edge, then the rest.
7. Place the slide in a Coplin jar with paraformaldehyde and SDS, if required (*see* **Note 4**), for 3 min.
8. Transfer the slide to a second Coplin jar containing only paraformaldehyde for an additional 3 min.
9. Wash 3 × 1 min each in Photo-Flo solution and air dry.
10. While the slides are in the fixative and washing solutions, additional spreads can be made: Discard the used hypotonic solution, rinse the spreading dish in soapy water, hot water, and distilled H_2O. Add fresh hypotonic solution and spread the next 5 μL of nuclei.
11. The slides can be used when dried or else stored dry at –70°C.
12. Nuclei and chromosome cores are visible with phase-contrast microscopy.

3.2. Immunostaining

1. When the slides are dry, wash in three changes of wash buffer for 10 min per wash. Add 1% Triton X-100 to the middle wash. It is important that the buffer is gently stirring during the washing process, so do not forget the stir bar.
2. Remove the slides one at a time. To perform several treatments on the same slide, wipe dry the edges of the multiwell slide. Use the edge of a glass slide wrapped with a lint-free tissue to dry the interior partitions. To each well, add 20–30 μL of antibody diluted in ADB. Otherwise, coat the slide with 50–100 μL of primary antibody and invert the slide onto the Parafilm-lined floor of a humid chamber.

3. Place the slide in a small humid chamber located inside a plastic bag containing a PBS-soaked tissue. Incubate in the dark at room temperature overnight or at 37°C for 1 h.
4. Repeat the washing procedure described in **step 1**. Add the secondary antibody as described in **step 2** and incubate at 37°C for 1 h.
5. Wash 3 × 10 min each in PBS with 1% Photo-Flo, the middle wash containing 1% Triton X-100. Rinse 2 × 1 min each in distilled water with 1% Photo-Flo. Air dry or dry in a desiccator. The slides can now be screened or immediately processed for *in situ* hybridization.
6. To screen the slides, add a few drops of antifade reagent, with or without DAPI or PI, add a coverslip, and view under the microscope. We use ProLong Antifade, which inhibits quenching of fluorescence and enhances signal from fluorochromes such as rhodamine. Vials of antifade may be stored at 4°C for more than a week if tightly sealed and light protected. Because this reagent loses effectiveness in the presence of water, slides must be completely dry before applying.
7. Select the slides of interest for further processing. Wash off the coverslip by soaking the slide at a 45° angle in PBS with 1% Triton X-100 and 1% PhotoFlo. Wash as in **step 5** and air dry. Continue with the *in situ* procedures.

3.3. FISH Detection

3.3.1. Probe Labeling

Detailed protocols for probe purification can be found in Heng and Tsui *(5)*.

3.3.1.1. BIOTIN LABELING

1. Purified DNA (1 µg) is labeled with a BRL BioNick kit according to the supplier's instructions (15°C for 60 min for probes of 1–40 kb; 120 min for probe >100 kb). Optional digoxigenin (DIG) labeling of DNA (1 µg) can be done with Boehringer's DIG labeling kit.
2. After labeling, the unincorporated nucleotides are removed using a Nick column (Pharmacia). Precipitate 6 µg of salmon sperm DNA with the labeled probes in 40 µL of 3 *M* NaOAc and 880 µL of ethanol. After washing with 70% ethanol, resuspend the probes in 20 µL of TE buffer.

3.3.2. Probe Treatment Before Hybridization

3.3.2.1. REPETITIVE PROBES

Denature labeled repetitive probes in hybridization solution II. Add 20–50 ng of labeled probes to 15 µL of denaturing solution and denature at 75°C for 5 min. Place the tube containing the denatured probe on ice immediately after denaturation.

3.3.2.2. GENOMIC PROBES (PHAGE, COSMID, BAC, AND YAC)

For genomic sequence detection, potential signal from repetitive elements within the probe itself must be suppressed by adding genomic DNA or Cot-I DNA. This is done as follows:

1. For phage, cosmid, and BAC probes: 20–50 ng probes + 13 µL of hybridization solution I + 2 µg of mouse Cot-I DNA or total mouse DNA.
2. For YAC probes isolated with total yeast DNA: 200–250 ng labeled probes + 13 µL of hybridization solution I + 2 µg of mouse Cot-I DNA + 2 µg of total yeast DNA.
3. After 5 min of denaturation at 75°C, transfer the tube to a 37°C water bath for another 15–30 min for prehybridization (prehybridization for YAC probe may be slightly longer: 20–60 min).

3.3.3. Prehybridization Treatment

3.3.3.1. SLIDE DENATURATION

1. Make fresh denaturation solution of 70% formamide by mixing 28 mL of formamide, 8 mL of distilled H_2O, and 4 mL of 20× SSC. Put the jar filled with denaturing solution into a water bath and bring the temperature to 70°C. Meanwhile, heat the slides at 50°C to avoid lowering the temperature of the denaturing solution.
2. Immerse one to four warmed slides in the denaturation solution for 1–1.5 min.
3. Quickly transfer the slides to a plastic jar with cold 70% ethanol for 2 min. Dehydrate the slides in 95% and 100% ethanol for 3 min each. Air dry and perform hybridization immediately.

3.3.4. Hybridization

1. Load 15 µL of denatured probe in hybridization solution onto each slide and cover with a 22 × 22 mm coverslip. Gently remove air bubbles and seal the edges with rubber cement to minimize evaporation.
2. Hybridize at 37°C in a humid chamber containing a water-soaked tissue. For repetitive probes, hybridize for a few hours or overnight. For cosmid or YAC probes, 8–24 h are suggested.

3.3.5. Posthybridization Wash

Wash conditions vary according to the type of probe used in the hybridization.

3.3.5.1. COSMID, PHAGE, YAC, OR BAC PROBES CONTAINING REPETITIVE ELEMENTS

1. Prewarm wash solution A to 46°C in three plastic jars.
2. With forceps, carefully remove the rubber cement from the slides. Allow the coverslips to float off in 2× SSC solution and agitate the slides a few times.
3. Wash slides in wash solution A with gentle agitation 3 × 3 min each.
4. Wash with warmed 2× SSC at 46°C 3 × 3 min.
5. Place slides in wash solution C. If necessary, slides can be kept at 4°C in this solution for up to 2 d or used immediately for detection.

3.3.5.2. REPETITIVE DNA PROBES

1. Prewarm wash solution B to 43°C and 2× SSC to 37°C.
2. Remove rubber cement and coverslips as described in the previous section.
3. Wash slides in wash solution B for 20 min with agitation.
4. Wash twice with warmed 2× SSC at 37°C for 4 min each.
5. Place slides in wash solution C and the slides are ready for detection.

3.3.6. Detection and Amplification

3.3.6.1. DETECTION WITH AMPLIFICATION

1. Remove slides from the wash jar and blot off excess liquid.
2. Quickly apply 30 μL of blocking solution per slide. Place a plastic coverslip over the solution and incubate at room temperature for 5 min.
3. Gently peel off plastic coverslips, tilt the slide, and drain the fluid.
4. Apply 30 μL of FITC detection solution per slide and cover with a fresh plastic coverslip. Incubate for 20 min at 37°C in a dark humidified chamber. (From this step onwards, it is critical that the samples have minimal light exposure).
5. Remove coverslips and wash the slides in wash solution C 3× for 3 min each in a fresh solution.
6. Apply 30 μL of blocking solution per slide. Cover with a plastic coverslip and incubate at room temperature for 5 min.
7. Apply 30 μL of biotinylated goat anti-avidin antibody working solution per slide and cover with a plastic coverslip. Incubate at 37°C for 20 min.
8. Following removal of the coverslip, wash the slides 3× for 3 min each in wash solution C.
9. Apply 30 μL of blocking solution per slide. Cover with a plastic coverslip and incubate at room temperature for 5 min.
10. Peel off the plastic coverslip, tilt the slide, and drain the fluid.
11. Apply 30 μL of FITC detection solution to each slide and cover with a fresh plastic coverslip. Incubate for 20 min at 37°C.
12. Remove coverslips and wash the slides in wash solution C 3× for 3 min each.
13. Stain with DAPI by immersing the slides in 0.2 μg/mL of DAPI in 2× SSC for 5–10 min at room temperature. Rinse in 2× SSC 3× for 2 min each.
14. Mount the slides with 10 μL of antifade solution. Cover with a 22 × 40 mm glass coverslip. Apply gentle downward pressure to flatten coverslip before examination.

3.3.6.2. DETECTION WITHOUT AMPLIFICATION

Signal amplification may not be necessary when the BAC, YAC, or repetitive sequence probes are used for FISH detection.

1. Remove slides from wash jar and blot off excess liquid.
2. Apply 30 μL of blocking solution per slide. Cover with a plastic coverslip and incubate at room temperature for 5 min.
3. Gently peel off plastic coverslips, tilt the slide, and allow fluid to drain.

4. Apply 30 μL of FITC detection solution to each slide and cover with a fresh plastic coverslip. Incubate 20 min at 37°C in a dark humid chamber. (From this step onwards, it is critical that the samples have minimal light exposure.)

5. Remove coverslips and wash the slides in wash solution C 3× for 3 min each.

6. Stain with DAPI by immersing the slides in 0.2 μg/mL of DAPI in 2× SSC for 5–10 min at room temperature. Rinse in 2× SSC 3× for 2 min each.

7. Mount the slides with 20 μL of antifade solution. Cover with a 22 × 40 mm glass coverslip. Apply gentle downward pressure to flatten the coverslip before examining the slide. If the signals need further amplification, carefully remove the oil from the coverslip before gently detaching it and rinse the slide quickly in 2× SSC. Wash the slide 3× for 5 min each in wash solution C. The remaining steps are same as 6–14 in **Subheading 3.3.6.1**. (Detection with amplification).

3.3.6.3. Two Color Detection (Optional)

To visualize the chromosome core and DNA sequences in separate colors, use different fluorochromes conjugated to secondary antibodies and FISH probes. For instance, if the chromosome core is detected by FITC-conjugated antibody, then the DNA probe should be labeled with biotin or DIG and detected by avidin or anti-DIG antibody conjugated with rhodamine or Texas red. For multicolored detection, follow the protocol outlined in **Subheading 3.3.6.2.** but in **step 4**, add anti-digoxigenin–rhodamine fragments instead of FITC-avidin.

3.4. Microscopy and Photography

1. Unstained spread nuclei can be visualized with phase contrast microscopy.

2. Fluorescent probes are visualized with a fluorescent microscope with the appropriate filters. FITC is visualized with an epifluorescent exciter filter 450–490 nm, reflector filter 510 nm, and barrier short pass of 515–560 nm. DAPI filters are exciter filter 330–380 nm, reflector 420 nm, and barrier 420 nm. PI filters are exciter filter 450–495 nm, reflector 510 nm, and barrier 515 nm. The latter will show FITC as well as PI.

3. For immunocytology, use Kodak T-Max 400 film for photographs or Fuji Color 400 film for slides. With a 100× objective lens, expose for 15–30 s. We recommend commercial development. No filters are necessary when making prints with Cibachrome paper.

4. For photography of FISH following immunocytology, use Kodak Ectachrome P800/1600 film, and push the exposure to 3200 ASA (exposure times: DAPI about 4–5 s, PI and/or FITC 20 s).

5. For two-color detection, it is easiest to take photographs using duo- or tri-filters. However, if none are available, multiple exposure steps can be used. First expose the FITC image, followed by rhodamine or DAPI.

4. Notes

1. While regular glass microscope slides or coverslips can be used, we use multiwell slides to test different combinations or concentrations of reagents on a single

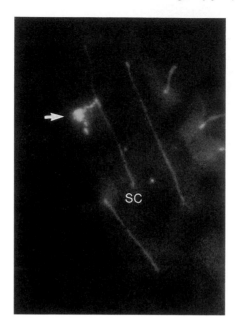

Fig. 1. Detection of insertion on meiotic chromosome. In this partial pachytene nucleus of a mouse, the protein core (SC) of the paired chromosomes are visualized green with FITC-conjugated secondary antibody attached to anti-synaptonemal complex antibody. The chromatin loops surrounding these protein cores are stained red with rhodamine. The heterochromatin surrounding the centromeres are particularly apparent. The heavy yellow signal (arrow) represents the inserted 1.7 Mb of bacterial *Lac*I-repeated sequence on chromosome no. 4. The packaging of this inserted sequence does not follow the pattern of the native sequence.

tissue preparation. The 15-nm plastic coating of these multiwell slides provides a solid containment for the reagents used on individual wells and protects the biological sample from physical damage when removing the coverslip for sequential treatments such as replacement of antibodies, probes, fluorescent tags, or antifade reagents *(19)*.

2. Variations of spreading technique for different species. In some species, a lack of material precludes surface spreading on a water bath as described previously. An alternate method that preserves all cells is to place a drop (50 μL minimum of liquid) of testicular material directly on a clean slide (regular or multiwell). Lower a coverslip or piece of Parafilm on top of the slide and let sit for 10 min. Peel off the Parafilm or remove the coverslip by immersing the slide in a Coplin jar of paraformaldehyde at a 45° angle until the coverslip falls off. Transfer the slide to a Coplin jar of fresh paraformaldehyde (with SDS, if necessary; *see* **Note 4**) and continue the fixation as described in **Subheading 3.1.3., steps 7–12**.

3. Spreading problems. When surface spreading, the cell suspension sometimes drops to the bottom of the dish rather than spreading out on the surface. This is

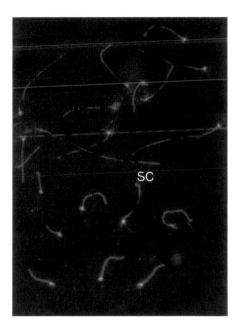

Fig. 2. Minor-satellite–sequence detection along the chromosome core. The chromosome cores of these mouse-pachytene-paired chromosomes (SC) are stained green with FITC-conjugated second antibody attached to antisynaptonemal complex antibody. The minor satellite at the centromeric end of these acrocentric chromosomes is visualized red with rhodamine-conjugated probe. Note that the size of the minor satellite signal is much smaller compared to the heterochromatin loops of the chromosomes in **Fig. 1**. Regulation of loop size is controlled by the position of the sequences on the chromosomes as well as the sequence itself.

due either to grease in the preparation or to a low cell concentration. Solution: change the spreading dish and/or pipet tips. Failing that, centrifuge the cells to concentrate them. In the worst case, the material can be resuspended in fresh MEM and left to stand for 1–2 min. Wipe the surface with lens paper to remove any grease and centrifuge to concentrate the nuclei once more. Often the situation is not as bad as it seems, so check the slides before aborting what initially appears to be a bad run.

4. Adjusting the degree of spreading of chromosomes with SDS. The degree of chromatin spreading can be controlled by adjusting the concentration of SDS in the first paraformaldehyde fixation from 0% to 0.06%. The more SDS used, the greater is the spreading. For rats and mice, use 0.03% in 50 mL paraformaldehyde; for hamsters, 0.015%; for humans, 0.06%.

5. Order of detection: protein or DNA? The order of detecting protein and DNA depends on the properties of each experiment. To minimize disruption of the protein structure and optimize detection with antibodies, we perform protein detec-

tion first. These signals survive the harsh conditions of denaturation in the FISH protocol, especially when the subject material is meiotic chromosomes owing to the copious amount of proteins in the meiotic chromosome core. However, it is also possible to perform the experiment in the reverse order.

6. Multiple-colored probes. One has the option of simultaneously using different colored probes such as FITC, Texas red, and Cy3 to visualize protein and DNA. Caution is needed to avoid interference from cross-reaction among the various secondary antibodies (**Figs. 1** and **2**).

7. New developments. Many interesting proteins including transcription factors are less abundant than meiotic chromosome core proteins. With improved sensitivity, we shall be able to visualize the transcription complex relative to the promoter and other regulatory regions of the gene and the chromatin. This may be achieved in the future by combining FISH and green-fluorescent protein detection or FISH with nonfluorescent protein detection. DNA–protein codetection using released chromatin fibers will also generate enormous information regarding chromatin structure and gene regulation *(6–8,20–23)*.

Acknowledgments

We thank Dr. Lap Chee Tsui for his support. We also thank X. M. Shi for her excellent assistance. This work was supported by a grant to Peter Moens from the Natural Sciences and Engineering Council of Canada and by SeeDNA Biotech Inc.

References

1. Lawrence, B. J. (1990) A fluorescence *in situ* hybridization approach for gene mapping and the study of nuclear organization, in *Genome Analysis*, vol I: *Genetic and Physical Mapping* (Davies, K. E. and Tilghman, S. M., eds.), Cold Spring Harbor Laboratory Press, Cold Spring Harbor, New York, pp. 1–38.

2. Lichter, P., Boyle, A., Cremer, T., and Ward, D. C. (1991) Analysis of genes and chromosomes by nonisotopic *in situ* hybridization. *GATA* **8**, 24–35.

3. Trask, J. B. (1991) Fluorescence *in situ* hybridization. *TIG* **7**, 149–154.

4. Spyropoulos, B. and Moens, P. B. (1994) *In situ* hybridization of meiotic prophase chromosomes, in *Methods in Molecular Biology:* In Situ *Hybridization Protocols* (Choo, K. H., ed.), Humana Press, Totowa, NJ, pp. 131–139.

5. Heng, H. H. Q. and Tsui, L.-C. (1994) FISH detection on DAPI banded chromosomes, in *Methods in Molecular Biology: In Situ Hybridization Protocols* (Choo, K. H., ed.), Humana Press, Totowa, NJ, pp. 35–49.

6. Heng, H. Q. H., and Tsui, L.-C. (1994) Free chromatin mapping by FISH, in *Methods in Molecular Biology: In Situ Hybridization Protocols* (Choo, K. H., ed.), Humana Press, Totowa, NJ, pp. 109–122.

7. Heng H. H. Q., Tsui L. -C., Windle B. and Parra I. (1995). High-resolution FISH analysis, in *Current Protocols in Human Genetics* (Dracopoli, N. C., ed.), John Wiley and Sons, New York, pp. 4.5.1-4.5.26.

8. Heng, H. H. Q., Spyropoulos, B., and Moens, P. B. (1997) FISH technology in chromosome and genome research. *Bioessays* **19**, 75–84.

9. Heng H. H. Q., Tsui L.-C., and Moens P. B. (1994) Organization of heterologous DNA inserts on the mouse meiotic chromosome core. *Chromosoma* **103**, 401–407.

10. Heng, H. H. Q., Chamberlain, J. W., Shi, X.-M., Spyropoulos, B., Tsui, L.-C., and Moens, P. B. (1996) Regulation of meiotic chromatin loop size by chromosome position. *Proc. Natl. Acad. Sci. USA* **93**, 2795–2800.

11. Moens, P. B., Heng, H. H. Q., Pearlman, R. E., Rosonina, E., and Spyropoulos, B. (1995) Meiotic chromosomes visualized with antibodies and DNA probes, in *Kew Chromosome Conference IV* (Brandham, P. E. and Bennett, M. D., eds.), Royal Botanical Gardens, Kew, pp. 375–379.

12. Barlow, A. L. and Hulten, M. A. (1996) Combined immunocytogenetics and molecular cytogenetic analysis of meiosis I human spermatocytes. *Chromosome Res.* **4**, 562–573.

13. Barlow, A. L. and Hulten, M. A. (1997) Sequential immunocytogenetics, molecular cytogenetics and transmission electron microscopy of microspread meiosis I oocytes from a human fetal carrier of an unbalanced translocation. *Chromosoma* **106**, 293–303.

14. Moens, P. B., Heddle, J. A. M., Spyropoulos, B., and Heng, H. H. Q. (1997) Identical megabase transgenes on mouse chromosomes 3 and 4 do not promote ectopic pairing or synapsis at meiosis. *Genome* **40**, 770–773.

15. Moens, P., Pearlman, R., Traut, W., and Heng, H. H. Q. (1998) Chromosome cores and chromatin at meiotic prophase, in *Meiosis and Gametogenesis* (Handel, M., ed.), Academic, San Diego, pp. 241–262.

16. Haaf, T. and Ward, D.C. (1994) Structural analysis of α-satellite DNA and centromeric proteins using extended chromatin and chromosomes. *Hum. Mol. Genet.* **3**, 679–709.

17. Hunt, P., LeMaire, R., Emburg, P, Sheean, L., and Mroz, K. (1995) Analysis of chromosome behaviour in intact mammalian oocytes: monitoring the segregation of a univalent chromosome during female meiosis. *Hum. Mol. Genet.* **4**, 2007–2012.

18. van den Berg, H., Vossen, J. M., van den Bergh, L., Bayer, J., and van Tol, M. J. D. (1991) Detection of Y chromosome by in situ hybridization in combination with membrane antigens by two-color immunofluorescence. *Lab. Invest.* **64,** 623–628.

19. Spyropoulos, B., Heng, H. H. Q., and Moens, P. B. (1997) Recycling cells in FISH and immunocytology studies. *Technical Tips on Line*, 25 June, T01090.

20. Heng, H. H. Q., Squire, J., and Tsui, L.-C. (1991) Chromatin mapping — a strategy for physical characterization of the human genome by hybridization *in situ*. Proc. 8th Int. Cong. Human Genetics, *Am J. Hum. Genet. Suppl.* p. 368.

21. Heng, H. Q. H., Squire, J., and Tsui, L.-C. (1992) High resolution mapping of mammalian genes by *in situ* hybridization to free chromatin. *Proc. Natl. Acad. Sci. USA* **89**, 9509–9513.

22. Wiegant, J., Kalle, W., Mullenders, L., Brookes, S., Hoovers, J. M. N., Dawerse, J. G., van Ommen, G. J. B., and Raap, A. K. (1992) High resolution *in situ* hybridization using DNA halo preparations. *Hum. Mol. Genet.* **1**, 587–592.

23. Heng, H. H. Q. and Tsui, L.-C. (1998) High resolution free chromatin/DNA fiber fluorescent in situ hybridization. *J. Chromatogr.* A **806,** 219–229.

3

Radioactive *In Situ* Hybridization to Animal Chromosomes

Graham C. Webb

1. Introduction

Although largely replaced by the use of fluorescent *in situ* hybridization (FISH) in animal and human molecular cytogenetics, the technique of radioactive *in situ* hybridization (RISH) still has some uses. Using practicable exposure times for autoradiographs of 3–4 wk, RISH is approx 50 times more sensitive than FISH using biotin- or digoxygenin-labeled probes, and probably 10 times more sensitive than the recently introduced system of tyramide FISH (NEN Life Sciences Prods., Code NEL 730/730A). In addition, the sensitivity of RISH can be increased with longer exposures, in a roughly linear fashion until the silver bromide grains in the emulsion approach saturation over the target; by contrast, FISH requires instantaneous expression of the signal. Because of its high sensitivity, RISH can be used with short probes *(1)*, down to 200 bp, poorly labeled probes *(2)*, short target sequences, and old slides (**Table 1**). Bands on the chromosomes can be excellent with RISH *(3)*, requiring no enhancement by image analyzing systems, and observed with simple brightfield microscopy (**Fig. 1**); although finding a suitable batch of Giemsa stain can be difficult (*see* **Subheading 3.6.** and **Note 1**).

Disadvantages of RISH compared with FISH include the waiting time for exposure of autoradiographs and the imprecise localization due to the electrons from tritium having a range of up to 3 μm in emulsion. However, RISH can be made more accurate by scoring from numerous chromosome spreads, which is made easy by the use of brightfield optics; such scoring leads to a peak of signal grains located over a target chromosomal segment of approx 1 μm (**Fig. 1**). By contrast, FISH appears to have a resolution of approx 0.5 μm *(4)*, owing

From: *Methods in Molecular Biology*, Vol. 123: In Situ *Hybridization Protocols*
Edited by: I. A. Darby © Humana Press Inc., Totowa, NJ

Table 1
Comparison of Radioactive vs Fluorescent *In Situ* Hybridization (RISH vs FISH)

Characteristic	RISH	FISH
Minimal practical length of unique probe DNA	200 bp	1000 bp
Probable minimal length of target sequence	150 bp	500 bp
Minimal incorporation of label into unique DNA	2%	10%
Age of slides, stored dry @ –20°C	Up to 10 years	Fresh, or up to a few weeks.
Unenhanced chromosome bands	Good	Poor
Method of observation	Brightfield	Fluorescence
Precision of localization	1 µm	0.5 µm

largely to the position and size of fluorescent signal spots, which are often made deliberately large to distinguish them from background. If small-grained emulsions, such as those normally used for electron microscopy, are used for RISH, then resolution due to signal grain size may be superior to FISH.

2. Materials

2.1. Cell Culture Materials

1. AmnioMax culture medium (Gibco-BRL Code 27000-025), mix and store as per instructions.
2. Pokeweed mitogen (e.g., Sigma Code L 8777), stock 1 mg/mL in water. Final concentration is 5 µg/mL. Store at –20°C.
3. Phytohemagglutinin (PHA) (e.g., M form Gibco-BRL Code 10576-015), stock to 10 mL with water. Use at 2 mL/100 mL. Store at –20°C.
4. Decon 90 (Decon Laboratories Ltd.), fresh 5% v/v solution in water. **CAUTION:** this detergent is so alkaline that it will etch the surface of the glass if slides are left in it for prolonged periods.
5. 5-Bromodeoxyuridine (e.g., Sigma Code B 9285), stock of 20 mg/mL in water; variously diluted. Store at –20°C.
6. Dulbecco's phosphate-buffered saline (e.g., Commonwealth Serum Labs. Code 09832471).
7. Thymidine (e.g., Sigma Code T 9250), stock aliquots, for release of arrest: 10^{-3} M in phosphate-buffered saline (PBS); for arrest: 20 mg/mL in water, variously diluted. Can be thawed and refrozen. Store at –20°C.
8. Colchicine (e.g., Sigma Code C 9754), stock 100 µg/mL, use at preferred concentration, e.g., 1.5 µg/mL. Store at –20°C.

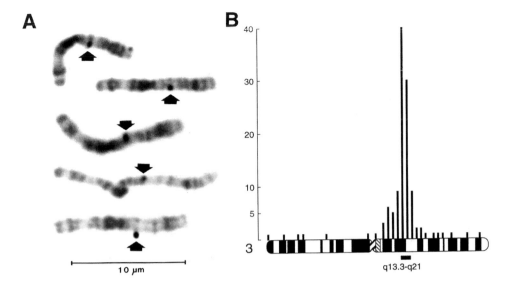

Fig. 1. Silver grain signals generated by *in situ* hybridization of a tritiated 778-bp probe for the gene encoding human urothelial tetraspan protein, UPK1B. **(A)** Grains, indicated by arrows, over photographs of chromosome 3. **(B)** Scores of grains over an idiogram of chromosome 3 with a bar linking the two tallest columns of grains over bands 3q13.3 and 3q21, the stated localization of the probe *(3)*.

9. 0.075 M KCl, stock 0.75 M, 55.91 g/L, diluted 1:10 (to 1/2 isotonic) for use.
10. Concanavalin A (e.g., Sigma C 5275), stock 1 mg/mL. Final concentration 3 µg/mL. Store at –20°C.

2.2. DNA Probes

1. Nick Translation Kit (e.g., Amersham Code N 5500). Store at –20°C.
2. Tritium-labeled nucleotide triphosphates (e.g., Amersham: [³H]dATP, Code TRK633; [³H]dCTP, Code TRK625; [³H]dTTP Code TRK576, all 1 mCi/mL). Store at –20°C.
3. DNA polymerase I, Klenow fragment (e.g., Sigma Code D 8276), stock of 5000 U/mL. Store at –20°C.
4. Sephadex G-50 (DNA grade, Pharmacia Code 17-0045-02). Gel preparation *(5)*.
5. $T_{10}E_1$: 10 mM Tris, 1 mM EDTA, pH 7.6 *(5)*.
6. Water-miscible scintillation fluid (e.g., Amersham, BCS, Code NBCS 104).
7. Scintillation Counter (e.g., Beckman LS 3801).
8. Oligo-labeling Kit (e.g., Bresatec, Code OLK-TAC) Store at –20°C.
9. Salmon sperm DNA (e.g., Sigma Code D 1626), stock about 10 mg/mL, for preparation *see* **ref. 5**. Store at –20°C.
10. 3 M Sodium acetate, 246.09 g/L of anhydrous CH_3COONa, adjust to pH 5.2 using acetic acid.

11. Cot-I DNA: human (Gibco-BRL Code 15279-011); mouse (Gibco-BRL Code 18440-016). For preparation of Cot-I DNA from other animals, consult the author.

2.3. Application of Probes to Slides

1. RNase (e.g., Sigma pancreatic RNase A, Code R 5125), stock aliquots of 10 mg/mL in 10 mM Tris-HCl, 15 mM NaCl, pH 7.5, immersed in boiling water for 10 min to kill contaminating DNase and cooled slowly. Dilute with 2× SSC (saline sodium citrate). The diluted solution can be frozen and thawed approx 10 times for repeated use. Store at –20°C.
2. 2× SSC, stock 20× SSC (12 M NaCl; 1.2 M Na acetate), dilute as required and adjust pH to 7.0.
3. Triethanolamine-HCl (e.g., Sigma Code T 9534), 0.1 M, 18.57 g/L in H$_2$O, adjust to pH 8.0 using solid NaOH.
4. 10× SSCP (1.2 M NaCl, 0.15 M Na citrate, 0.1 M Na$_2$HPO$_4$, 0.1 M NaH$_2$PO$_4$, pH 6.0; keeps indefinitely at –20°C and can be repeatedly thawed and refrozen).
5. 20% w/v Dextran sulfate (Pharmacia Code 17-0340-01): 0.2 g in 1.0 mL or proportionate. This mix can be kept at 4°C for a few days for reuse.
6. Deionized formamide: stir 5 g of AR-Grade mixed bed resin (e.g., Bio-Rad AG 501-X8D) with 100 mL of formamide for 1 h and then filter, store at –20°C (or –70°C *[5]*).
7. 70% Formamide/2× SSC. For one Coplin jar (50 mL): 5 mL 20× SSC, 10 mL of H$_2$O, 35 mL of deionized formamide, adjust to pH 7.0 with HCl. This mix can be kept at 4°C for a few days for reuse.

2.4. Posthybridization

1. 50% Deionized formamide/2× SSC: 50% v/v deionized formamide, 10% v/v 20× SSC, 40% v/v distilled water, adjust to pH 7.0 with HCl.

2.5. Autoradiography

1. Ilford Nuclear Research Emulsions from Ilford Australia, imported from Ilford, Mobberley, Cheshire, England.
2. Kodak Nuclear Research Emulsions only from Eastman Kodak, Rochester, 14650-0532, New York.
3. D19 Developer, made from powder (Kodak Code 146 4393). Store stock at 4°C.
4. Hypam Fixer (Ilford Code CP 4143): dilute with four parts of water to make to film strength.
5. Hardener (Ilford Code CP 4121): add 1 part to 39 parts of film-strength fixer.

2.6. Staining, Viewing, and Photographing

1. Hoechst 33258 (Sigma Code B 2883), stock 500 µg/mL in H$_2$O for up to 6 mo, in the dark, at 4°C; freshly dilute 1:50 with 2× SSC to 10 µg/mL.
2. Long-wavelength UV from a Sylvania Blacklite-blue, 15 W tube, 350 nm.
3. R 66 Giemsa stain (BDH Code 15427), store at 4°C.
4. Approx 0.01 M phosphate buffer, pH 6.8, from tablets (BDH tablets, Code 33199 2P), dissolve one tablet per liter of distilled water.

5. Depex (BDH Code 36125 4D): dispense from a rubber-free syringe.
6. Techpan 2415 35-mm film (Kodak Code 129 9916).
7. HC-110 developer (Kodak Code 140 8998), stock made by diluting the supplied 473 mL to 1890 mL with water. Dilution B made by freshly diluting 1 part of the stock with 7 parts of water.
8. Photoflo (Kodak Code 405 5786), dilute 1 mL in 600 mL of water, replace after about 40 films.

3. Methods

The methods described are an adaptation of those of Buckle and Craig *(6)* and Choo and Earle *(7)*. At all stages practice normal procedures of sterility and noncontamination for molecular biology.

3.1. Slide Preparations of Chromosomes

Methods of cell culture and chromosome preparation are too many and varied to be described, except to say that the author has found AmnioMax to be a most satisfactory general medium for the culture of human and other mammalian cells, with the addition of pokeweed mitogen as well as PHA as a useful step, especially for animals. It will be assumed that good-quality preparations of chromosomes are available from tissue culture, either short-term stimulated blood, splenic lymphocytes, or long-term fibroblasts. Direct preparations from uncultured cells can be used, including male meiosis *(8)*, but usually without the advantages of chromosome bands, unless the chromosomes are polytenic *(9)*.

If bands are required on the chromosomes they are best induced chemically by incorporation of 5-bromodeoxyuridine (5-BrdU) during the first half of S-phase (DNA synthesis) ultimately to produce GBG-bands (G-banding with 5-BrdU and Giemsa stain) after Giemsa staining; or into the second half of S-phase ultimately to produce RBG-bands (R-banding with 5-BrdU and Giemsa stain), the latter being the simpler method because rinsing out of the 5-BrdU is not required. Typical methodologies are as follows.

3.1.1. Slide Cleaning

1. Slides used for RISH should be precleaned and frosted at one end.
2. Load slides into racks that space them apart.
3. Soak the slides for 2 h, or overnight as a maximum, in a 5% solution of Decon 90.
4. Rinse with running tap water for at least 30 min, with some of the water leaving the bottom of the rinsing tank.
5. Rinse briefly in distilled water
6. Pass through three changes of ethanol, only the last of which needs to be new.
7. Stand vertically to dry, preferably with the ethanol running down to the frosted end.

3.1.2. Early S-Phase Incorporation of 5-BrdU into Cultured Human Lymphocytes

This method yields a high number of prophasic chromosome spreads, most of which have incorporated 5-BrdU during early S-phase, and which will show GBG-bands after appropriate staining; however, a few cells will be RBG-banded owing to the intervals of the cell cycle having a chaotic distribution.

1. Culture stimulated lymphocytes from whole blood at 37°C for 2–3 d by a preferred method.
2. Add 5-BrdU to a final concentration of 200 µg/mL and leave overnight. In most cells the 5-BrdU will be incorporated into the DNA of the faint G-bands and cause the cell cycle to stop in the middle of the S-phase.
3. Next day, rinse the cells 3× with Dulbecco's PBS, by centrifuging and discarding the supernatant.
4. Rinse once with the culture medium.
5. Resuspend the cells in culture medium with 10^{-5} *M* additional thymidine (Tdr).
6. Culture the cells for a further 6–7 h, to span the average mid-S to mitosis (mS-mit) interval.
7. Add colchicine solution for 10–20 min, only to disrupt the mitotic spindle.
8. Harvest in the common manner, using 0.075 *M* KCl hypotonic treatment, methanol:acetic acid fixation and air drying the cells onto cleaned slides (*see* **Subheading 3.1.1.**).

3.1.3. Late S-Phase Incorporation of 5-BrdU into Cultured Human Lymphocytes

This method will yield only RBG-banded chromosomes; there is no confusing admixture of both R- and G-bands, and no rinsing if **step 2** is omitted.

1. Culture stimulated lymphocytes from whole blood for 2–3 d by the preferred method.
2. Optional step: synchronize the cells in mid-S with an overnight treatment with 300–400 µg/mL of Tdr, which must be rinsed out as for a 5-BrdU arrest (*see* **Subheading 3.1.2.**).
3. Add 5-BrdU to a final concentration of 20 µg/mL and leave for a 6–7 h mS-mit interval.
4. Add colchicine solution (*see* **Subheading 3.1.2.**), 10–20 min if **step 2** is taken; otherwise 30–60 min.
5. Harvest in the common manner (*see* **Subheading 3.1.2.**).

3.1.4. Incorporation of 5-BrdU into Cultured Bovine Lymphocytes

Cattle, sheep, and deer seem to be sensitive to high levels of 5-BrdU or thymidine.

1. To produce mainly GBG-bands follow **Subheading 3.1.2.** but use an arresting dose of less than 70 µg/mL of 5-BrdU.

2. To produce RBG-bands follow **Subheading 3.1.3.** but use an arresting dose of less than 100 µg/mL of Tdr if the cells are to be synchronized in mid-S.

3.1.5. Incorporation of 5-BrdU into Cultured Mouse Splenic Lymphocytes

Mouse splenic lymphocytes seem to be tolerant to high arresting doses of 5-BrdU or Tdr, up to at least 800 µg/mL, and there appear to be differences between laboratory strains: e.g., a 200 µg/mL arresting dose of 5-BrdU produced mainly G-banded chromosomes in BALB/c mice, but in C57BL mice more than half of the cells are harlequin banded, either because they have a very short G_2 phase or because they have escaped a mid-S block. The cells do not produce any confusing R-banded chromosome sets in a GBG-banding experiment, so perhaps the cell cycle times in them are less chaotic than in larger mammals.

1. Culture mouse splenic lymphocytes stimulated with 3 µg/mL of Concanavalin A for 2–3 d.
2. Label overnight with 5-BrdU and Tdr as for human lymphocytes to produce GBG-bands (**Subheading 3.1.2.**) or RBG-bands (**Subheading 3.1.3.**), but in C57BL mice arresting 5-BrdU or Tdr doses should be about 500 µg/mL.
3. Rinse out the arresting agent and substitute with 10^{-5} *M* Tdr or 20 µg/mL of 5-BrdU as required.
4. Culture for only a further 4.5 h.
5. Add colchicine solution (**Subheading 3.1.2.**) for 10–20 min if the cells are synchronized, otherwise 30–60 min.
6. Harvest in the common manner (**Subheading 3.1.2.**).

3.1.6. Incorporation of 5-BrdU into Cultured Fibroblasts

The advantage of using attached fibroblasts is that they do not have to be centrifuged between changes of treatment. The m S-mit interval tends to be longer in fibroblasts than in lymphocytes, but it can be observed by noting the appearance of cells in mitosis after an arresting treatment, and these rounded-up cells can be shaken off the surface to give slides with a high mitotic index.

1. Establish fibroblast cultures by a preferred method.
2. Before the cells reach confluence add overnight arresting doses of 5-BrdU or Tdr as for lymphocytes of the animal in culture.
3. The next day pour off the medium and rinse twice with PBS.
4. Continue culturing with medium that is 10^{-5} *M* Tdr or 20 µg/mL of 5-BrdU as required. Check for the appearance of rounded-up cells in mitosis.
5. Add colchicine solution (*see* **Subheading 3.1.2.**) to 1.5 µg/mL and either:
6a. After 30–60 min, use trypsin to strip off the cells, or
   b. After 15 min, shake the rounded-up metaphase cells off the base of the culture flasks using a vortex mixer, centrifuge, and return medium to flask to repeat the step at least once.

c. After 15 min, add hypotonic 0.075 *M* KCl to the flask and shake the rounded-up metaphase cells off the base of the culture flasks using a vortex mixer, repeat with new hypotonic KCl, pool the suspensions, and continue hypotonic treatment for a total of 25 min.
7. Continue the harvest and make slides as for lymphocytes (*see* **Subheading 3.1.2.**).

3.1.7. Storage of Fixed Cells

Unused fixed cells can be stored in methanol–acetic fixative at –20°C, but the keeping qualities are very variable, a few being much deteriorated on the day after harvest; for this reason the author always makes up a useful supply of slides on the day of harvest. Storage of cells in pure methanol, as practiced by the chromosome microdissecters, seems to be an advantageous alternative.

1. Store fixed cells in at least 5 mL of 3:1 methanol:acetic acid fixative or methanol at –20°C.
2. Recover the cells to make further slides by centrifuging at about 3100*g* (1800 rpm, radius 165 mm). This may seem high, there are always some cells left in suspension, so the faster the better.
3. Replace the fixative or methanol with fresh fixative, and repeat.
4. Reduce to a suitable volume and make slides.

3.1.8. Storage of Prepared Slides

Mouse slides so stored have been used for RISH after 10 yr (*10*); however, human slides stored for long periods often do not band as well as comparatively new slides.

1. Label the slides with pencil on the frosted surface.
2. Pack in boxes without hinges (Kartell), with silica gel drying agent in a punctured plastic tube.
3. Seal the box with electrical tape and place at –20°C.

3.1.9. Exposure of Slides to Light

To minimize random nicking of 5BrdU-substituted DNA, the slides should be protected from strong light whenever possible, at least until after hybridization.

3.2. DNA Probes

For RISH, DNA probes should be at least 200 bp long. Such short probes should be excised from cloning vectors. If the probe sequence is left in the vector, groups of silver grains may be present in the final autoradiograph instead of single grains, as both signal and background noise, resulting in a very messy slide which is usually difficult or impossible to interpret. This clustering of silver grains seems to be related to electrostatic charge and sometimes can be controlled by using an acetylation step *(11)*. Longer probes, larger than 1 kb, can usually be left in the cloning vector, but, in this condition, some may still cause problems with clustering of grains *(12)*.

The DNA probe can be labeled by incorporating tritiated nucleotides by nick-translation, oligonucleotide primer extension, or polymerase chain reaction (PCR). A detailed method is shown for a well-tested variant of schedule D of the Amersham Nick Translation Kit, adapted for tritium labeling. An outline of a random priming method is also given.

3.2.1. Nick-Translation of Probe DNA

1. To label 200 ng of probe DNA (scale up if desired), add 12.5 µCi of each of the [^{3}H]dATP, [^{3}H]dCTP, [^{3}H]dTTP to an Eppendorf tube, wrap the top with plastic film, and pierce it several times (optional step: freeze to –70°C). The author has found it satisfactory to use equal amounts in microCuries of Amersham nucleotides with specific activities ranging from 50 to 130 Ci/mol. For more exactly equimolar proportions, use: y µCi = y ÷ SA nmol (thus, in this protocol, 0.1–0.25 nmol is used).

2. Dry down the tritiated nucleotides in a vacuum desiccator or SpinVac. This is a particularly hazardous step, with a high level of tritium emitted. Take necessary precautions.

3. Add to the dry tube:

Dried dNTPs	3×12.5 µCi
probe	200 ng
H$_2$O# to:	10.0 µL
300 µ*M* dGTP#	1.25 µL (0.375 nmol)
enzyme mix#	1.25 µL (# Amersham Nick Kit)
Total	12.5 µL

4. Incubate at 15°C for 2 h. (Optional step, designed to rejoin at least some of the breaks in the probe DNA: Add 7.5 U of DNA polymerase I to each 12.5-mL nick and continue to incubate at 15°C for a further 30 min).

5. During incubation, lightly block the neck of a Pasteur pipet with sterile nonabsorbent cotton wool and fill with a suspension of Sephadex G-50 gel in T$_{10}$E$_1$. Allow the gel to settle until a packed column is formed to within about 15 mm of the top. Rinse the column with two changes of T$_{10}$E$_1$. (Such a column will not suck air into the gel if left to drain to a stop.)

6. Layer the nick mix onto the G-50 column and allow it to run into the gel. Rinse the nick Eppendorf tube with 300–600 µL of T$_{10}$E$_1$ (depending on probe) and drive the probe into the gel by adding a small volume of this T$_{10}$E$_1$ and waiting for it to penetrate, then add the rest of the T$_{10}$E$_1$ to the column. Add measured volumes of T$_{10}$E$_1$ to collect the fractions: Eppendorf 1, 300–600 + nick µL; Eppendorfs 2–12, 100 µL; Eppendorfs 13–16, 500 µL.

7. Eppendorf 1 should contain only background counts. Add 2-µL samples of fractions 2–16 to 200 µL of water in scintillation counting tubes. Add 2 mL of a

water-miscible scintillation fluid and mix well. Count, using a tritium channel for a maximum of 1 min.

There should be two peaks in the readout from the counter: the first is the probe DNA, which runs out in the void volume of the column; the second is the unincorporated bases, which are retarded in the pores of the gel beads. The low point between the peaks can be either, a single tube, which is eliminated from calculations, or two tubes that count roughly equally, one being included in the probe counts and the other with the unused bases.

8. Determine which tubes contain the labeled probe and pool them along with one or two more tubes containing the tail of the probe peak under the first of the bases.

9. For tubes 13–16, correct for volume by multiplying by 5 and then add up the counts of the probe and the total of all single and corrected counts. Calculate (a) the fraction of the original radioactivity incorporated in the probe, and (b) the specific activity of the probe, thus:

a. Fraction incorporated $= \dfrac{\text{Total Counts in Probe}}{\text{Total counts (probe plus unincorporated bases)}}$

b. $\dfrac{\text{Specific activity}}{\text{(DPM/μg)}} = \dfrac{\text{Total μCi in reagent bases} \times \text{fraction incorporated} \times 2.22 \times 10^6}{\text{Mass of probe in μg}}$

where 1 μCi $= 2.22 \times 10^6$ disintegrations/minute (DPM).

A fraction incorporated of 0.25 giving specific activity of 1.0×10^8 dpm is satisfactory label incorporation, but do not be discouraged if the specific activity is less than 10^7 DPM, the author has published results from similar poor incorporation *(13)*.

3.2.2. Oligo-Labeling of Probes

1. Using a suitable kit, follow the instructions for first mixing the entire contents of the tubes containing lyophilized random primer to the dGTP/buffer cocktail.

2. Dry down 5 μCi (or scaled-up amounts) of each of the labeled nucleotide triphosphates, [^{3}H]dATP, [^{3}H]dCTP, [^{3}H]dTTP to an Eppendorf tube (Epp), or equimolar amounts, as for a nick-translation (**Subheading 3.2.1.**).

3. Dilute the probe DNA to be labeled to 7 μg/mL (7 ng/μL) with H_2O.

4. Warm 11.5 μL of the DNA solution to 95°C for 2 min. Chill on ice thoroughly and centrifuge to settle condensation.

5. Add the 11.5 μL of DNA solution, containing 80.5 ng of the probe DNA.

6. Add 12.5 μL of random primer/dGTP/buffer cocktail. Vortex gently.

7. Add 1.0 μL of the DNA polymerase I from the kit and gently tap the tube to mix.

8. Incubate at room temperature overnight for best incorpation, or at 40°C for 20 min.

9. Run a G-50 column, collect fractions and count them, as for a nick-translation (**Subheading 3.2.1.**).

10. Pool the tubes containing the probe, plus one or two more (**Subheading 3.2.1.**).
11. Calculate the specific activity of the probe DNA (**Subheading 3.2.1.**).

3.2.3. Precipitation of Probes

3.2.3.1. PROBES WITH NO REPEAT SEQUENCES AND PROBES THAT ARE REPEATED SEQUENCES

1. To the solution of the probe, add sheared, denatured herring or salmon sperm carrier DNA, usually to a final 1000× the concentration of the probe DNA, although 2000×–3000× can be used to control background *(14)*.
2. If required by the experimental design, fractionate the sample at this stage.
3. Alcohol-precipitation of the carrier and probe (mixed) DNA. For volume V:
   a. Add 1/10 vol V of 3 *M* sodium acetate, pH 5.2.
   b. Either add 2.75 vol V ethanol, preferably at –20°C, or add 0.6 vol V isopropanol, at –20°C.
   c. Place at –20°C for at least 1 h. The mixed DNA can be stored at at –20°C this stage.

3.2.3.2. UNIQUE-SEQUENCE PROBES CONTAINING REPEAT SEQUENCES

Generally, probes of this type will be greater than 1000 bp long, and, being suitable for FISH, it would now be unusual to use them for RISH.

1. To the solution of the probe, add sheared, denatured herring or salmon sperm carrier DNA to a final concentration of 500× the mass of the probe DNA. The making of an unsuppressed slide is recommended; for this, keep enough of this mixed DNA.
2. To most of the mixed DNA from the previous step, add Cot-I DNA to a final 50–300× the concentration of the probe. The amount of Cot-I DNA should be increased stepwise until suppression of signal from repeated DNA sequences in the probe is observed to be effective.
3. If required by the experimental design, fractionate the sample at this stage.
4. Alcohol-precipitate and store the mixed DNA, as in **Subheading 3.2.3.1**.

3.3. Application of Probes to Slides

All treatments and rinses are carried out in Coplin jars, the slides transferred with fine forceps. For large-scale experiments larger dishes of reagent or rinsing solution can be used to treat slides carried along in racks. Agitation of rinses is usually continuous on a mechanical shaker. Solutions are at pH 7.0 unless otherwise specified. It is usual with RISH to process about five times more slides than for an equivalent FISH experiment; the extra slides are not expensive to produce and, after hybridization, there is much less slide handling for RISH than for FISH. The extra slides are useful for exploring the exposure time of autoradiography and the vagaries of Giemsa staining.

3.3.1. Labeling Slides

1. Warm the unopened box of slides to room temperature. Replace the drying agent before repacking unused slides.
2. Label each slide, using a diamond pencil, with the date and a number.
3. Make the slide numbers visible by overwriting with a carbon pencil.

3.3.2. RNase Treatment of Slides

This step seems to be traditional rather than necessary, because most genes do not transcribe amounts of RNA that are readily detectable. It certainly can be omitted if the probe is nontranscribing. It was becoming fashionable to omit RNase treatment for RISH when kits for FISH reintroduced it. For RISH, DNase treatment may slightly improve the final GBG- and RBG-banding (**Subheading 3.6.**).

1. Treat slides with RNase, 100 µg/mL in 2× SSC for 1 h at 37°C.
2. Rinse the slides in four changes of 2× SSC, 2 min each, at room temperature, with agitation.
3. Dehydrate slides through an ethanol series: approx 2 min each, with agitation, in 35%, 70%, 95%, and 100% ethanol.
4. Air dry by draining toward the frosted handling end.
5. Slides may be stored at this stage (**Subheading 3.1.8.**).

3.3.2. Acetylation of Slides

See **ref. 11**. Acetylation is not required if insert is used as a probe, but is very useful if plasmid is used as a probe. Results are variable; treat only a portion of the available slides.

1. Place the slides in 500 mL of 0.1 *M* triethanolamine-HCl. Add 2.5 mL of acetic anhydride drop by drop, stirring continuously. Leave still for 10 min.
2. Rinse slides in four changes of 2× SSC, dehydrate through an ethanol series, air dry, and perhaps store the slides (**Subheading 3.1.8.**).

3.3.4. Preliminary to Denaturation and Hybridization

Make up to 20% w/v dextran sulfate in deionized formamide and heat to 70°C. Mix frequently over the next few hours, using a vortex mixer.

3.3.5. Denaturation of Chromosomal DNA

1. Make up a denaturation solution of 70% formamide/2× SSC in a plastic Coplin jar. Heat in a water bath to 70°C.
2. Preheat the slides to about 70°C, in an empty plastic Coplin jar.
3. Denature the DNA on the slides by immersing them in the denaturation solution for 2 min, with frequent agitation.
4. Stop snapback of the DNA by immersing the slides in cold 70% EtOH for 2 min, with frequent agitation. The 70% ethanol is kept at −20°C and the Coplin jars

containing it are surrounded by ice. Use only one jar of cold 70% ethanol per five slides.

5. Dehydrate the slides through 70% ethanol (room temperature), 80, 95, and 100%, and air dry (**Subheading 3.3.2.**). Some protocols allow storage of the slides at this stage (**Subheading 3.1.8.**).

3.3.6. Preparation of Hybridization Mix

Most of the author's RISH has been done with probes at 200 ng/mL, because of early warnings about background produced by higher probe concentrations. However, after some experience with the much higher probe concentrations used for FISH, probes have recently been applied for RISH at 400 ng/mL, and they could probably go much higher. Rather than risk increased background, the author has to date usually preferred to accumulate the signal from RISH with the appropriate passage of time.

For the steps that follow, "mixed DNA" refers to the labeled probe plus carrier and Cot-I DNA; the "probe" is referred to as such.

1. Centrifuge the alcohol-precipitated mixed DNA(s) for 15 min at 17,000g (maximum speed of a microfuge).
2. Pour off the supernatant and check that a pellet has been left behind. Mop as much alcohol as possible from the mouth of the Eppendorf tube.
3. Rinse the pellet with 70% ethanol at –20°C, centrifuge for 15 min at 17,000g, pour off the supernatant, and mop off alcohol as in the previous two steps. (This step is advisable, to remove excessive salt from the DNA, but the author has found it to be optional.)
4. Dry the mixed DNA at room temperature or briefly at 37°C. If the tube contains the exact amount of mixed DNA for hybridization then leave it dry.
5. If required by the experimental design, resuspend the mixed DNA in H_2O at a concentration of 5 ng/mL of the probe. Preferably leave overnight on a shaker, or heat to 70°C for a few minutes and vortex mix.
6. The hybridization mix should contain the following, added in the order A–C:
   A. 3 vol, mixed DNA from **Subheading 2.3.5.**, **step 4**, to yield final concentration of the probe DNA in the hybridization mix as:
      200 ng/mL: 0.4 vol of 5 ng/μL probe in mix plus 2.6 vol pure water
      400 ng/mL: 0.8 vol of 5 ng/μL probe in mix plus 2.2 vol pure water
   B. 2 vol, 10X SSCP.
   C. 5 vol, 20% dextran sulfate in deionized formamide at 70°C.
7. The actual volumes used can be calculated from the amount required to be placed on each slide under coverslips of different sizes (*see* **Note 2**).
8. If a measured amount of the probe DNA has been left dry (**Subheading 3.2.4.**), then simply add the necessary volume of water at **step 6A**, heat briefly to 70°C, and mix well; e.g., for 20 ng of probe DNA to be used in 100 μL of hybridization mix, add 30 μL of water.

3.3.7. Denaturation of the Probe DNA

1. Homogenize the hybridization mix by heating to 75°C for 1 min in a water bath. Vortex-mix and centrifuge.
2. To denature the probe DNA heat at 75°C for a further 10 min.
3. For probe mixes with no Cot-I DNA: cool quickly on ice.
4. For probe mixes containing Cot-I: transfer to 37°C for 60 min, then to ice. (Most protocols recommend only 20 min annealing of the Cot-I, but 60 min is the theoretical optimum.)

3.3.8. Hybridization

1. Clean the required number of coverslips by dipping them in 70% ethanol and wiping them dry.
2. Place the required volume of hybridization mix to suit the size of coverslip (see **Note 2**) on each slide and add the coverslip, leaving a 3–4-mm gap at the nonfrosted end (where high-power objective lenses cannot safely reach and the emulsion will be too thick).
3. Seal the coverslip with generous quantities of a rubber cement, containing a hydrocarbon solvent that is not miscible with water. Allow the cement to dry, for swiftness at 37°C.
4. Incubate the slides flat in a slightly humid environment (e.g., plastic slide box with lightly dampened paper on the bottom) overnight for a maximum of 16 h, at 37°C for short unique probes, 42°C for long (+1500 bp) or repetitive probes.

3.4. Posthybridization

3.4.1. Stringency Rinses

For most purposes these are best done at a stringency similar to that used for hybridization. The steps below are made slightly more stringent than for hybridization only by use of a slightly elevated temperature. For probes consisting of repetitious DNA the stringency of rinsing might be raised, by use of a higher temperature or by diluting some of the final rinses to 0.1–1.0× SSC; for very short probes it might be lowered. However, the author has only rarely varied the following steps:

1. Make up 200 mL, or more, of 50% deionized formamide/2× SSC and add 40–50 mL to at least four Coplin jars.
2. Prewarm, to 2°C above the hybridization temperature, all but one of the three Coplin jars containing 50% deionized formamide/2× SSC, plus five jars containing 2× SSC. Prepare enough jars to renew the first one of each type every 20 slides.
3. Remove the rubber cement from the slides using fine forceps and lift off the coverslips.
4. Rinse briefly in the Coplin jar containing 50% deionized formamide/2× SSC at room temperature, from **Subheading 3.4.1**.

5. Rinse the slides for 3 min, with agitation, in each of the eight Coplin jars from **Subheading 3.4.1**.
6. Dehydrate through 35%, 70%, 95%, 100% ethanol, 2 min each, and dry. Store, if required, at –20°C with silica gel drying agent. Warm to room temperature before opening.

3.5. Autoradiography

An excellent general reference for autoradiographic techniques is Rogers *(15)*; much of the autoradiographic methodology is taken from this book.

3.5.1. Darkroom

Autoradiography should preferably be performed in a clean darkroom. If the room is used for general photography, take steps to protect the work from chemical contamination. Safelighting should be with an Ilford 914 filter in front of a 15-W globe, at least a meter away from bulk emulsions, and preferably further from coated slides. (However, the author has exposed a coated slide taped to the safelight filter for about 30 min without detectable fogging.) The steps to be done under safelighting are shown below with bold/underlined numbers.

3.5.2. Nuclear Research Emulsions in Liquid Form

Ilford L4 emulsion is the best for chromosomal *in situ* hybridization. It has an average undeveloped AgBr crystal diameter of 0.14 µm and a sensitivity rated as S11, i.e., 11 silver grains formed per 100 tritium disintegrations *(16)*. Ilford actually recommend emulsion K2 for recording tritium, but it has a crystal diameter of 0.20 µm and sensitivity of only S1, so would require 11 times the exposure time of the autoradiographs compared with L4. K5, with a similar grain size to K2 but sensitivity S7, would be preferred if larger grains are required without a large increase in exposure time. Larger grains result in a small fall in resolution *(15)*, but the author is considering changing to K5 as he gets older, with consequent reduction of visual acuity. Eastman Kodak emulsions are not as versatile as the Ilford ones. The Kodak NTB series, for light microscopy, have a mean crystal diameter of 0.26 µm, but with unfortunate wide variation of this size, in the author's experience. The sensitivity of NTB is only S4. NTB2 would be more sensitive and is recommended for recording tritium. The Kodak equivalent of L4, NTE, has a crystal diameter of 0.07 µm: impossibly small for light microscopy.

Bulk emulsion must be carefully stored at 4°C, the formation of ice crystals will spoil it. Occasionally new batches of emulsion have unacceptable background, so they should be tested for this using a blank slide. L4 can be stored for many months but should be tested for background before use if older than about 3 mo. Allowance should also be made for a drop in sensitivity of old emulsion, as with all old photographic materials.

3.5.3. Preparing Emulsion

1. Warm the emulsion package to room temperature and mark a 50-mL plastic screw-capped tube at the 15- and 30-mL levels.
2. Using a stirring water bath that has all illuminated lights and dials covered, preheat 15 mL of 1% glycerol in H_2O to 46°C in the marked tube. Glycerol is the plasticizer used in the emulsion and the concentration of it should be maintained.
3. Place a five-slide plastic mailer in a totally immersed Coplin jar in the water bath. For preheating the slides, place a glass Coplin jar at a higher level so that the water in the bath cannot enter it.
4. Under safe-lighting, using a small plastic spoon, carefully remove shreds of emulsion from the brown glass bottle in which it is delivered and add it to the warm glycerol solution until the volume reaches the 30-mL mark.
5. Allow the shreds of emulsion to melt into the glycerol solution at 46°C. Gently mix the diluted emulsion until it looks homogeneous and coats the inside of the tube evenly.
6. When fully melted and mixed, pour the 1:1 diluted emulsion into the slide mailer.
7. The diluted emulsion may be stored in the 50-mL container, protected from light, at 4°C. It can be safely remelted for use about five times.

3.5.4. Dipping and Exposing Slides

1. If required, dip some slides to test the background in the emulsion. Process these as described in **Subheading 3.5.5.** and check them under the microscope before proceeding. If available, test slides should have some material on them, e.g., old chromosome slides, this helps to find the focal plane.
2. Preheat two pairs of slides back-to-back in the dry Coplin jar.
3. Dip two of the slides, back-to-back, in the emulsion, by inserting the slides into the mailer diagonally to its square mouth, avoiding the internal flutes.
4. Hold the slide in the emulsion for about 5 s, then remove it slowly and evenly, allowing the last drips of emulsion to return to the mailer.
5. Drain the slides briefly on multiple layers of paper towel; wipe emulsion from the back of the last slide if an odd number is dipped.
6. Place each slide on two glass rods on a level rack that preferably can be covered and chilled to 10–15°C below room temperature with iced water; but if the room has a high humidity, avoid the formation of condensation droplets on the slides. A suitable rack is illustrated by Rogers (*15*), it allows the emulsion to set before the slides are hung vertically, avoiding background due to physical damage that occurs if half-set emulsion is allowed to run.
7. Hang the slides vertically using pegs or clips to transfer them to a rod. Make sure that the last slides on the rack have set.
8. Transfer the slides on the rod a drying box to dry in a forced air flow for at least 30 min. For convenience it is best if this dryer is light tight.

9. Transfer slides to small black plastic slide boxes (Clay Adams) equipped with silica gel drying agent in a pierced plastic tube. Seal the boxes with black electrical tape. Rogers *(15)* suggests that the humidity for exposure should be 45%, but this is difficult to organize and total dryness seems to be suitable.

10. Expose at 4°C, to assist the preservation of latent images. Usually expose for 2–6 wk, but often much longer, depending on the specific activity of the probe, age of the emulsion, and length of the probe (and target) DNA. Use trial slides to determine a satisfactory time.

3.5.5. Developing and Fixing Slides

1. Allow the box of slides to warm to room temperature and transfer those to be processed to a suitable rack, preferably double-spaced.
2. Dilute Kodak D19 Developer with an equal volume of water, mix, and warm to exactly 20°C.
3. Immerse the slides in the diluted developer and agitate briefly, to disperse bubbles adhering to the slides
4. Cover so that slides are in total darkness and develop for 5 min, without further agitation.
5. Drain off developer and immediately fix the emulsion with Ilford Hypam Fixer, film strength with hardener, for 5 min, with agitation. Clearing time is only a few seconds, so this step is light safe after 1 min. Avoid the common step of rinsing slides after developing them, it can cause the emulsion to lift off the surface. Because the fixer is contaminated, do not reuse it.
6. Rinse five times in distilled water, 2 min or more each, with agitation. Do not allow the slides to dry before staining.

3.6. Staining, Viewing, and Photographing

The basis of GBG- and RBG-banding is that DNA that has incorporated 5-BrdU is more sensitive to damage by UV light than unsubstituted DNA. If the Hoescht dye, 33258, is intercalated into the DNA it becomes sensitive to long-wavelength (350 nm) UV and differentiation by the subsequent staining with Giemsa stain is improved (*see* **Note 1**).

3.6.1. Staining for GBG- and RBG-Banding

1. Sensitize slides in a Coplin jar for 30 min at room temperature in 10 μg/mL Hoescht 33258 in 2× SSC.
2. Rinse briefly, twice in 2× SSC.
3. Lay slides in a flat container, just cover them with 2× SSC and expose to 350 nm UV light for 1 h at a distance of 10 cm. Make sure that the slides do not dry.
4. Dilute R 66 Giemsa stain to 8-15% v/v in phosphate buffer, pH 6.8, and filter immediately before use through a Whatman No. 4 paper.
5. Stain the slides in a Coplin jar for 20 min.
6. Rinse twice, very briefly, in the pH 6.8 buffer.

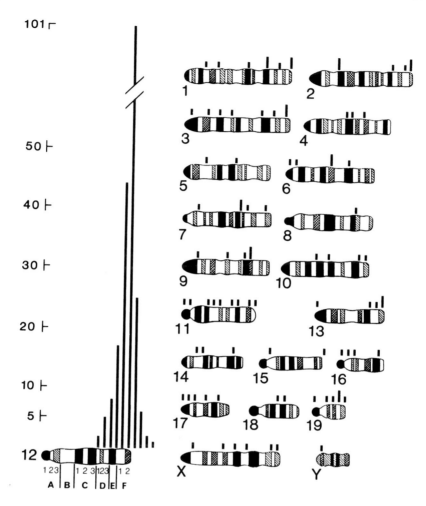

Fig. 2. Scores of silver signal grains over an idiogram of the mouse karyotype generated by *in situ* hybridization of a 10.5-kb probe from a flanking region of the IgH genes *(8)*. The long probe is very specific for the distal bands of chromosome 12, with a peak over band 12F; note the low background.

7. Stand vertically to drain to the frosted end, and dry. The slides may be kept in this state at room temperature until ready to view.

3.6.2. Mounting and Restaining

1. To check the success of staining, use microscopal immersion oil to mount a coverslip, for an indefinite length of time.
2. If the staining or banding is unsatisfactory, lift off the coverslip and rinse the slides in two changes of 100% ethanol, 5 min each, with agitation.

3. Soak the slides in 80% ethanol overnight to remove the stain.
4. Return the slides to 100% ethanol and dry.
5. Restain as in **Subheading 3.6.1.**, making adjustments to improve density of staining. Generally the contrast of the banding will be improved by restaining and a tendency to produce annoying stained flecks in some batches of emulsion will be eliminated by the alcohol treatments. However it is difficult to beat initial good GBG-banding (**Fig. 1**).
6. For permanent mounting, soak off the immersion oil with xylenes and mount a coverslip with Depex.

3.6.3. Photography

Black-and-white photography in use for general cytogenetics can be used to capture images of RISH, provided that contrast of the final image is not too high, e.g.,

1. Load the camera with Kodak Techpan and set the camera control box for ASA 100.
2. Adjust the microscope condensor to Kohler illumination and its iris to the correct numerical aperture, usually wide open for an oil immersion objective.
3. Close down the field stop to limit stray light.
4. Depending on the intensity of staining, choose a blue or pale green filter. If in doubt try both.
5. Focus mainly at the level of the signal grain(s). Take several shots, going from the chromosomes in focus, through average focus, to grains only in focus
6. Develop the film with Kodak HC-110 developer at dilution B, for 8 min at 20°C, shaking by inversion continually for the first 30 s then two inversions every 30 s.
7. Rinse twice with tap water.
8. Fix for 3 min, with agitation, with Ilford Rapid Fixer (**Subheading 3.5.5.**).
9. Rinse in running water for 10 min, briefly in Kodak Photoflo 1:600, hang, sponge off excess water, and dry.
10. Print at contrast 2–4.

3.7. Scoring

The author is in the habit of scoring to high signal peaks, and may hold the record, of 289 grains *(17)*; high scoring can authenticate subpeaks where other workers have failed to do so *(18,19)*.

1. To restrict observer bias, metaphase spreads selected under low power (10× objective lens) should be scored under high power for all silver grains present unless this is impossible.
2. First enter the grains onto an idiogram of the chromosomes that shows the subbands and a few of the larger sub-subbands of the karyotype in question. This will establish a picture of signal to noise (**Fig. 2**); usually a figure of this plot is not published. Score to at least 30 grains in the tallest column.

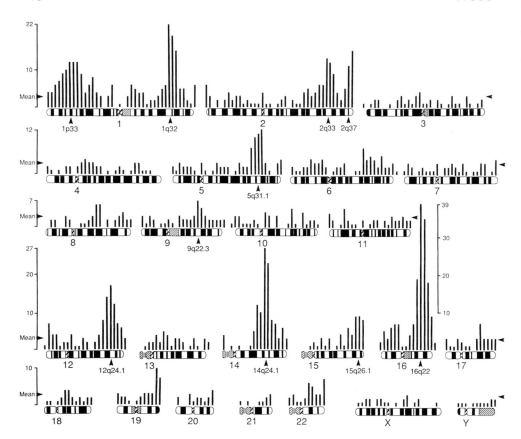

Fig. 3. The pattern of silver grains produced by *in situ* hybridization of a short (274 bp) cDNA of the gene encoding human early pregnancy factor/chaperonin 10 *(1)*. Although 1517 grains have been plotted, the mean grain density is only 2.9. The plot is one of the most complex on record, but scoring to this plot, and a control one, was feasible because of the use of RISH, which was also essential for mapping the short probe.

3. Having established the whereabouts of the peak(s) of signal, score to more detailed idiograms of the target chromosome(s), preferably from prophases. The target chromosome(s) in prophase are often identifiable, initially from their labeling, and there is the luxury of avoiding any further scoring for background. Score to at least 50 grains in the tallest collumn.

4. Draw the idiogram(s) required for publication using a suitable program, e.g., Claris McDraw II (*[3]* and **Fig. 1**). The idiogram of the main peak of signal is usually the only one published, but sometimes it is advisable to show the idiogram of all of the chromosomes where multiple peaks due to the presence of multiple genes and pseudogenes have been established (*[1,19]* and **Fig. 3**).

4. Notes

1. Unfortunately only the successor of the British Gurr's R66 Giemsa stain works through emulsion to produce GBG-bands. After three takeovers, R66 is now made by BDH/Merck and unfortunately not all batches of it can be used for RISH. The author has worked since 1994 with Australian Batch 15427, purchased with the advice from BDH England that it had been found to produce good GBG-banding by American workers.
2. Suitable volumes of probe for different coverslips are:

Coverslip (mm^2)	Volume hybridization mix, µL
22×22	10
22×40	25
22×50	40
24×50	50

References

1. Summers, K. M., Murphy, R. M., Webb, G. C., Peters, G. B., Morton, H., Cassidy, A. I., and Cavanagh, A. C. (1996) The human early pregnancy factor/chaperonin 10 gene family. *Biochem. Mol. Med.* **58**, 52–58.
2. Webb, G. C., Earle, M. E., Merritt, C., and Board, P. G. (1988) Localization of human alpha-one acid glycoprotein genes to 9q31 to 9q34.1. *Cytogenet. Cell Genet.* **47**, 18–27.
3. Finch, J. L., Webb, G. C., Evdokiou, A., and Cowled, P. A. (1997) Chromosomal localization of the human urothelial "tetraspan" gene, UPK1B, to 3q13.3–21 and detection of a *Taq*1 polymorphism. *Genomics* **40**, 501–503.
4. Kamei, M., Campbell, H. D., Webb, G. C., and Young, I. G. (1998) SOL, a human homologue of the *Drosophila melanogaster* small optic lobes gene, is a member of the Calpain and zinc-finger gene families and maps to human chromosome 6p13. 3 near CATM (cataract with micropthalmia). *Genomics* **51,** 197–206.
5. Sambrook, J., Fritsch, E. F., and Maniatis, T. (1989) *Molecular Cloning: A Laboratory Manual*, 2nd ed., Cold Spring Harbor Laboratory Press, Cold Spring Harbor, NY.
6. Buckle, V. J. and Craig, I. W. (1986) *In situ* hybridization, in *Human Genetic Diseases: A Practical Approach* (Davies, K., ed.), IRL, Oxford, England, pp. 85–100.
7. Choo, K. H. A. and Earle, E. (1994) Radioactive *in situ* hybridization to replication-banded chromosomes, in *Methods in Molecular Biology*, vol. 33: *In Situ Hybridization Protocols* (Choo, K. H. A., ed.), Humana Press, Totowa, NJ, pp. 147–158.
8. Webb, G. C. and Fabb, S. A. (1985) Probing murine male meiosis using unique DNA flanking the immunoglobulin heavy chain genes. *Cytobios* **43**, 159–165.
9. Bedo, D. and Webb, G. C. (1990) Localization of 5S RNA genes in polytene chromosomes of *Lucilia cuprina* (Diptera: Calliphoridae). *Genome* **33**, 941–943.
10. Chapman, G., Remiszewski, J. L., Webb, G. C., Schulz, T. C., Bottema, C. D. K., and Rathgen, P. D. (1997) The mouse homeobox gene, *Gbx-2*: genomic organization and expression in pluripotent cells in vitro and in vivo. *Genomics* **46**, 223–233.
11. Pardue, M. L. (1985) *In situ* hybridization, in *Nucleic Acid Hybridisation: A Practical Approach* (Hames, B. D. and Higgins, S. J., eds.), IRL, Oxford, England, pp. 179–202.

12. Webb, G. C., Coggan, M., Ichinose, A., and Board, P. G. (1989) Localization of the coagulation factor XIII B subunit gene (F13B) to chromosome bands 1q31–32.1 and restriction fragment length polymorphism at the locus. *Hum. Genet.* **81**, 157–160.

13. Evdokiou, A., Webb, G. C., Peters, G. B., Dobrovic, A. N., O'Keefe, D. S., Forbes, I. J., and Cowled, P. A. (1993) Localization of the human growth arrest-specific gene (GAS1) to chromosome bands 9q21.3–q22, a region frequently deleted in myeloid malignancies. *Genomics* **18**, 731–733.

14. Webb, G. C., Baker, R. T., Fagan, K., and Board, P. G. (1990) Localization of the human UbB polyubiquitin gene to chromosome bands 17p11.1–17p12. *Am. J. Hum. Genet.* **46**, 308–315.

15. Rogers, A. W. (1979) *Techniques of Autoradiography*. Elsevier, Amsterdam.

16. Ron, A. and Prescott, D. M. (1970) The efficiency of tritium counting with seven radioautographic emulsions, in *Methods in Cell Physiology*, vol. IV (Prescott, D. M., ed.), Academic Press, New York, Chap. 11.

17. Board, P. G. and Webb, G. C. (1987) Isolation of a cDNA clone and localisation of human glutathione *S*-transferase-2 genes to chromosome band 6p12. *Proc. Natl. Acad. Sci. USA* **84**, 2377–2381.

18. Board, P. G., Webb, G. C., and Coggan, M. (1989) Isolation of a cDNA clone and localization of the human glutathione S-transferase 3 genes to chromosome bands 11q13 and 12q13–14. *Ann. Hum. Genet.* **53**, 205–213.

19. Webb, G. C., Parsons, P. A., and Chenevix-Trench, G. (1990) Localization of the gene for human proliferating cell nuclear antigen/cyclin by *in situ* hybridization. *Hum. Genet.* **86**, 84–86.

4

Native Polytene Chromosomes
of *Drosophila melanogaster*
for Light and Electron Microscopic Observation
of the Conformation and Distribution of Molecules

Ronald J. Hill and Margaret R. Mott

1. Introduction

In the 1930s, the discovery of a simple method for the isolation and detailed microscopic observation of the banded structures that lie within the nuclei of salivary gland cells of *Drosophila melanogaster* was soon followed by the realization that these structures were in fact a highly amplified form of interphase chromosomes *(1)*. The method involved squashing the salivary gland, immersed in 45% acetic acid, between a coverslip and a microscope slide. Aqueous acetic acid dissolves the cell and nuclear membranes and generally disperses cellular contents except for the chromosomes which are toughened by "acid fixation." The acid-squashing procedure has, in general, served as the basis for the isolation of *D. melanogaster* polytene chromosomes for cytological study since that time. It is ideal for the rapid preparation of salivary gland chromosomes for morphological observation in the light microscope and as targets for *in situ* hybridization (see, however, **Note 1**).

However, when it comes to studies that require more than morphological integrity at the level of chromomeres (bands) in the light or electron microscope, or that require integrity of higher levels and orders of macromolecular organization, exposure to 45% acetic acid (apparent pH 1.6) can obviously lead to unwanted complications. For example, at such a low pH nucleosomes are disrupted *(2)*, the strands of DNA separate *(3)*, chromosomal proteins are extracted *(4–6)*, and chemical changes can occur in both proteins *(7)* and DNA *(8)*. While protein extraction can be countered by crosslinking with aldehydes

From: *Methods in Molecular Biology*, Vol. 123: In Situ *Hybridization Protocols*
Edited by: I. A. Darby © Humana Press Inc., Totowa, NJ

(9,10), it may not always be a simple matter to balance two opposing chemical and physical influences to maintain the protein distribution in vivo without, eg., perturbing equilibrium distributions between nucleoplasm and chromosomes *(11)*. We have found that aspects of chromosome ultrastructure involving the organization of nascent ribonucleoprotein particles appear to be damaged in acid-fixed chromosomes *(12,13)*. Moreover, it is clearly advantageous to avoid extremes of pH if one wishes to detect biological activity of native proteins or examine such processes as chromatin decondensation and gene activation in vitro.

This chapter describes in detail a microsurgical procedure that has been under development for some years *(14)*. It now allows the routine isolation of cytologically mappable spreads of *D. melanogaster* salivary gland polytene chromosomes as native chromatin, in which higher order molecular conformations have not been disrupted by exposure to low pH. Some examples of the use of such native polytene chromosome preparations are given and possible future applications considered.

2. Materials
2.1. Culturing of D. melanogaster
1. Ethanol/acetic acid agar: 400 mL of water, 9 g of agar, 15 g of treacle, 23 mL of absolute ethanol, 2.7 mL of glacial acetic acid.
2. Yeast glucose medium: 130 g of compressed yeast, 100 g of glucose, 17 g of agar, 30 mL of 10% nipagen in ethanol and 1000 mL of water tinted faintly green with food coloring.
3. Fresh yeast paste: Slowly stir approx 10 mL of water into 30 g of fresh compressed baker's yeast and allow to equilibrate at room temperature. The consistency of the paste should be firm enough to not run but be at the point where an additional drop of water noticeably lowers the viscosity.

2.2. Equipment, Tools, and Solutions for Microsurgery
1. Tungsten needles. Prepare these by sealing a 5 cm length of 0.4 mm diameter tungsten wire into a length of 7 mm diameter Pyrex glass capillary tubing by briefly melting the glass in an oxygen–natural gas flame. Sharpen the needles by dipping repeatedly into molten sodium nitrite in a nickel crucible until observation under the dissecting microscope at ×16 magnification reveals a fine needle-shaped tip as seen in **Fig. 1**.
2. Glass needles. Prepare these on a David Kopf needle puller or equivalent from Pyrex melting-point capillaries. Affix the needles to pulled out Pyrex-rod "handles" (diameter 5 mm attenuated to 2 mm near the tip) by heating the tip to a red heat in a Bunsen burner, removing it for a few seconds, and then touching it to the needle approx 8 mm from the point to produce the configuration shown in **Fig. 2**. Needles employed for transport of nuclei are prepared in the usual fashion

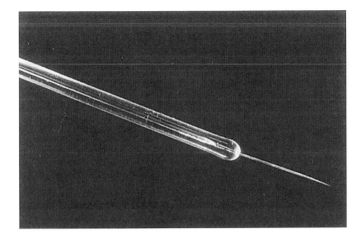

Fig. 1. Sharpened tungsten wire needle set in a Pyrex glass handle.

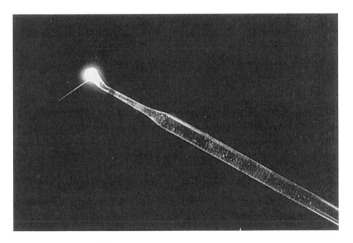

Fig. 2. Fine glass needle affixed to attenuated Pyrex rod for mounting in micro-manipulator.

except that their points are subsequently rounded on a microforge or by brief exposure to the proximity of a microflame.
3. "Well-slide" for microsurgery (**Fig. 3**). This is a modification of a chamber origi-nally designed by J. G. Gall for observation of lampbrush chromosomes as de-scribed in **ref. *16*** and consists of a 76 mm × 25 mm microscope slide through which a central 6 mm or 10 mm diameter hole has been bored. Such slides are cleaned in chromic acid and siliconized lightly. Seal a 22 mm diameter No. 1 round coverslip, previously cleaned in 95% ethanol, across the central hole on

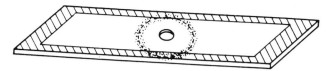

Fig. 3. Oblique view of well-slide modified for dissection of salivary gland nuclei. The floor of the well is provided by a round coverslip attached to the lower surface of the slide with paraffin wax and sealed around the circumference with rubber cement (*stippling*). A Parafilm border on the upper surface of the slide is indicated by *hatching. See* **Subheading 2.2.** for details.

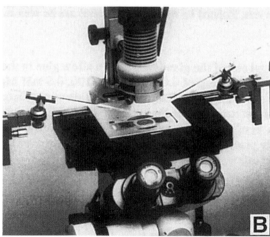

Fig. 4. (**A**) Overall view of the apparatus for nuclear microsurgery. Left- and right-hand operated Leitz micromanipulators are mounted on a vibration-free table to access the stage of a Zeiss inverted microscope. (**B**) Close-up view of the configuration of instrument-holders, needles, and well-slide on the stage of the microscope.

the bottom of the slide with paraffin wax to provide a floor to the well. Seal the edge of the coverslip with rubber cement to prevent leakage during washing and long-term storage of chromosome preparations. Heat-seal with the aid of a small flame a border of parafilm, to the top surface of the slide to prevent spillage of fluid during microdissection.

4. Micromanipulation apparatus (**Fig. 4**). Mount two Leitz mechanical microman-ipulators on blocks on a vibration resistant table to access a Zeiss inverted micro-scope with ×16, ×25, ×40, and ×60 phase-contrast objectives, a long working distance condenser as used for examination of tissue-culture flasks, and both ×10 and ×16 eyepieces. The configuration of needle-holders and needles relative to the well-slide in place on the microscope stage is illustrated in **Fig. 4**.

5. Medium of Cohen and Gotchel *(17)*: 25 mM Na glycerophosphate, 10 mM KH$_2$ PO$_4$, 3 mM KCl, 10 mM MgCl$_2$, 3 mM CaCl$_2$, 160 mM sucrose, pH 6.8.

6. Saline D (based on D'Angelo's saline, **ref. *18***): 90 mM KCl, 60 mM NaCl, 1 mM CaCl$_2$, 5 mM sodium phosphate, pH 7.0.

7. Buffer A (Hewish and Burgoyne, **ref. *19***): 60 mM KCl, 15 mM NaCl, 0.15 mM spermine, 0.15 mM spermidine, 15 mM Tris-HCl, pH 7.4

2.3. Equipment, Tools, and Solutions for Electron Microscopy

1. 3% Glutaraldehyde in 0.1 M phosphate buffer, pH 7.2, made up from E. M. grade 50% in water in an ampoule under inert gas (Fluka).

2. Holders for examining the surface of the specimen block to identify chromosomes and for trimming of the block to give a pyramid and for subsequent sectioning.

   For hand trimming with fine files and single-sided razor blades, an LKB Ultratome chuck holding the peg with the chromosomes on the top surface is held by a snap lock in a circular trimming holder, fitting into a rectangular clamping plate that fits across the stages of the dissecting microscope and Ultratome.

   For ultramicrotome trimming with 55° glass knives, the chuck with the speci-men is held in the removable LKB Ultratome orienting head which also fits the rectangular clamping plate. The head can be tilted so that the minute 0.1 mm^2 chromosome spread can be seen by either focused or diffused light.

   With these devices and the photographs of the chromosomes, the block can be trimmed so that the area just surrounds the whole spread and the chromosomes and their landmarks identified to ensure that the whole face of the block and therefore the entire chromosome spread are sectioned at first stroke.

3. Epoxy-propane for removing last traces of alcohol when embedding the chromo-somes.

4. Embedding medium. Polarbed (Polaron, UK) made up fresh from the four com-ponents according to the manufacturer's instructions.

5. D'Angelo's saline D + 0.02 M Tris, pH 8.1 (DT).

6. Monoclonal antibodies against chromosomal proteins.

7. Immunogold reagent — goat anti-mouse IgG coupled to 10 nm gold particles. Auroprobe E M GA M IgG G10. (Janssen Pharmaceutica Life Sciences Belgium) — now handled by Amersham International plc.

3. Methods

3.1. Culturing of D. melanogaster

1. Maintain Oregon R flies in 300 mL of well-yeasted cornmeal-agar culture bottles.
2. Transfer approx 100 3- to 4-d old flies to fresh bottles on a Monday and leave there until the following Wednesday, when they are transferred to a "mini-population cage" consisting of a disposable plastic bottle whose bottom has been cut off to allow attachment of a 6 cm diameter culture dish half-filled with freshly yeasted ethanol:acetic acid agar (400 mL of water, 9 g of agar, 15 g of treacle, 23 mL of absolute ethanol and 2.7 mL of acetic acid) to stimulate egg production. The original fly bottles, at this point, contain sufficient eggs to produce a new population of flies suitable for continuing the cycle on the Monday 2 wk later.
3. Change the minicage laying plates twice on the Thursday .
4. Transfer the flies into glass bottles containing yeast–glucose medium on the Friday. Allow the flies to lay eggs in the dark for 1 h and tip the flies out.
5. Count a total of 100 eggs distributed between four quadrants on the central part of the agar surface and remove the remainder carefully .
6. Apply a layer of fresh baker's yeast, approx 6 mm high, around the periphery of the bottle. Place the bottle in an incubator at 23.5°C.
7. On the following Monday feed the larvae with a paste of fresh baker's yeast to build up a layer some 8 mm thick covering the agar surface except for four symmetrically positioned holes approx 6 mm in diameter. Add the paste to the bottle in three portions at 10-min intervals to prevent smothering the young larvae.
8. Place the bottles back in the incubator until the following Wednesday.
9. Take the bottles out in a small styrofoam box containing 100 mL of water to maintain humidity.
10. Generally take larvae that have crawled to near the top of the bottle for microsurgery.

3.2. The Microsurgical Procedure

1. Dissect salivary glands out of third instar larvae into the medium of Cohen and Gotchel (*see* **Note 2**).
2. Perform subsequent microdissection steps in solutions based on saline D.
3. Place pairs of glands in a drop of 1% Triton X-100, saline D for 25–30 s.
4. Pass them through three drops of 0.25% Triton X-100, saline D and allow to remain in a fourth drop of 0.25% Triton X-100, saline D for 5–6 min before transfer into a drop of 0.01% Triton X-100, saline D.
5. Tear the distal end of the glands slightly to allow glue in the lumens to leak out.
6. Transfer to a 3-µL drop of 0.01% Triton X-100, 0.5 mM MgCl$_2$, saline D on the lightly siliconized surface of a well-slide some distance from the well.
7. Employ a sharp tungsten needle freehand to make an incision in the gland, usually approximately one third of a gland's length from the duct to release nuclei.
8. Immediately place a 5 mm × 5 mm siliconized coverslip over the drop to aid microscopic observation and the release of nuclei.
9. Gently add another 3 µL of 0.01% Triton X-100, 0.5 mM MgCl$_2$, saline D under the edge of the coverslip.

Fig. 5. The transport of a nucleus on a blunted glass needle. *See* **Subheading 2.2.** for details. Two chromosome sets, isolated by micromanipulation, can be seen in the background. The bar represents 50 μm.

10. Examine very briefly under the microscope and immediately add 40 μL of 0.05% formaldehyde in saline D or of buffer A of Hewish and Burgoyne *(19)*.

11. Flood the well and surface of the slide with 0.05% (w/v) formaldehyde in saline D or with buffer A.

12. Pick up individual nuclei on a glass needle whose tip has been rounded in a microforge as illustrated in **Fig. 5**, transport on the needle to the well, and transfer to the well floor by gently touching to the coverslip and withdrawal of the needle (*see* **Note 3**).

13. Puncture the nuclear envelope with two sharp glass needles whose tips are maintained in close proximity and make an incision in the envelope by drawing the needles apart.

14. Withdraw the chromosome mass through the incision with one needle while the other serves to hold the incision open (*see* **Note 4**).

15. Use both needles to unravel the chromosomes and spread them out in an optical plane immediately above the floor of the well (*see* **Note 5**).

16. Light microscopy. After completion of micromanipulation, strip the parafilm border from the slide, draw off excess fluid, and cover the well with a coverslip, taking care to avoid trapping a bubble. Invert the slide and examine the preparation under phase contrast on a Zeiss Universal microscope. Phase-contrast micrographs are recorded on Ilford Pan F film. Immunofluorescence is photographed on Ilford HP5[+] or Fujichrome 400 film.

3.3. Preparation of Spread Chromosomes for Electron Microscopy

The spread chromosomes in saline D, attached to the glass base of the well-slide, were treated by the following method described in **refs.** *12, 13*, and *20*.

The slide is kept in a moist atmosphere in a Petri dish, supported by match sticks above moist filter paper and the coverslip over the well removed.

1. Fix with 3% glutaraldehyde in 0.1 *M* sodium phosphate buffer, pH 7.2, for 30 min, repeating with fresh glutaraldehyde in phosphate buffer for another 30 min.
2. Wash in several changes of phosphate buffer.
3. Post-fix in 1% osmium tetroxide in phosphate buffer for 30 min, repeating with fresh osmium tetroxide for another 30 min.
4. Wash in several changes of phosphate buffer, place a coverslip over the well, examine the slide under inverted phase contrast, and photograph the chromosomes at various focal planes; the chromosomes are not flattened to the coverslip along their whole lengths. Make a drawing on tracing paper. The negatives are printed normally and also reversed to monitor the chromosomes using the dissecting and the inverted phase microscopes. These help immensely to monitor the trimming of the block, the cutting of the sections, and the identification of chromosomes and their landmarks.
5. Dehydrate by passing through an ethanol series to absolute alcohol (50%, 75%, 95%, 100% × 2) in a Coplin Jar.
6. To pass the slide through epoxy-propane to embedding medium, three glass Petri dishes are prepared, with rolled aluminum foil "match stick" strips to keep the slide from the bottom of the Petri dish. In the first Petri dish pour over the slide epoxy-propane only; in the second pour on epoxy-propane/Polarbed embedding medium 2:1; in the third apply epoxy/embedding medium 1:2. The slide is in each Petri dish for about 1 min.
7. Place the well-slide in a glass Petri dish lined with aluminum foil and flood with approx 2 mL of embedding medium using a very wide bore pipet, e.g., a plastic disposable pipet with the tip cut off. Do not pipet directly over the well. Add sufficient embedding medium to ensure displacement of epoxy-propane as indicated by the absence of Schleirin lines.
8. Place the Petri dish containing the slide in a 60°C oven for approx 1 h to evaporate any remaining epoxy-propane.
9. Flood the well with additional embedding medium and incubate at 60°C overnight for polymerization.
10. With a permanent waterproof marker, circle the position of the well on the underside of the bottom coverslip and end-embed by inverting a Beem capsule of embedding medium over the slide, centering on the marked circle.
11. Polymerize at 60°C for several days.
12. Split the resulting peg with the end-embedded chromosomes from the slide and bottom coverslip by placing in a slurry of pulverized solid CO_2 pellets in liquid nitrogen.
13. File the peg down to about 1 cm length and stand upright on a glass slide supported by plasticine with the chromosome end uppermost. Observe the chromosomes just below the surface that had been in contact with the coverslip using the dissecting microscope. Tilting the peg so that the incident illumination gives a

clear silver effect aids detection of the chromosomes. Using a mounted needle, scratch a small square with orientation marks on the surface to surround the barely visible chromosome group, which generally occupies an area of about 0.1 mm^2.

14. Wet this surface containing the chromosome group with immersion oil and invert the peg onto a slide (chromosome end now down).
15. Wet the filed upper end of the peg with immersion oil and lay a 13-mm coverslip over it.
16. Examine with the inverted phase-contrast microscope to check the chromosomes against the photographs taken earlier and mark these photographs with appropriate orientation marks.
17. Lightly remove the immersion oil with alcohol and trim the block using the holders described in **Subheading 2.3.**, first by hand with fine files and single sided razor blades and finally using the LKB Ultramicrotome with 55° glass knives to a pyramid block face of approx 0.2 mm × 0.1 mm, just a little larger than the area of the chromosome group, monitoring all the time with the photographs.
18. Orientating the block with great care so that the entire face of the block is cut at first stroke, cut 50-nm silver sections with a diamond knife (Freidrich Dehmer) mounted on the LKB Ultramicrotome and pick up the ribbon of sections on carbon-coated Formvar grids.
19. Stain the sections with 4% aqueous uranyl acetate and bismuth subnitrate as described by Ainsworth and Karnovsky *(21)* and Riva *(22)*.
20. Examine in the transmission electron microscope at 40 and 60 kV. Calibrate the magnification with a carbon-grating replica grid (Balzer's Union).

3.4. Immunolocalization of Chromosomal Proteins at the Ultrastructural Level by Gold-Labeled Antibodies

The spread chromosomes in the well-slide are treated for immunolocalization with colloidal gold-labeled antibodies *(23,24)*.

1. Wash with D'Angelo's saline D.
2. Incubate in 0.1 *M* glycine, pH 7.1, for 30 min.
3. Wash with saline D.
4. Treat with 0.1% bovine serum albumin (BSA) in saline D.
5. Wash with saline D.
6. Incubate successively for 1 h at room temperature with anti-chromatin monoclonal antibodies in saline D.
7. Wash with 0.1% BSA in saline DT.
8. Incubate for 1 h at room temperature with goat antibodies against mouse IgG coupled to colloidal gold beads of 10 nm diameter, diluted 1:20 in 1% BSA in saline DT.
9. Wash in 0.1% BSA in DT.
10. Wash with 0.1 *M* sodium phosphate, pH 7.1.
11. The immunolabeled chromosomes were prepared and sectioned as described in **Subheading 3.3.**

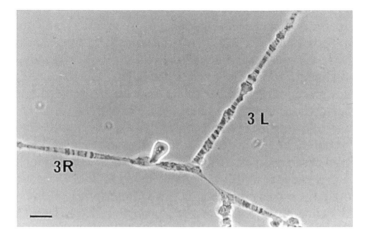

Fig. 6. Phase-contrast micrograph depicting portion of a native chromosome preparation isolated by microsurgery from a salivary gland at puffing stage 6 of the ecdysone response. Early ecdysone puffs are apparent on chromosome arm 3L and a general high degree of preservation of the banding pattern may be discerned. The bar represents 10 μm.

12. To give sufficient staining to the ultrastructure without obscuring the electron-dense label, the 50-nm silver sections were lightly counterstained by a modification of the method of Ainsworth and Karnovsky *(21)* with 1 h in bismuth subnitrate and 15 min in 4% aqueous uranyl acetate.
13. Examine in the calibrated transmission electron microscope at 60 kV, as previously.

3.5. Applications of Native Preparations of D. melanogaster Polytene Chromosomes

Figure 6 depicts part of a preparation of salivary gland chromosomes isolated as native chromatin by microsurgery. This phase-contrast micrograph of unstained native chromosomes records the proximal region of chromosome arm 3L bearing two obvious early ecdysone puffs, the base of 3R and the chromocenter. It can be seen that fine details of the banding pattern are maintained in the isolated gel. **Figure 7** illustrates the degree of preservation of morphology observed in the electron microscope. Ultrastructure is apparent that has not been seen in classical acid squash preparations of *D. melanogaster* salivary chromosomes, e.g., within puffs nascent ribonucleoprotein (RNP) particles may be observed in linear array sometimes clearly connected by stalks to an axial filament. Such preparations are suitable for a number of applications including the localization of proteins by immunofluorescence in the light microscope and by gold-labeled antibodies at an ultrastructural level in the electron micro-

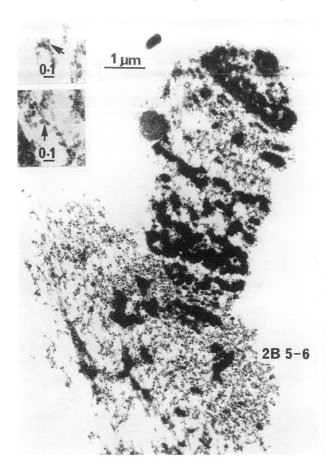

Fig. 7. Electron micrograph of the tip of the X chromosome and the large puff at 2B5-6. Inserts show parts of the puff 2B5-6 at higher magnification with linear arrays of ribonucleoprotein (RNP) particles some of which display stalks attaching them to an axial filament. (Reproduced from Mott et al. *[12]* with permission.)

scope, the examination of molecular conformations *in situ*, the localization of biological activities of native proteins, and possibly the future activation of specific chromosomal loci in vitro. Here we shall illustrate such applications with just two specific examples.

3.5.1. The Investigation of DNA Conformation In Situ

Pohl and Jovin *(25)* detected a structural change occurring in a synthetic alternating dG–dC copolymer as the ionic strength is raised. The remarkable structure of the high-salt form of the polymer, termed Z-DNA, was determined

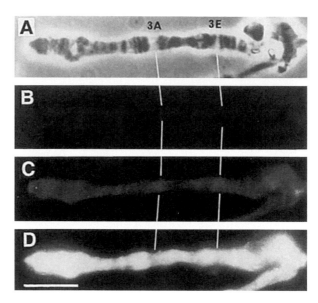

Fig. 8. Phase-contrast (**A**) and Z-DNA antibody-stained fluorescence micrographs (**B–D**) of the X-chromosome of *D. melanogaster*. The chromosome was isolated by the microsurgical procedure described in this chapter. In (**B**) it was treated directly by Z-DNA antibodies. In (**C**) it was treated with antibody after a 5-s exposure to 45% acetic acid and return to pH 7.0. In (**D**) it was antibody treated after a 30-s exposure to acid. The bar represents 10 μm. Reproduced from Hill and Stollar *(30)*, with permission.

by groups at the Massachusetts Institute of Technology and the California Institute of Technology *(26,27)*. It is a left-handed double helix, in contrast with the classical right-handed double-helical forms as originally modeled by Watson and Crick in 1953. The pitch of the Z-DNA double helix is 45 Å and the phosphate groups along the backbone trace out a zigzag path, in contrast to the smoother contour along the phosphates of B-DNA. Z-DNA is highly immunogenic, allowing the production of highly specific antibodies *(28)*. Such antibodies exhibit strong immunofluorescence with classically isolated and fixed chromosomes. These findings initially caused much excitement because they suggested that readily detectable levels of Z-DNA occurred, not just in crystals of synthetic copolymers, but in chromosomes. However, on comparison of the immunofluorescence patterns obtained by different laboratories, paradoxical qualitative differences became apparent. Some investigators found immunofluorescence primarily over bands *(29)* and others observed that the major fluorescence occurred over interbands and puffs *(28)*.

On treatment of chromosome preparations isolated by microsurgery at neutral pH with Z-DNA antibodies only low levels of fluorescence, close to background, were observed (**Fig. 8B**). However, if such preparations were exposed to the classical acid fixative, 45% acetic acid, for 5 s then immunoreactivity started to appear, primarily over the less condensed interbands and puffs (**Fig. 8C**). Exposure to 45% acetic acid for 30 s led to massive enhancement of immunoreactivity, which now followed the mass distribution of the chromosomes, i.e., resided predominantly over bands (**Fig. 8D**). Clearly, the solvent was having a major effect on the immunoreactivity and this effect occurred first in regions that are most accessible and subsequently in more highly compacted structures. Experiments suggested that a major driving force for the unwinding of the right-handed double helix and rewinding into a left-handed double helix derives from the liberation of the free energy of deformation of the DNA associated with its left-handed super-helical path on the surface of nucleosomes *(30)*. Nucleosomes are disrupted on exposure to acid fixatives *(2)*. When the DNA is freed from the constraints imposed by interactions with the nucleosome core, the molecules extend out and writhing into a left-handed superhelix transduces into a left-handed twist. It is likely that the underwinding of DNA associated with nucleosome disruption in classical acid-fixed and denatured chromosome preparations tends to prevent the "snap-back" renaturation of strand-separated regions, keeping them available for *in situ* hybridization.

3.5.2. Observation of In Situ *Protein Distribution*

The apparent distribution of proteins on polytene chromosomes can be influenced by the method of isolation. Exposure to acid can lead to extraction of proteins, an effect that may be countered by crosslinking with formaldehyde or glutaraldehyde. However, such crosslinking agents may also cause accretion by the chromosomes of components of the nucleoplasm and even the precipitation of nucleoplasmic material. The apparent chromosomal distribution of the protein D1 as a function of the parameters of chromosome preparation was carefully studied by Alfageme *et al. (31)*. These investigators found that the distribution of immunofluorescence produced by antiserum against D1 varied from background to general strong fluorescence depending upon the relative exposure of the chromosomes to 50% acetic acid and formaldehyde during isolation. An intermediate distribution was passed through with antibody binding at just a small number of discrete sites. Treatment of native chromosome preparations isolated by microsurgery with D1 antiserum immediately gave the discrete distribution as illustrated in **Fig. 9**.

At the ultrastructural level in the electron microscope, specific antigens can be localized on native chromosome preparations using gold-labeled antibodies as illustrated in **Fig. 10**. Here gold labeling reveals the binding of monoclonal antibodies specific for a major chromatin antigen at transcriptionally active chromosomal loci bearing putative RNP particles. However, the absence of

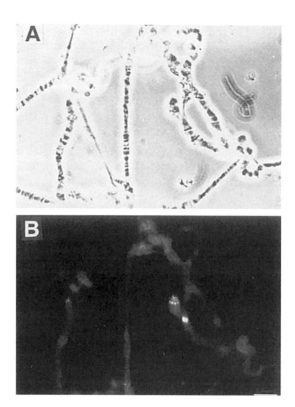

Fig. 9. Phase-contrast **(A)** and **(B)** immunofluorescence pattern induced by antiserum against chromosomal protein D1. The native chromosome preparation was isolated by microsurgery. The bar represents 10 μm.

gold labeling on RNP particles at one locus in the region depicted suggests the existence of different proteins at different active loci. A range of antigens can be localized at different active loci using specific monoclonal antibodies and gold labeling.

4. Notes

1. There have been reports of *in situ* hybridization to acid-squashed chromosome preparations that have been treated with RNase but not subjected to a defined "denaturation" step *(32,33)*. Such hybridization appears to correlate with intense transcriptional activity and may directly reflect differences in DNA underwinding and/or conformation in vivo or simply a more open chromatin structure, permitting some denaturation by the acid fixatives employed. *In situ* hybridization to native chromosomes isolated without exposure to low pH fixation could distinguish between these two possibilities.

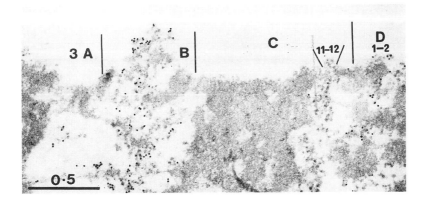

Fig. 10. Ultrastructural localization of monoclonal antibody to a major 74-kDa chromatin antigen in divisions 3A–D of the distal region of the X chromosome. Gold spheres (10 nm diameter) indicate location of antibody molecules. The diffuse band at 3B consists predominantly of ribonucleoprotein (RNP) particles with antibody binding. There is relative absence of antibody binding to RNP at locus 3C 11–12, in contrast to that indicated by the presence of gold spheres at 3B and 3D 1–2. (Reproduced from Hill et al. *[24].*)

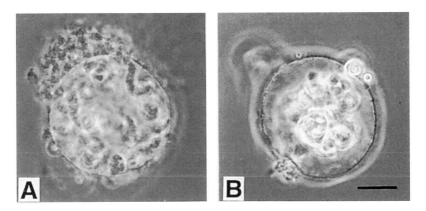

Fig. 11. Nuclei dissected out of third instar *D. melanogaster* salivary glands. (**A**) Nucleus containing chromosomes dispersed through the nuclear space. (**B**) Nucleus in which the chromosomes are condensed into a central mass. Bar = 20 μm.

2. In some instances it is useful to culture isolated glands in Schneider's medium diluted with a one fifth volume of 10% ethanol in water *(15)* to which 20-hydroxyecdysone may be added if it is desired to initiate a puffing response in vitro.

3. The overall distribution of the chromosomes within nuclei isolated from third instar salivary glands is found to vary between the two extremes shown in **Fig. 11**. These observed chromosomal configurations within nuclei are stage- or physiology-related properties of the glands. It is possible to dissect chromosomal material out of nuclei exhibiting configurations throughout this range. However, complete spreads can most readily be obtained from nuclei exhibiting intermediate distributions, i.e., when the chromosomes are not too tightly bunched together on the one hand or too closely associated with the nuclear envelope on the other. Glands in which the chromosomes are clumped into a tight ball in the center of the nucleus can be converted into those with chromosomes more dispersed through the nucleus by simply incubating in Schneider's medium for 5–15 min.

4. In some glands the chromosomes are found to be too "soft" and fragile for micromanipulation. These almost certainly correspond to glands that do not give good squashes by classical techniques. Usually the same culture will provide glands suitable for microsurgery a few hours later.

5. Once attached to the coverslip in this manner, their configuration is usually quite stable. Alternatively, nuclei or isolated chromosomes may be transferred to the surface of an electron microscope grid for examination of nuclear envelopes or chromosomes, respectively.

Acknowledgments

We wish to thank Professor H. G. Callan FRS for introducing us to nuclear microsurgery, Garth Watson, Zhong Yuan Jiang, and Kerrie Medveczky for technical assistance and Louise Lockley for help with photography.

References

1. Painter, T. S. (1934) A new method for the study of chromosome aberrations and the plotting of chromosomal maps in *Drosophila melanogaster. Genetics* **19**, 175–188.

2. Cole, R. D. and Lawson, G. M. (1979) Selective displacement of histone H1 from whole HeLa nuclei: effect on chromatin structure *in situ* as probed by micrococcal nuclease. *Biochemistry* **18**, 2160–2166.

3. Peacocke, A. R. (1957) The dissociation and molecular structure of the nucleic acids. *Spec. Publ. Chem. Soc. Lond.* **8**, 139–164.

4. Baker, J. R. (1958) *Principles of Biological Microtechnique*. Methuen, London.

5. Hancock, J. M. and Summer, A. T. (1982) The role of proteins in the production of different types of chromosome bands. *Cytobios* **35**, 37–46.

6. Hill, R. J., Watt, F., and Stollar, B. D. (1984) Z-DNA immunoreactivity of *Drosophila* polytene chromosomes. Major effects of 45% acetic acid, 95% ethanol and DNAse I nicking. *Exp. Cell Res.* **153**, 469–482.

7. Chen, C. C., Bruegger, B. B., Kern, C. W., Lin, Y. C., Halpern, R. M., and Smith, R. A. (1977) Phosphorylation of nuclear proteins in regenerating liver. *Biochemistry* **16**, 4852–4855.

8. Chargaff, E. (1955) Isolation and composition of the deoxypentose nucleic acids and of the corresponding nucleoproteins, in *The Nucleic Acids* (Chargaff, E. and Davidson, J. N., eds.), Academic Press, New York, NY, pp. 307–371.

9. Silver, L. M. and Elgin, S. C. R. (1976) A method for determination of the *in situ* distribution of chromosomal proteins. *Proc. Natl. Acad. Sci. USA* **73**, 423–427.

10. Zink, B. and Paro, R. (1989) *In vivo* binding of a trans-regulator of homoeotic genes in *Drosophila melanogaster*. *Nature* **337**, 468–471.

11. Alfageme, C. R., Rudkin, G. T., and Cohen, L. H. (1980) Isolation, properties and cellular distribution of D1, a chromosomal protein of *Drosophila*. *Chromosoma* **78**, 1–31.

12. Mott, M. R., Burnett, E. J., and Hill, R. J. (1980) Ultrastructure of polytene chromosomes of *Drosophila* isolated by microdissection. *J. Cell Sci.* **45**, 15–30.

13. Mott, M. R. and Hill, R. J. (1986) The ultrastructural morphology of native salivary gland chromosomes of *Drosophila melanogaster*. The band-interband question. *Chromosoma* **94**, 403–411.

14. Hill, R. J. and Watt, F. (1978) Native salivary chromosomes of *Drosophila melanogaster*. *Cold Spring Harb. Symp. Quant. Biol.* **XLII**, 859–865.

15. Ashburner, M. (1972) Patterns of puffing activity in the salivary gland chromosomes of *Drosophila*. VI. Induction by ecdysone in salivary glands of *D. melanogaster* cultured *in vitro*. *Chromosoma* **38**, 255–281.

16. Callan, H and Lloyd, L. (1960) Lampbrush chromosomes of crested newts *Triturus cristatus* (*Laurenti*). *Philos. Trans. R. Soc. Lond. B* **243**, 135–219.

17. Cohen, L. H. and Gotchel, B. V. (1971) Histones of polytene and non-polytene nuclei of *Drosophila melanogaster*. *J. Biol. Chem.* **246**, 1841–1848.

18. D'Angelo, E. G. (1946). Microsurgical studies on *Chironomus* salivary gland chromosomes. *Biol. Bull.* **90**, 71–87.

19. Hewish, D. R. and Burgoyne, L. A. (1973) Chromatin sub-structure. The digestion of chromatin DNA at regularly spaced sites by a nuclear deoxyribonuclease. *Biochem. Biophys. Res. Commun.* **52**, 504–510.

20. Mott, M. R. and Callan, H. G. (1975) An electron microscope study of the lampbrush chromosomes of the newt *Triturus cristatus*. *J. Cell Sci.* **17**, 241–261.

21. Ainsworth, S. K. and Karnovsky, M. J. (1972) An ultrastructural staining method for enhancing the size and electron opacity of ferritin in thin sections. *J. Histochem. Cytochem.* **20**, 225–229.

22. Riva, A. (1974) A simple and rapid staining method for enhancing the contrast of tissues previously treated with uranyl acetate. *J. Microscopie* **19**, 105–108.

23. Hill, R. J., Mott, M. R., Watt, F., Fifis, T., and Underwood, P. A. (1986) The localization of an Mr 74,000 major chromatin antigen on native salivary chromosomes of *Drosophila melanogaster*. *Chromosoma (Berl.)* **94**, 441–448.

24. Hill, R. J., Mott, M. R., and Steffensen, D. M. (1987) The preparation of polytene chromosomes for localization of nucleic acid sequences, proteins and chromatin conformation. *Int. Rev. Cytol.* **108:** 61–118.

25. Pohl, F. M. and Jovin, T. M. (1972) Salt-induced cooperative conformational change of a synthetic DNA: equilibrium and kinetic studies with poly (dG-dC). *J. Mol. Biol.* **67**, 375–396.

26. Wang, A. H.-J., Quigley, G. J., Kolpak, F. J., Crawford, J. L., van Boom, J. H., van der Marel, G., and Rich, A. (1979) Molecular structure of a left-handed double-helical DNA fragment at atomic resolution. *Nature* **282**, 680–686.

27. Drew, H., Takano, T., Tanaka, S., Itakura, K., and Dickerson, R. E. (1980) High-salt d(CpGpCpG), a left-handed Z-DNA double helix. *Nature* **299,** 312–316.

28. Nordheim, A., Pardue, M. L., Lafer, E. M., Moller, A., Stollar, B. D., and Rich, A. (1981) Antibodies to left-handed Z-DNA bind to interband regions of *Drosophila* polytene chromosomes. *Nature* **294,** 417–422.

29. Arndt-Jovin, D. J., Robert-Nicoud, M., Zarling, D. A., Greider, C., Weimer, E., and Jovin, T. M. (1983) Left-handed Z-DNA in bands of acid fixed chromosomes. *Proc. Natl. Acad. Sci. USA* **80,** 4344–4348.

30. Hill, R. J. and Stollar, B. D. (1983) Dependence of Z-DNA antibody binding to polytene chromosomes on acid fixation and DNA torsional strain. *Nature* **305,** 338–340.

31. Alfageme, C. R., Rudkin, G. T., and Cohen, L. H. (1980) Isolation, properties and cellular distribution of D1, a chromosomal protein of *Drosophila*. *Chromosoma* **78,** 1–31.

32. Artavanis-Tsakonas, S., Schedl, P., Mirault, M.-E., and Moran, L. (1979) Genes for the 70,000 dalton heat shock protein in two cloned *D. melanogaster* DNA segments. *Cell* **17,** 9–18.

33. Henikoff, S. (1981) Position-effect variegation and chromosome structure of a heat shock puff in *Drosophila*. *Chromosoma* **83,** 381–393.

5

Released Chromatin or DNA Fiber Preparations for High-Resolution Fiber FISH

Henry H. Q. Heng

1. Introduction

When combined with molecular analysis and immunocytological localization, fluorescence *in situ* hybridization (FISH) represents one of the most direct and precise experimental tools in current biological research *(1–4)*. Direct visualization and *in situ* detection fill the gap between molecular analysis and cytological description and provide a new avenue for in vitro and in vivo comparison. Such a system facilitates the understanding of cellular events not only with respect to what and how, but also where and when.

A major breakthrough in FISH methodology was the concept and the development of protocols for high-resolution fiber FISH *(5)*. By hybridizing probes on released, less condensed, and somewhat linearized chromatin or DNA fibers, ultra-high-resolution FISH data can be generated after specific probe–target interaction and detection *(6–7)*. Because the map resolution of fiber-FISH is much higher than chromosome-based FISH and interphase FISH, this approach has contributed to the Human Genome Project by generating high-resolution physical mapping data that have uncovered many details of chromosome and genome structure *(8–10)*. Applications of fiber FISH to study the DNA–protein *in situ* interactions are expected to elucidate the function aspects of chromosomes and the genome *(1,11,12)*.

In this chapter, we describe fiber FISH techniques for physical mapping and genome structure analysis, emphasizing details for generating chromatin or DNA fibers and FISH detection using these released fibers.

From: *Methods in Molecular Biology*, Vol. 123: In Situ *Hybridization Protocols*
Edited by: I. A. Darby © Humana Press Inc., Totowa, NJ

1.1. Brief Development of Fiber FISH

Prophase chromosomes (13) and interphase nuclei (2) and pronuclei (14) have been used for high-resolution FISH detection. However, the resolution can be dramatically improved and the data simplified if the spatial arrangement of the chromatin is removed by artificially releasing the chromatin or DNA fiber from within the nucleus. The notion of performing in situ hybridization on released chromatin fiber was proposed in the 1980s (15) but because of the lack of awareness of the released chromatin fiber, the scientific community showed no interest until the first report of free chromatin fiber FISH in 1991 (5,16,17). This successful use of free chromatin fiber for performing high-resolution FISH suggested the possibility of using even less condensed DNA fiber to establish the order of genes or DNA segments on a chromosome.

Further modifications immediately followed that simplified the protocols of fiber generation and improved the resolution. These included FISH detection on chromatin released by alkaline treatment (6); on DNA "halo" preparations (7); on DNA fiber released by alkaline solution, nonionic detergents, or sodium dodecyl sulfate (SDS) with or without mechanical force; on released DNA strands confined in agarose gel; and on DNA molecules attached to the surface of a solid phase (18–23). Although all these protocols, as well as FISH detection using "molecular combing" (24), generate linearized chromatin or DNA fibers ready for FISH, their key difference is the method of fiber preparation and the corresponding resolution.

Many terms have been introduced to describe the systems of FISH detection on released chromatin/DNA fiber, including free chromatin/DNA FISH, DNA halo FISH, extended chromatin/DNA fiber FISH, elongated chromatin/DNA FISH, individual stretch DNA molecule FISH, visual mapping, direct visual hybridization (DIRVSH), quantitative DNA fiber mapping, and dynamic molecular combing (6,18–25). Recently, it was suggested that all high-resolution chromatin and DNA fiber FISH methods should be simply referred to as fiber FISH (26).

1.2. DNA Fiber or Chromatin Fiber?

It is essential to determine what type of fiber is best suited to a particular FISH experimental situation: chromatin fiber or DNA fiber. The selection is based on the degree of resolution required by the nature of research. For physical mapping, the lower degree of condensation of the DNA fiber provides higher mapping resolution than chromatin fibers. However, the coverage of DNA fiber is much less than that of chromatin fiber (Note 1). In chromatin structure studies, chromatin fiber FISH must be used, because few if any chromatin proteins remain associated with the DNA fiber. To perform DNA–protein in situ codetection experiments to analyze the DNA–protein interactions and to study the high-order organization of chromosomes, the chromatin fiber FISH is the method of choice.

2. Materials

2.1. Cells and Cell Culture

1. Lymphocytes isolated from healthy donors or cord blood (**Note 2**).
2. Human fibroblastoid cell line.
3. Lymphocyte culture medium: RPMI 1640 (Gibco-BRL) supplemented with: 10% (v/v) fetal calf serum (Gibco-BRL); 2% phytohemagglutinin (PHA) (Gibco-BRL); and 5 U/mL of heparin (Hepalean) (Organon Canada Ltd.).
4. Cell line culture medium: α-minimum essential medium (α-MEM) supplemented with 10% fetal calf serum.

2.2. Free Chromatin Fiber Preparation

1. N-[4-(9-acridinylamino)-3-methoxyphenyl]methanesulfonamide (m-AMSA, Drug Synthesis Branch, National Cancer Institute, Bethesda, MD, USA). Prepare stock solution by dissolving *m*-AMSA in dimethyl sulfoxide at 10 mg/mL, diluting with an equal volume of distilled water and filter-sterilizing (**Note 3**).
2. 10 mg/mL Ethidium bromide (EtBr) (*see* **Note 3**).
3. 5-Bromodeoxyuridine (BrdU) (*see* **Note 3**).
4. Potassium chloride (KCl) (Fisher Scientific) (*see* **Note 4**).
5. Alkaline lysis buffer: 1 m*M* sodium borate solution adjusted to pH 10–11 with NaOH, mixed with KCl solution (0.4%) in a 1:1 ratio before use.
6. Methanol (JT Baker).
7. Acetic acid (Fisher Scientific).
8. Microscope slides (VWR).

2.3. DNA Fiber Preparation

1. PBS solution.
2. Alkaline solution: 0.07 *M* NaOH: ethanol (5:2) (*see* **Note 5**).
3. Potassium chloride (KCl) (Fisher Scientific).
4. Methanol (JT Baker).
5. Acetic acid, Glacial (Fisher Scientific).
6. 70%, 95%, and 100% ethanol.
7. Microscope slides (VWR).

2.4. Probe Labeling

1. Biotin labeling nick-translation kit (BRL) (**Note 6**).
2. Digoxigenin (DIG)-nick-translation mix (Boehringer Mannheim) (**Note 6**).
3. Nick column (Pharmacia).
4. Salmon sperm DNA (100–500-bp fragments obtained by sonication).
5. 3 *M* NaOAc.
6. 70% and 100% ethanol.

2.5. Hybridization

1. Hot plate (37–70°C); water baths at 37°C, 43°C, 46°C, and 70°C; 37°C incubator.
2. 25-mL Plastic slide mailers (Surgipath).
3. Plastic slides chamber (slide holder) (VWR CanLab).
4. RNase A (Boehringer Mannheim).
5. Ethanol: 75%, 90%, and 100%.
6. Denaturation solution: 70% deionized formamide (IBI) in 2× SSC (saline sodium citrate) (20× SSC stock solution: 3 M NaCl, 300 mM Na citrate).
7. Hybridization solution I (for use with genomic probes): 50% deionized formamide (IBI) and 10% dextran sulfate in 2× SSC.
8. Hybridization solution II (for use with repetitive DNA probes): 65% formamide and 10% Dextran sulfate in 2× SSC.
9. Human Cot-I DNA (BRL) or "Super Blocker" (SeeDNA Inc.) (**Note 7**)
10. Wash solution A (for nonrepetitive DNA probe): 50% formamide in 2× SSC.
11. Wash solution B (for repetitive DNA clones): 65% formamide in 2× SSC.
12. Wash solution C: 0.1 M phosphate buffer, pH 8.0, with 0.1% Nonidet P-40 (Boehringer Mannheim).
13. 2× SSC (before and after 4',6-diamidino-2-phenylindole [DAPI] staining wash).

2.6. Detection and Amplification

1. Blocking solution: 3% BSA (bovine serum albumin, Sigma fraction V) in 4× SSC with 0.1% Tween-20.
2. Detection solution: 1% BSA and 0.1% Tween-20 in 4× SSC. Store at 4°C.
3. Avidin–FITC (fluorescein isothiocyanate, Vector): 500 µg/mL (stock solution). FITC detection working solution: 10 µL of avidin–FITC stock solution to 990 µL detection mixture. Store in the dark at 4°C. Good for up to 6 mo.
4. Biotinylated goat anti-avidin antibody (Vector): 500 µg/mL (stock solution). Aliquots (50 µL each) can be kept at –20°C. Working solution: dilute stock solution with detection solution. The final concentration of the solution with detection mixture is 5 µg/mL.
5. Antidigoxgenin–rhodamine Fab fragments (20 µg/mL).

2.7. Counterstaining and Antifade

1. DAPI (Sigma): 0.2 mg/mL of stock solution in H_2O. Store in the dark at 4°C.
2. Propidium iodide (PI, Sigma): 0.1 mg/mL of stock solution in PBS (phosphate-buffered saline) (Gibco-BRL).
3. Antifade solution: Dissolve 100 mg of p-phenylendiamine in 10 mL of PBS, then add 90 mL of glycerol. Adjust pH to 9.0 by using Litmus paper. Store at –20°C.

2.8. Microscopy and Photography

1. Epifluorescent microscope (Leitz Aristoplan) with DAPI, rhodamine and FITC filters.
2. Kodak Ektachrome P800/1600.

3. Methods

3.1. Free Chromatin Fiber Preparation

3.1.1. Preparation of Free Chromatin from Lymphocytes by Drug Treatment

1. Cell culture. Isolate lymphocytes from 5–10 mL of fresh human peripheral or cord blood by low-speed centrifugation (500 rpm for 5 min) or unit-gravity sedimentation. Collect the white cells suspension. Transfer 0.5–0.8 mL of the blood cell suspension to 20 mL of RPMI 1640 in a 50–mL culture flask (Falcon 3082 [Los Angeles, CA, USA]). Culture lymphocytes in CO_2 incubator at 37°C for 48–52 h.
2. Drug treatment. Incubate cultured lymphocyte with mAMSA, or EtBr, or BrdU (10 μg/mL) for 2–4 h before harvesting (*see* **Note 3**).
3. Cell harvest and slide preparation.
   a. Transfer 4 mL of the culture to a 15-mL disposable test tube and collect the cells by centrifugation at 1000 rpm for 7 min. Resuspend the cell pellet in 0.3 mL of culture medium, add 5 mL of 0.4% KCl solution, mix well, and incubate in a 37°C water bath for 10 min (**Note 4**).
   b. At the end of the hypotonic treatment, add 0.1–0.2 mL of freshly prepared fixing solution (3:1 methanol:acetic acid) for prefixation at room temperature. Mix the contents gently by inverting the tube, then pellet the cells again by centrifugation (1000 rpm for 7 min).
   c. Discard the supernatant. Loosen the cell pellet by gently tapping at the bottom of the tube. Resuspend the cells in 5 mL of the fixing solution. Fix at room temperature for 20 min.
   d. Centrifuge the cells and resuspend in 0.5 mL of fixing solution.
   e. Allow 1 drop of the fixed cell suspension to fall onto the surface of a prechilled microscope slide (left on ice for at least 10 min). Air dry the slide.
4. Storage of slides. Slides containing free chromatin preparations may be stored for several weeks at –20°C. Once a good batch of slides is obtained, they should be dried at room temperature for 1 d and then sealed in slide containers with parafilm before transferring to the freezer.

3.1.2. Release of Free Chromatin from Cultured Cells with Alkaline Buffer (*Note 4*)

The rationale behind the use of alkaline buffer in the releasing of chromatin fibers from cultured cells is based on the observation that nuclear lamins can be disrupted by high pH treatment. Prolonged alkaline treatment, however, will generate over-releasing, resulting in a mixture of chromatin and DNA fibers. For chromatin fiber FISH, it is important to avoid producing DNA fibers to simplify the data analyzes. The following protocol is designed for fibroblastoid cell cultures. It may also be used for other cell types with modifications.

1. Cell culture. Grow fibroblasts in α-MEM for 2–4 d after reaching confluency to accumulate cells arrested at the G_1 phase.

2. Cell harvesting. Add 1 mL of trypsin-EDTA (1×) solution to each 60-mm Petri dish. Incubate at 37°C for 30–60 s. Once the majority of the cells become detached from the dish, stop the reaction by adding 1 mL of culture medium with serum. The next step should follow quickly.

3. Alkaline treatment. Because the optimal condition for each particular cell line has to be obtained empirically, a brief screening test is recommended:

    a. Prepare a series of tubes each containing 1 mL of alkaline buffer. Transfer dropwise an aliquot (0.2–0.3 mL) of the trypsinized cell suspension into each of these tubes and mix the contents gently by tapping the tube. At various time intervals (3–5 min), terminate the alkaline treatment by adding 5 mL of the fixing solution (methanol:acetic acid, 3:1).

    b. Collect the alkaline-treated cell suspension by centrifugation (1000 rpm). Resuspend the cells in 5 mL of new fixing solution. The remaining steps of slide preparation are as described in **Subheading 3.1.1., steps 3c-e**.

4. Use the optimal conditions determined by the screening test to maximize chromatin fiber production.

3.2. DNA Fiber Preparation (see Note 8)

1. The cell culture protocol for lymphocytes cells is described in **Subheading 3.1.1., step 1**.

2. Harvest and fix by routine methods:

    a. Transfer cultured cells into 15-mL centrifuge tube. Collect cells by centrifuging at 1000 rpm for 7 min.

    b. Resuspend the cell pellet in 0.3 mL of medium, and add 4 mL of 0.4% KCl for 10 min at 37°C.

    c. Collect cells at the end of hypotonic treatment by centrifuging the cells.

    d. Discard the supernatant. Add 5 mL of fixative (3:1 methanol:acetic acid), mix well.

    e. Fix once more in fresh fixative (20 min).

3. Drop fixed cells onto the slide.

4. Soak the slide in PBS for 1 min at room temperature. (Do not let the slide dry out.)

5. Add 200 μL of alkaline solution (0.07 *M* NaOH:ethanol = 5:2) to the slide.

6. Put one edge of the cover slide onto the slide and pull it along the slide from one end to another.

7. Add one drop of methanol to the slide.

8. Soak the slide in 70%, 95%, 100% series of ethanol.

9. Air dry.

3.3. In Situ Hybridization

3.3.1. Probe Labeling

Detailed protocols for probe purification can be found in Heng and Tsui *(29)*.

1. Plasmid, phage and cosmid probes. Purified DNA (1 μg) is labeled using BRL BioNick kit or Boehringer DIG-nick-translation mix according to the supplier's instructions (15°C for 60 min) (**Note 6**). Check the size of the products of nick-translation with a 1% agarose minigel. The optimal size of labeled fragments is approx 150–250 bp.
2. YAC or BAC probes. Longer nick-translation times are usually necessary. Generally, 2 h are sufficient. The size of nick-translation product can be checked by a minigel assay.
3. After labeling, the unincorporated nucleotides are removed using a Nick column (Pharmacia). Then 60 μg of salmon sperm DNA are precipitated together with the labeled probes in 40 μL of 3 *M* NaOAc and 880 μL of ethanol. After washing with 70% ethanol, the probes are resuspended in 20 μL of TE buffer.

3.3.2. Probe Treatment Before Hybridization

3.3.2.1. REPETITIVE PROBES

1. Labeled repetitive probes are denatured in hybridization solution II. Twenty to fifty nanograms labeled probes are added in 15 μL of denaturing solution and denatured at 75°C for 5 min.
2. No prehybridization is needed for these probes. The tube with the denatured probe should be put into ice immediately after the denaturation.

3.3.2.2. PHAGE, COSMID, YAC, AND BAC PROBES

For single-copy sequence detection, it is necessary to suppress the potential signal of repetitive elements in the probe by adding Cot-I DNA or "Super Blocker DNA" (**Note 7**).

1. For phage or cosmid probes: 20–50 ng of probes + 13 μL of hybridization solution I + 2 μg of Cot-I DNA or total human DNA.
2. For YAC probes isolated with total yeast DNA, 200–250 ng labeled probes + 13 μL of hybridization solution I + 2 μg of human Cot-I DNA + 2 μg of total yeast DNA.
3. After 5 min denaturation at 75°C, transfer the tube into a 37°C water bath for another 15–30 min for prehybridization (prehybridization for YAC probe may be slightly longer, e.g., 20–60 min).

3.3.3. Slide treatment before hybridization

3.3.3.1 SLIDE BAKING

Slides stored at –20°C are baked at 55°C for 2 h.

3.3.3.2. RNASE TREATMENT

1. Incubate slides in the 25-mL plastic jars containing RNase (100 μg/mL in 2× SSC) at 37°C for 1 h.
2. Wash the slides in 2× SSC solution for 2 min.
3. Dehydrate the slides by incubating for 3 min in 75%, 90%, and 100% ethanol and air dry.

3.3.3.3. Slide Denaturation

1. Make fresh denaturation solution of 70% formamide by mixing 14 mL of formamide, 5 mL of distilled H_2O, and 2 mL of 20× SSC. Put the jar filled with denaturing solution into a water bath and adjust the temperature to let the solution reach 70°C. Meanwhile, heat the slides at 50°C to avoid dropping the temperature of the denaturing solution.
2. Immerse one to four warmed slides in the denaturation solution for 3–5 min (*see* **Note 9**).
3. Quickly transfer the slides into a plastic jar with cold 70% ethanol for 2 min, dehydrate the slides in 95% and 100% ethanol for 3 min each, air dry, and perform hybridization immediately.

3.3.4. Hybridization

1. Load 15 µL of denatured probe with hybridization solution on each slide, and cover the hybridization slides with a 22 × 22 mm coverslip. Gently remove air bubbles and seal the edges with rubber cement to minimize evaporation.
2. Hybridize at 37°C in a moist chamber containing absorbent paper soaked in water. For repetitive probes, a few hours or overnight hybridization is suggested. For cosmid, YAC, or BAC probes, overnight is suggested.

3.3.5. Posthybridization Wash

Wash conditions depend on the type of probe used in the hybridization.

3.3.5.1 Cosmid, Phage, or YAC Probes Containing Repetitive Elements

1. Prewarm wash solution A to 46°C in three plastic jars.
2. With forceps, carefully remove the rubber cement from the slides. Allow the coverslips to float off in 2× SSC solution. Wash the slides a few times by agitation.
3. Wash slides in wash solution A by gently agitating 3× for 3 min each.
4. Further wash with warmed 2× SSC at 46°C, 3×.
5. Place slides in wash solution C; the slides are now ready for detection. If necessary, slides can be kept at 4°C in wash solution C for up to 2 d.

3.3.5.2. Repetitive DNA Probes

1. Prewarm wash solution B to 43°C and 2× SSC to 37°C.
2. Remove rubber cement and coverslips as described in **Subheading 3.2.6.1., step 2**.
3. Wash slides in wash solution B for 20 min by agitating.
4. Further wash 2× with warmed 2× SSC at 37°C for 4 min each.
5. Place slides in wash solution C; the slides are now ready for detection.

3.3.6. Detection and Amplification

3.3.6.1. Detection with Amplification

1. Remove slides from the wash jar and blot excess liquid from the edge.

2. Quickly apply 30 µL of blocking solution to each slide. Place a plastic coverslip over the solution and incubate at room temperature for 5 min.
3. Gently peel off plastic coverslips, tilt the slide, and allow the fluid to drain.
4. Apply 30 µL of FITC detection solution to each slide and cover with a fresh plastic coverslip. Incubate 20 min at 37°C in a dark humidified chamber (from this step onward, it is very important to minimize exposure to light).
5. Remove coverslips and wash the slides in wash solution C 3× for 3 min each in a new solution.
6. Apply 30 µL of blocking solution to each slide, place a plastic coverslip on top of the solution, and incubate at room temperature for 5 min.
7. Apply 30 µL of biotinylated goat anti-avidin antibody working solution to each slide and cover with a plastic coverslip. Incubate at 37°C for 20 min.
8. Wash the slides 3× for 3 min each in wash solution C after removing the coverslips.
9. Apply 30 µL of blocking solution to each slide. Place a plastic coverslip over the solution and incubate at room temperature for 5 min.
10. Peel off plastic coverslips, tilt the slide, and allow the fluid to drain.
11. Apply 30 µL of FITC detection solution to each slide and cover with a fresh plastic coverslip. Incubate for 20 min at 37°C.
12. Remove coverslips and wash the slides in wash solution C 3× for 3 min each.
13. Stain with DAPI by immersing the slides in 0.2 µg/mL of DAPI in 2× SSC for 5–10 min at room temperature. Rinse in 2× SSC 3× for 2 min each.
14. Mount the slides with 10 µL of antifade solution. Place a 22 × 40 cm glass coverslip on the slide. Apply gentle downward pressure to flatten coverslip before examining the slide. PI (propidium iodide) staining may be applied to visualize the FISH signals on the counterstained chromosome.

3.3.6.2. DETECTION WITHOUT AMPLIFICATION

Signal amplification may not be necessary when YAC or BAC probes or α-satellite probes are used for FISH detection.

1. Remove slides from the wash jar and blot excess liquid from the edge.
2. Apply 30 µL of blocking solution to each slide. Place a plastic coverslip over the solution and incubate at room temperature for 5 min.
3. Gently peel off plastic coverslips, tilt the slide, and allow fluid to drain.
4. Apply 30 µL of FITC detection solution to each slide and cover with a fresh plastic coverslip. Incubate for 20 min at 37°C in a dark humidified chamber (from this step onwards, it is very important to minimize light exposure).
5. Remove coverslips and wash the slides in wash solution C 3× for 3 min each.
6. Stain with DAPI by immersing the slides in 0.2 µg/mL of DAPI in 2× SSC for 5–10 min at room temperature. Rinse in 2× SSC 3×for 2 min each.
7. Mount the slides with 10 µL of antifade solution or 10 µL of PI solution. Place a 22 × 40 cm glass coverslip on the slide. Apply gentle downward pressure to flatten the coverslip before examining slide.

If the signals need further amplification after the slide has been examined, carefully clean all the oil from the coverslip, remove it, and place the slide in 2× SSC for a quick wash, then wash slides 3× for 5 min each in wash solution C. The remaining steps are the same as **steps 6–14** in **Subheading 3.3.6.1**.

3.3.6.3. TWO-COLOR DETECTION

1. Follow all the steps as in **Subheading 3.3.6.2.**, except that in **step 4**, add anti-digoxigenin–rhodomine fragments together with FITC–avidin (**Note 10**).
2. Mix 100 μL of anti-digoxigenin–rhodamine fragments, 10 μL of FITC–avidin, and 890 μL of detection solution.
3. Apply 30 μL of the above mixture to each slide. Incubate the slides at 37°C for 20 min with light protection. Wash them in washing solution C and 2× SSC at room temperature.
4. Stain with DAPI by immersing the slides in 0.2 μg/mL of DAPI in 2× SSC for 5-10 min at room temperature (do not use propidium iodide). Rinse in 2× SSC 3× for 2 min each.
5. Mount the slides with 10 μL of antifade solution.

3.4. Photography

The slides may be examined immediately. Use the DAPI filter to identify DAPI-stained chromatin or DNA fibers first, then switch to the FITC or rhodamine or dual filter to localize the FISH signals. Specialized equipment such as sensitive charged-couple device (CCD) camera is needed for fiber FISH. In many cases, quantitative analysis and image processing is required. If those systems are not available, high-speed film should be used to record the images. Kodak Ektachrome P800/1600 can be push-developed up to 3200.

4. Notes

1. Although high resolution normally precludes wide coverage of the physical mapping, careful selection of the FISH target can achieve an acceptable compromise. Since less condensation results in higher mapping resolution and vise versa, the resolution of FISH detection increases from metaphase chromosomes to interphase nuclei, to free chromatin fiber to DNA fiber. The combination of multiple-level targets is required to produce the right resolution and coverage.
2. According to our experience, cord blood is ideal for the generation of free chromatin fiber because of its sensitivity to drug treatment and the large volume of sample.
3. EtBr, BrdU, and *m*-AMSA are toxic reagents. Handle with care. To induce the free chromatin fiber, *m*-AMSA is the most effective, but if unavailable, it can be replaced by either BrdU or EtBr.
4. For lymphocyte cells, drug treatment plus hypotonic shock usually generates good results for chromatin fibers. An optional step to promote fiber production is to increase the pH of hypotonic solution to 9–10 by mixing with 1 m*M* sodium

borate solution *(19)*. For fibroblastoid cell lines, however, drug treatment is much less effective and the alkaline lysis buffer has to be used. Because releasing conditions vary among different cell lines, the concentration of KCl must be determined empirically, from 0.4% to 1–2%. Additional protocols for preparation of free chromatin fibers from lymphocyte cells can be found in our previous publication *(27)*.

5. Alkaline treatment can be replaced by formamide *(20)* or SDS lysis buffer: 2.5 mL of 20% SDS (0.5% final concentration), 10 mL of 0.5 M EDTA (50 mM final concentration), 20 mL of 1 M Tris-HCl, pH 7.4 (200 mM final concentration), and 67.5 mL of H_2O *(18,28)*.

6. Biotin and digoxigenin nick-translation labeling kit can be prepared as follows: DNA polymerase I, DNase I, 10× dNTP mix (0.2 mM dATP, 0.2 mM dCTP, 0.1 mM dGTP, and 0.1 mM dTTP), 10× reaction buffer (500 mM Tris-HCl, pH 7.8, 50 mM $MgCl_2$, 100 mM β-mercaptoethanol, 100 μL/mL bovine serum albumin), and biotin-16-dUTP or DIG-11-dUTP. For labeling 1 μg of probe DNA, mix 5 μL of 10× reaction buffer, 2.5 μL of biotin-16-dUTP, 20 U of DNA polymerase I, and diluted DNase I (1:1000). Bring the total reaction volume to 50 μL with dH$_2$O. Enzyme solution is the last ingredient to add. For labeling probe with digoxigenin, use 5 μL of DIG-11-dUTP instead of biotin-16-dUTP *(27)*.

7. "Super Blocker" is a cocktail of DNA developed to compete out signals from repetitive elements during FISH analysis and Southern blotting. The effective blocker reagent can be used as Cot-I DNA. Information can be obtained from SeeDNA Biotech Inc. (http://www.seedna.com).

8. Recently, FISH detection on DNA fibers prepared from "water–air phase interaction," so-called "molecular combing" (a type of fiber FISH), showed its great potential for deletion detection and physical mapping. Similarly, lysis of nuclei in gel blocks and stretching the DNA fiber by mechanical or electronic force are effective protocols for DNA fiber FISH *(22)*. The protocol we suggest, however, is the simplest one *(20)*. It should be useful for most of cases and can serve as a technical gate for more complicated, yet powerful protocols.

9. Different from the chromosomal FISH mapping in which the overdenaturation should be avoided to keep the chromosomal morphological features including banding pattern *(29,30)*, longer denaturation time is needed for fiber FISH.

10. The different combinations of color detection should be used according to designed experiments. For example, if biotin labeling is used for one probe and detected by avidin-FITC, it can be amplified by using biotinylated-goat-anti-avidine, and then detected by one more layer of avidin-FITC. If digoxigenin labeling is used for another probe, antidigoxigenin-rhodamine Fab fragments can be used for detection. Alternatively, the amplification and detection can be achieved by using mouse-antidigoxigenin, then rabbit-antimouse-rhodamine (TRITC), or goat-antirabbit-TRITC.

Acknowledgments

Free chromatin fiber FISH was originally developed in Dr. Lap Chee Tsui's Laboratory of Hospital for Sick Children, Toronto. I thank him for his encouragement and support. Thanks also go to Dr. Peter Moens of York University, Toronto, for his advice on this project. A special thanks belongs to Barbara Spyropoulos for her valuable help including editing this manuscript. This work was supported by SeeDNA Biotech Inc.

References

1. Heng, H. H. Q., Spyropoulos, B., and Moens, P. B. (1997) FISH technology in chromosome and genome research. *Bioessays* **19,** 75–84.

2. Lawrence, B. J. (1990) A fluorescence *in situ* hybridization approach for gene mapping and the study of nuclear organization. in *Genome Analysis*, vol. I: *Genetic and Physical Mapping* (Davies, E. K. and Tilghman, M. S. eds.), Cold Spring Harbor Laboratory Press, Cold Spring Harbor, NY, pp. 1–38.

3. Trask, J. B. (1991) Fluorescence *in situ* hybridization. *Trends Genet.* **7,** 149–154.

4. Lichter, P., Boyle, A. L., Cremer, T., and Ward, D. C. (1991) Analysis of genes and chromosomes by nonisotopic *in situ* hybridization. *GATA* **8,** 24–35.

5. Heng, H. H. Q., Squire, J., and Tsui, L.-C. (1991) Chromatin mapping — a strategy for physical characterization of the human genome by hybridization *in situ*. Proc. 8th. Int. Cong. Hum. Gen. *Am. J. Hum. Genet.* **49 (Suppl),** 368.

6. Heng, H. H. Q., Squire, J., and Tsui, L.-C. (1992) High resolution mapping of mammalian genes by *in situ* hybridization to free chromatin. *Proc. Natl. Acad. Sci. USA* **89,** 9509–9513.

7. Wiegant, J., Kalle, W., Mullenders, L., Brookes, S., Hovers, J. M. N., Dauwerse, J. G., van Ommen, G. J. B., and Raap, A. K. (1992) High resolution *in situ* hybridization using DNA halo preparation. *Hum. Mol. Genet.* **1,** 587–592.

8. Heiskanen, M., Peltonen, L., and Palotie, A. (1996) Visual mapping by high resolution FISH. *Trends Genet.* **12,** 379–382.

9. Raap, A. K., Florijn, R. J., Blonden, L. A. J., Wiegant, J., Vaandrager, J.-W., Vrolijk, H., den Dunnen, J., Tanke, H. J., and van Ommen, G.-J. (1996) Fiber FISH as a DNA mapping tool. *Methods: A Companion to Methods in Enzymology* **9,** 67–73.

10. Heng, H. H. Q. and Tsui, L.-C. (1998) High resolution free chromatin/DNA fiber FISH. *J. Chromatogr. A* **806,** 219–229.

11. Heng, H. H. Q., Spyropoulos, B., and Moens, P. B. (1999) DNA–protein *in situ* covisualization for chromosome analysis, in *Methods in Molecular Biology,* In Situ *Hybridization Protocols* (Darby, I. A., ed.), Humana Press, Totowa, NJ, Chap. 2.

12. Heng, H. H. Q., Chamberlain, J. W., Shi, X.-M., Spyropoulos, B., Tsui, L.-C., and Moens, P. B. (1996) Regulation of meiotic chromatin loop size by chromosomal position. *Proc. Natl. Acad. Sci. USA* **93,** 2795–2800.

13. Lichter, P., Tang, C.-C., Gall, K., Hermanson, G., Evans, G., Housman, D., and Ward, D. (1990) High-resolution mapping of human chromosome 11 by *in situ* hybridization with cosmid clones. *Science* **247,** 64–69.

14. Brandriff, B., Gordon, L., and Trask, B. (1991) A new system for high-resolution DNA sequence mapping in interphase pronuclei. *Genomics* **10,** 75–82.

15. Heng, H. H. Q. and Zhao X. L. (1987) Studies of the free chromatin structure. *J. Sichuan Univ. Nat. Sci. Edi.* **24,** 479–485.

16. Heng, H. H. Q. and Shi, X-M. (1997) From free chromatin analysis to high resolution fiber FISH. *Cell Res.* **7,** 119–124.
17. Heng, H. H. Q. and Chen, W. Y. (1985) The study of the chromatin and the chromosome structure for Bufo gargarizans by light microscope. *J. Sichuan Univ. Nat. Sci.* **22,** 105–109.
18. Parra, I. and Windle, B. (1993) High resolution visual mapping of stretched DNA by fluorescent hybridization. *Nat. Genet.* **5,** 17–21.
19. Haaf, T. and Ward, D. C. (1994) Structural analysis of α-satellite DNA and centromere proteins using extended chromatin and chromosomes. *Hum. Mol. Genet.* **3,** 697–709.
20. Fidlerova, H., Senger, G., Kost, M., Sanseau, P., and Sheer, D. (1994) Two simple procedures for releasing chromatin from routinely fixed cells for fluorescence *in situ* hybridization. *Cytogenet. Cell Genet.* **65,** 203–205.
21. Houseal, T. W., Dackowski, W. R., Landes, G. M., and Klinger, K. W. (1994) High resolution mapping of overlapping cosmids by fluorescence *in situ* hybridization. *Cytometry* **15,** 193–198.
22. Heiskanen, M., Karhu, R., Hellsten, E., Peltonen, L., Kallioniemi, O. P., and Palotie, A. (1994) High resolution mapping using fluorescence *in situ* hybridization to extended DNA fibers prepared from agarose-embedded cells. *BioTechniques* **17,** 928–934.
23. Weier, H.-U. G., Wang, M., Mullikin, J. C., Zhu, Y., Cheng, J.-F., Greulich, K. M., Bensimon, A., and Gray, J. W. (1995) Quantitative DNA fiber mapping. *Hum. Mol. Genet.* **4,** 1903–1910.
24. Michalet, X., Ekong, R., Fougerousse, F., Rousseaux, S., Schurra, C., Hornigold, N., van Slegtenhorst, M., Wolfe, J., Povey, S., Beckmann, J, and Bensimon, A. (1997) Dynamic molecular combing: stretching the whole human genome for high-resolution studies. *Science* **277,** 1518–1522.
25. Samad, A., Huff, J. E., Cai, W., and Schwartz, D. C. (1995) Optical mapping: a novel, single-molecule approach to genomic analysis. *Genome Res.* **5,** 1–4.
26. Florijn, R. J., Bonden, A. J., Vrolijk, H., Wiegant, J., Vaandrager, J.-W., Baas, F., den Dunnen, J. T., Tanke, H. J., van Ommen, G. J. B., and Raap, A. K. (1995) High-resolution DNA fiber-FISH for genomic DNA mapping and color bar-coding of large genes. *Hum. Mol. Genet.* **4,** 831–836.
27. Heng, H. H. Q. and Tsui, L.-C. (1994) Free chromatin mapping by FISH, in *Methods in Molecular Biology*, vol. 33: In Situ *Hybridization Protocols* (Choo, K. H. A., ed.), Humana Press, Totowa, NJ, pp. 109–122.
28. Heng, H. H. Q., Tsui, L.-C., Windle, B., and Parra I. (1995) High resolution FISH analysis, in *Current Protocols in Human Genetics* (Dracopoli, N., Haines, J., Korf, B., Moir, D., Morton, C., Seidman, C., Seidman, J., and Smith, D., eds.), John Wiley and Sons, New York, pp. 4.5.1–4.5.25.
29. Heng, H. H. Q. and Tsui, L.-C. (1994) FISH detection on DAPI-banded chromosomes, in *Methods in Molecular Biology*, vol. 33: In Situ *Hybridization Protocols* (Choo, K. H. A., ed.), Humana Press, Totowa, NJ, pp. 35–49.
30. Heng, H. H. Q. and Tsui, L.-C. (1993) Modes of DAPI banding and simultaneous *in situ* hybridization. *Chromosoma* **102,** 325–332.

6

In Situ Hybridization to Polytene Chromosomes of *Drosophila melanogaster* and Other Dipteran Species

A. Marie Phillips, Jon Martin, and Daniel G. Bedo

1. Introduction

In situ hybridization of nucleotide sequence to *Drosophila melanogaster* interphase polytene chromosomes was initially developed by Pardue, who has published an extensive account of *Drosophila* polytene chromosomes and hybridization to these chromosomes *(1)*. The procedure, which used radioactively labeled probes, was later adapted to allow the use of more sensitive and safer nonradioactive labeling methods *(2)*. The method has been used extensively for gene mapping and identification of transposable element insertion sites. *In situ* hybridization to chromosomes has also been performed in cross-species studies within Diptera to give valuable evolutionary information, including conservation of linkage groups, and to identify homologous genes in related species *(3–5)*. In addition, there have been recent advances that extend this technique to salivary gland preparations from stored fixed specimens *(6)*.

In this chapter we describe in detail a method (which is derived from that of Engels et al. *[2]*) for the *in situ* hybridization of biotinylated DNA probes to *D. melanogaster* salivary gland polytene chromosomes (*see* **Fig. 1** and **Subheading 3.**).

This method can be simplified by experienced researchers *(7)* but, as described, has been found to be suitable for use at all levels of expertise, from undergraduate teaching to the research laboratory. We have successfully used this method for gene mapping *(8,9)*, orientation of chromosomal walks *(10)*, location of transposable elements *(11)*, and investigations of structural abnormalities in mutant flies *(12)*. The probes used include single-locus probes of 400 bp to 14 kb and repeat sequence probes including RNA genes. *In situ*

From: *Methods in Molecular Biology*, Vol. 123: In Situ *Hybridization Protocols*
Edited by: I. A. Darby © Humana Press Inc., Totowa, NJ

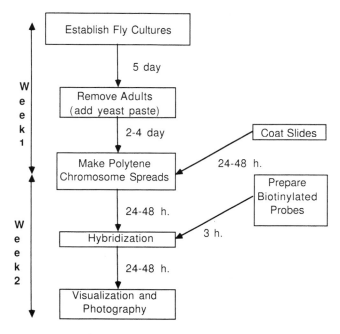

Fig. 1. Flow diagram of method.

hybridization has been valuable during gene cloning in the identification of chimeric clones. This may be particularly advantageous where yeast artificial chromosome (YAC) libraries containing a high proportion of chimeric clones must be screened. The method is reliable but we discuss potential problems and how they may be avoided. Considerable information is given on the preparation of good chromosome spreads as the success of the experiment is largely dependent on this step. The polymerase chain reaction (PCR) has replaced nick-translation as a method of generating probes in many laboratories, and this method of probe labeling is outlined in **Subheading 3.2.1**.

As there are other Dipteran species that are less well characterized genetically but are of scientific interest and/or great importance as pest species, we also present information on *in situ* hybridization to polytene chromosomes from *Chironomus* species, *Lucilia cuprina*, *Chrysomya bezziana*, and *Ceritatis capitata*.

2. Materials

2.1. Chromosome Spreads

1. Microscope slides. These must be of good quality. Check under phase contrast for microscopic chips in the glass, which can make analysis of data and photography difficult when phase optics are used. Clean slides are essential. A conve-

nient and safe cleaning method is to soak racks of slides overnight in a 5% solution of Decon 90 or similar laboratory cleaning agent. This is followed by a 1-h rinse in running tap water. To ensure that there is good circulation between the slides during rinsing, the rack is suspended above the bottom of the container. The slides are then rinsed twice, for 5 min each rinse, in distilled water and air dried in a dust-free environment. The cleaned slides are then coated in Denhardt's solution (*[13]* 1% polyvinylpyrrolidone [Kodak], 1% Ficoll, 1% bovine serum albumin — nuclease free [Sigma A 7906]) diluted 1:50 with 3× saline sodium citrate (SSC) (*see* **Subheading 2.1., step 14**). Incubate the slides in this solution at 65°C for 2.5 h, rinse briefly in distilled water at room temperature, and fix in acetic acid:ethanol (3 vol glacial acetic acid:1 vol ethanol) for 30 min. The slides can then be air dried in a dust-free environment (a laminar flow cabinet is useful, if available) and stored in a closed container at 4°C until required.

Alternatively, chromic acid can be used to clean the slides. As chromic acid is very corrosive, care must be taken to protect skin and clothing from exposure, and eye protection (safety glasses or head shield) is essential. Gelatin/chrome alum can be used to coat slides (*1*).

2. Coverslips. 22 mm × 22 mm glass coverslips. These should be siliconized by dipping in a commercial compound e.g., Siliclad (Sigma). We dip 100 coverslips at a time using a slotted Perspex grid. The coverslips are immersed briefly in the siliconizing solution, then allowed to air dry. Immediately before use wipe the coverslips with ethanol using a lint-free tissue.

3. Larvae. Wandering third instar larvae are obtained from well-yeasted stocks established at low density (12 females, 6 males per 250-mL cream bottle) on a sugar/semolina food. We maintain these stocks at 22°C with a 12 h light/dark cycle. Five days after the cross is established the adult flies are removed and a teaspoon of thick yeast paste is added to the bottle. Moistening the food with water after 7 d will induce the larvae to seek sites for pupation on the sides of the bottle, and increases the ease of their capture (*see* **Note 1**).

4. Dissecting needles and fine forceps.

5. 150 mm diameter plastic Petri dish with lid.

6. Small soft paint brush.

7. Paper tissues, lint-free lens type.

8. Dissecting solution: The larvae are dissected in a 1:5 dilution in water of the fixing solution (*see* **step 9**). Tap water or Ringer's solution can also be used as dissecting medium. Experienced workers can dissect the larvae in the fixing solution.

9. 1:2:3 Fixing solution: 1 part lactic acid, 2 parts water, 3 parts glacial acetic acid.

10. One-kilogram brass weights for flattening chromosome spreads. There are many alternative methods that can be used to apply constant even vertical pressure to the chromosome preparations. Included in these are the use of "C" clamps, or of glass/metal plates weighted to give the desired pressure. The chromosome squashes must be flat, well adhered to the slide, and ribbonlike under phase contrast for optimal results. The weighting added should be determined empirically.

11. Liquid nitrogen in a wide-necked flask or dry ice in the form of a flat block is used as a freezing agent.
12. 6× Coplin jars, plastic slide mailers, or other containers suitable for submerging slides while keeping them apart. If more than five slides are to be prepared at a time, use a slide tank.
13. Ethanol (reagent grade): Keep 100 mL of ethanol at –20°C and 100 mL at 4°C, in addition to a supply at room temperature.
14. 20× SSC stock: 3 M NaCl, 300 mM Na citrate, pH 7.4. Dilute to desired concentration as required.
15. 1 M Sodium hydroxide: Prepare fresh on day of use.

2.2. Probes and Hybridization

1. DNA: 400–500 ng of the probe DNA to be hybridized.
2. Commercial nick-translation kit (BRL, Gaithersburg, MD, USA).
3. Biotin-11-dUTP (Sigma).
4. 3 M Sodium acetate, pH 4.8.
5. Absolute ethanol (analytical grade).
6. 10% Sodium dodecyl sulfate (SDS) solution in water.
7. 20 mg/mL Salmon/herring sperm DNA in sterile water. The DNA can be sheared by sonication at 20 µm peak-to-peak, in 30-s bursts with 30-s rests on ice, until the solution is nonviscous. Alternatively, shear DNA by repeatedly passing the solution through an 18-gage syringe needle.
8. Plastic coverslips, 22 mm × 22 mm.
9. 50% Dextran sulfate solution in water.
10. Formamide (analytical grade): Deionize the formamide prior to use by adding mixed bed resin beads, and stirring for 30 min. Filter before use.
11. Rubber cement for sealing coverslip to slide, e.g., vulcanizing solution, Tip Top Stahlgruber, München, Germany.
12. Carrier mix: to make 500 µL, mix 150 µL of 50% dextran sulfate (warm to reduce the viscosity and aliquot using a sterile disposable plastic tip cut to increase the bore size), 250 µL of 20× SSC, and 100 µL of 20 mg/mL sheared salmon sperm DNA.
13. Microliter pipets (e.g., Gilson, Paris, France) 20 µL, 200 µL, and 1000 µL.
14. 1.5-mL Microfuge tubes, sterilized by autoclaving.
15. Plastic tips for the micropipets, sterilized by autoclaving.
16. TE buffer: 10 mM Tris-HCl, 1 mM EDTA, pH 8.0.

2.3. Staining

1. Commercially available staining kit using conjugated alkaline phosphatase–streptavadin (BRL BluGene kit).
2. Buffers. These buffers are as specified in the notes accompanying the staining kit:
   a. Buffer 1: 150 mM NaCl, 100 mM Tris-HCl, pH 7.5. This is the same as a 1:10 dilution of the neutralizing solution used in preparing gels for Southern blotting.

    b. Buffer 2: 2% solution of nuclease-free bovine serum albumin (BSA) in buffer 1. Prepare this buffer on the day it is to be used.
    c. Buffer 3: 100 mM Tris-HCl, pH 9.5; 100 mM NaCl; 10 mM MgCl$_2$. This buffer can be made in advance and stored at 4°C.

3. Methods

3.1. Preparing the Chromosome Spreads

1. Wipe siliconized coverslips free of dust with lint-free tissues and arrange about six coverslips in a single layer on the bottom of a large plastic Petri dish. Cover to keep them dust-free.
2. Dissect salivary glands from third instar larvae in a small drop of a 1:5 dilution of fixative (dissecting solution) in water on a clean noncoated slide on a dark background. Visualize the dissection using a binocular microscope.
3. Clean the slide of all nonsalivary gland material and check the size of the dissected glands. Discard all except large glands with clearly visible polytene cells.
4. Remove any fat adhering to the glands using the point of a syringe needle as a knife. A dissecting needle can be used if a syringe needle is not available. Avoid rupturing the glands during fat removal. Keep the glands well moistened with dissecting solution until transfer.
5. Transfer one gland from each pair to a 10-µL drop of 1:2:3 fixative on a siliconized coverslip.
6. Check, using the dissecting microscope, that no nonglandular material has been transferred and remove any contaminating material.
7. Leave the gland to fix for 2–3 min.
8. Pick up the coverslip using a dust-free Denhardt's treated slide. Position the coverslip on the lower half of the slide but with several millimeters of uncovered slide left at the bottom to allow adequate room for sealing later.
9. Using the blunt end of dissecting forceps, "write" gently across the coverslip to disperse the tissue and break open the nuclei (the word "write" gives an indication of the pressure that should be applied).
10. View the preparation under phase contrast with magnification ×40.
11. If the preparation is not well spread repeat the "writing" and/or move the coverslip a millimeter sideways and up and down.
12. If the chromosome arms are well spread, press to flatten, using even vertical thumb pressure or pressure applied using a "C" clamp (the chromosomes when adequately flattened should be almost transparent and ribbonlike, i.e., well adhered to the coated slide [*see* **Note 2**]). Place the slides under weights for approx 15 min, then store flat at 4°C until all slides have been prepared. The slides may be stored at 4°C overnight and this is our normal practice. The slides are stored uncovered on a flat surface. It has been suggested that further flattening occurs during this period as the fixative evaporates (*2*).
13. Using forceps dip the slides into liquid nitrogen until boiling of the N$_2$ stops, then hold in the air and allow the slide to warm until clouded with condensation (a few seconds).

14. Lift one corner of the coverslip and flip it off the slide using a razor blade or scalpel blade. The coverslip comes off intact and there is no loss of chromosomes.

15. Plunge the slides into a Coplin jar of cold ethanol (–20°C) and allow to come to room temperature. If treating slides in batches, place the slides into –20°C ethanol for 15 min, 4°C ethanol for 15 min, and then in room temperature ethanol for 15 min.

16. Air dry in a dust-free environment.

17. Dried slides can be kept, at 4°C in a slide storage box with desiccant, for weeks before use. It is our usual practice, however, to heat pretreat and dehydrate the slides for storage (*see* **steps 18–23**) within 1 wk.

18. Heat one slide tank containing 2× SSC, two slide tanks containing 70% ethanol, and one containing 95% ethanol to 65°C, in a water bath.

19. Leave all four of the tanks in the water bath at 65°C and incubate the slides in the 2× SSC tank for 30 min.

20. Remove all four tanks to the bench (room temperature) and rinse the slides in one 70% ethanol tank.

21. Transfer the slides to the second 70% ethanol tank and let stand 20 min.

22. Place the slides in the 95% ethanol tank for 10 min.

23. Remove the slides and allow to air dry. The chromosome preparations can now be stored, dry, in a slide box at 4°C for at least 6 mo. Slides can also be stored long term in 100% propanol-2.

3.2. Preparing the Biotinylated Probe

Biotinylated probes are routinely made using a commercial nick-translation kit with biotin-11-dUTP (or other label of choice; *see* **Subheading 3.2.1.** and **Note 3**) replacing the radioactively labeled nucleotide. Alternatively, DNA for probes can be labeled using a PCR method (*see* **Subheading 3.2.1.**). All buffers, microfuge tubes, and pipet tips used should be sterilized by autoclaving. It is essential to wear gloves for steps involving the use of formamide and dimethylformamide.

1. Nick-translate 1 µg of DNA from a genomic clone (including λ arms) or 500 ng of purified insert DNA in a microfuge tube in a 50 µL reaction containing 5 µL of dATP, dCTP, dGTP nucleotide mix; 2.5 µL of biotin-11-dUTP (4 mM); and 5 µL of enzyme. The reaction is carried out for 90 min at 14°C. Stop with 5 µL of stop buffer (in kit).

2. Add 0.75 µL of 10% SDS, 3 M Na acetate (pH 5.2) to 0.3 M in final reaction volume, 1 µL of sheared salmon sperm DNA, 10 mg/mL, and 2 vol of cold absolute ethanol.

3. Precipitate the probe at –70°C, 15 min or longer.

4. Centrifuge the microfuge tube at 10,000g for 15 min to pellet the DNA.

5. Wash the pellet with 70% ethanol.

6. Dry the DNA.

7. Resuspend in 20 μL of TE buffer, pH 8.0.
8. Reprecipitate the DNA with sodium acetate and ethanol, and repeat **steps 3–7**.
9. Store the probe at 4°C. We have successfully used probes stored for more than 2 years.

3.2.1. PCR as a Method of Labeling the Probe

PCR is a convenient alternative to nick-translation as a method of labeling probes. The template for the PCR reaction may be genomic DNA or a clone containing the desired region. A single PCR amplification, using appropriate forward and reverse oligonucleotide primers, both produces and labels the required fragment. An advantage of this system is that essentially only the sequences required as the probe are labeled. Also smaller probes, readily generated by PCR, may be better able to penetrate the chromosomal matrix and bind to target DNA. The technique described here can be adapted for biotin-11-dUTP or DIG-11-dUTP. Digoxigenin labeled probes, generated by PCR, are now routinely used by one of us (J. M.), *(14)* (*see* **Fig. 2**).

1. PCR and labeling the probe: Quantities and final concentrations are given for 100 μL of PCR mix. Add 1 ng of template DNA, 200 nM of each primer; 250 μM of each nucleotide (10 μL of 2.5 mM dNTP mix containing DIG-11-dUTP [Boehringer DIG Labeling Mix], or biotin-11-dUTP); 10 μL of 10× PCR buffer (containing 1.5–5.0 mM Mg^{2+}): 2.5 U of polymerase:water to 100 μL.
   Thermocycler program: 95°C, 5 min: 55°C, 2 min; followed by 30 cycles of denaturation (95°C, 1 min), annealing (55°C, 2 min), and extension (68°C, 10 min). Thermocycler times and temperatures and component concentrations, particularly Mg^{2+} concentrations, may need to be optimized for the particular probe.

3.3. Denaturation and Prehybridization

(An alternative, heat denaturation, method of denaturation and hybridization of chromosomes is described in **Subheading 3.9.4.**).
Carry out all dipping and washing steps in Coplin jars or slide tanks.

1. Denature the DNA *in situ* by dipping the slides in 0.07 M NaOH (made from a freshly prepared 1 M NaOH stock) for 2 min. The time should be accurately measured. Two minutes is usually optimal for *Drosophila melanogaster* chromosomes probed with DNA from this species (a series of time points between 1.75 and 2.5 min can be tested) but a longer denaturation time is necessary for cross-species studies *(3)*.
2. Immediately rinse the slides twice in 2× SSC, 5 min each rinse.
3. Dehydrate the spreads by immersing in 70% ethanol twice for 5 min, and then in 95% ethanol for 5 min.
4. Air dry the slides in a dust-free environment.
5. Check the slides by viewing under phase contrast (×40) and mark the sides of the slide with a pencil to indicate the position of the chromosomes.

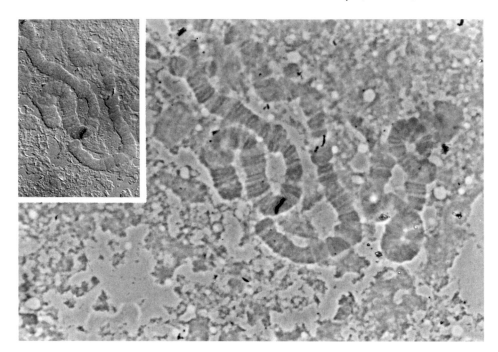

Fig. 2. A cDNA sequence for a single copy gene was labeled with biotin by nick-translation and is shown hybridized to a polytene chromosome from *Drosophila melanogaster* salivary gland. Phase-contrast optics (×40) were used to visualize the stained biotin probe/chromosomal DNA hybrid (larger photograph). The bands on the chromosome are clear and the chromosome is identified as chromosome 1 by comparison with published photographic maps, with the probe hybridizing to band 4F. In the inset, the same chromosome has been photographed using Nomarski optics, and illustrates the usefulness of this method of chromosome visualization for distinguishing the hybrid signal from a background of dark bands.

3.4. Hybridization

1. Freshly prepare a hybridization mixture of 270 μL of deionized formamide and 258 μL of carrier mix (**Subheading 2.2., step 12**). The carrier mix may have to be warmed for accurate pipetting and to ensure a homogeneous mixture with the formamide.
2. Place 26.4 μL of hybridization mixture and 3.6 μL of biotinylated probe in a microfuge tube. Mix by pipetting. This is enough probe solution for four or five slides. The formamide concentration in the probe solution is 45%. Smaller volumes of probe solution can be made providing that aliquots can be measured with sufficient accuracy to ensure that the formamide concentration is not significantly changed. Lower formamide concentrations are used for cross-species studies (*4*) (*see* also **Subheading 3.9.2.**).

3. Denature the probe by heating to 80°C for 10 min.
4. Immediately place the tube on ice, cool, and touch spin in a microcentrifuge to bring down condensation. The probe should be kept ice-cold until used to prevent reannealing.
5. Place clean plastic coverslips on a flat surface and pipet 6 µL of the denatured probe onto the center of each coverslip
6. Lower the region of the slide containing the denatured chromosomes (as indicated by the pencil markings) onto the coverslip. It is important to avoid introducing air bubbles. Ensure that sufficient room is left around the coverslip to enable adequate sealing. The probe solution should spread to completely cover the area beneath the coverslip. If necessary apply a little gentle vertical pressure.
7. Turn the slide over so that the coverslip is now facing upwards and seal the edges of the coverslip to the slide with rubber cement.
8. Place the slides in a box, or other flat container, and hybridize overnight at 37°C.

3.5. Posthybridization

1. Place two Coplin jars or slide tanks containing 2× SSC at 37°C.
2. Set another water bath at 60°C.
3. Take the hybridized slides from the container and peel off the rubber cement by holding one corner with fine forceps and lifting. The rubber cement and the coverslip should come off together.
4. Rub off any remaining cement, being careful not to touch the chromosomes.
5. Place the slides in one Coplin jar of 2× SSC at 37°C for 10 min.
6. Transfer to the second Coplin jar of 2× SSC at 37°C for a further 10 min.
7. Wash the slides twice in 2× SSC at room temperature.
8. Briefly rinse the slides in buffer 1.
9. Transfer the slides to buffer 2 and incubate at 60°C for 1 h.
10. Rinse the slides briefly in buffer 2 at room temperature.

3.6. Staining the Biotin Probe/Chromosomal DNA Hybrid

This is an adaptation of the BRL BluGene Kit method used for staining biotin labeled probes applied to Southern DNA blots.

1. Place the slides on a flat surface.
2. Dilute the commercial stock of streptavadin–alkaline phosphate (sv–ap) solution 1:1000 in buffer 1 (1 µL in 1 mL).
3. Pipet this solution over the hybridized chromosomes on the slide, using approx 100 µL/slide. After 10 min add an additional 100 µL of sv–ap solution. The chromosomes should be exposed to the conjugate for 15–20 min.
4. Wash the slides twice in buffer 1, for 15 min each wash.
5. Wash the slide in buffer 3 for 15 min.
6. During the buffer 3 wash make up a staining mix of 4.4 µL of nitroblue tetrazolium (NBT) and 3.3 µL BCIP (in kit) in 1 mL of buffer 3.
7. Clean a plastic coverslip for each hybridized slide.
8. Place 50 µL of fresh NBT/BCIP dye mix on each plastic coverslip.

9. Pick up each coverslip with a slide, using the pencil marks to ensure that the dye mix covers the chromosome spreads. Do not seal.
10. Place the slides in a tightly closed box in the dark for at least 2 h (2–4 h is usually sufficient for complete staining).
11. Monitor the staining by viewing the chromosomes under phase-contrast optics.
12. When a blue band is clearly visible stop the reaction by rinsing the slide with water.
13. Allow the slides to dry.

3.7. Visualization

The stained dry slides will store for several years if kept in a slide storage box at room temperature. Care must be taken to exclude dust and to prevent scratching. It is wise to obtain a permanent photographic record as soon as possible and we routinely photograph slides within 24 h of staining.

To visualize the stained band, place a drop of water on the slide and add a clean glass coverslip. View under phase contrast. A clear signal should be visible at the same chromosomal location/s in all labeled cells (*see* **Note 3** for possible explanations of lack of signal).

In our experience the chromosome banding pattern is clearly distinguishable in most cells and the chromosomes can be photographed under phase contrast without Giemsa or Orcein staining. If you wish to stain the chromosomes it is important to understain rather than overstain. Distinguishing the hybrid signal from dark chromosomal bands can be readily achieved by switching from phase to light optics without moving the microscope stage. The hybrid band can be clearly distinguished from the background using Nomarski optics and we find that photographing the preparations under both phase and Nomarski gives excellent results (**Fig. 2**).

3.8. Reprobing Chromosomes

It is sometimes desirable to reprobe chromosomes with an additional probe. This is particularly useful when orientating chromosome walks or when attempting to identify the nature of chromosomal abnormalities *(12)*. The chromosomes should be photographed first, and then denatured for a second time (for 2 min) before reprobing. There will be some loss of band resolution; however, the chromosomal position of the signal can be determined using the photographs taken prior to the denaturation step.

3.9. In Situ *hybridization to Other Diptera and to Stored, Fixed Samples*

3.9.1. In situ *Hybridization to Polytene Chromosomes from Fresh and Stored Fixed Samples of* Chironomus

The basic technique outlined for preparing *Drosophila* chromosomes can be used for salivary gland chromosomes of *Chironomus* but problems may be encountered with the large amounts of salivary secretion in the glands of the late fourth instar larvae required for *in situ* hybridization. This secretion can cover the chromosomes in a squash preparation and prevent entry of the probe or the chemicals of the staining reaction. Therefore it is preferable to use a preparative technique that reduces the secretion, or to use glands from prefixed larvae in which the hardened salivary secretion can be easily separated from the cells. These procedures are described in **Subheading 3.9.2**.

3.9.2. Reduction of Salivary Secretion in Living Larvae

Reduction of salivary secretion in living larvae can be achieved by pretreatment with Pilocarpine (modified from Kato et al. *[15]*) (0.04% pilocarpinium chloride [Merck] made up in Walter's solution: 5 mg of $NaHCO_3$, 35 mg of NaCl, 27 mg of $CaCl_2$, 2 mg of KH_2PO_4, 30 mg of $MgSO_4$, 10 μL of 1% $FeCl_3$ solution, make up to 1 L in distilled water *[16]*). Larvae may be left at room temperature for about 3–5 h or at 4°C overnight. Australian species appear relatively refractive to the effects of pilocarpine so that dissection of the cells away from the secretion while in the 1:2:3 fixing solution (*see* **Subheading 2.1., step 9**) is a better option.

Following this pretreatment, salivary chromosome squashes may be prepared using the technique outlined for *Drosophila* larvae. Using the modification of the Engels et al. technique suggested by Whiting et al. *(4)* it is possible to use probes on different species and different genera (**Fig. 3**). The strength of the signal that can be obtained across genera will of course be dependent upon the extent to which the probe sequence is conserved and the phylogenetic separation of the taxa. We routinely use a formamide concentration of 42% to facilitate hybridization of heterologous probes.

3.9.3. Use of Prefixed Chromosomes

It has been standard practice in our laboratory to place larvae that were not required for immediate cytological analysis into modified Carnoy fixative (3 parts absolute ethanol : 1 part glacial acetic acid), or absolute ethanol, for storage at –20°C. We have not found it necessary to open the thorax of these larvae, as the fixative readily enters through the larval cuticle. The fixative should be replaced after 24 h and, although it may be advisable to replace it again following prolonged storage, this does not appear to be essential. Larvae stored in this way for over 15 years have been found to produce good *in situ* preparations of both multiple *(6)* and single-copy probe DNA (**Fig. 3**).

When chromosomes are required, the larvae are transferred to a drop of 45% acetic acid on a slide, slit along the dorsal thorax, and the salivary glands lifted out. Next, move the glands to a fresh drop of 45% acetic acid and leave there for about 2 min while a freshly siliconized coverglass is prepared. Place a drop

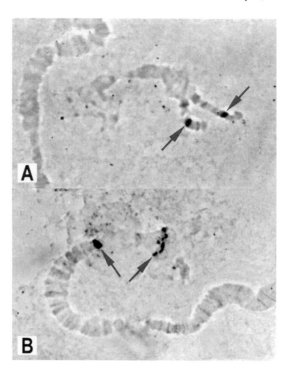

Fig. 3. *In situ* hybridization of the ssp160 gene (clone λ160.1) from the tissue-specific Balbiani ring, BRa, of European *Chironomus thummi* (= *riparius*) *(26)* to prefixed polytene chromosomes of: (**A**) North American *Chironomus anonymus*, (B) Australian *Chironomus occidentalis* (*arrows* point to hybridization signals). The probe was DIG-labeled by PCR amplification and stained as detailed in **Subheadings 3.2.1.** and **3.9.3**. The chromosomes are unstained and have been photographed under phase contrast (×40). Note that in *C. occidentalis* (**B**) there are two clear sites of hybridization

of 1:2:3 fixing solution (*see* **Subheading 2.1.**, **step 9**) on the coverglass and transfer the glands into this drop. Using fine needles, tease the cells away from the salivary secretion and remove the mass of secretion. Squash the cells, remove the coverslip and heat pretreat as outlined for *Drosophila* (*see* **Subheading 3.1.8.**, **step 23**).

It is recommended that the chromosomes be photographed before denaturation as prefixed chromosomes tend to be subject to greater loss of band resolution (**Fig. 3A**) than do fresh chromosomes.

3.9.4. Simultaneous Heat Denaturation of Chromosomes and Probes

Prepared chromosome spreads (*see* **Subheading 3.1.**) and probes (*see* **Subheading 3.2.** and **Note 3**) can be simultaneously denatured in a simple procedure that uses heat as the denaturing agent. Heat denaturation has been found to be gentler than alkali denaturation and can assist in band preservation (*see also* **Note 3**). Hybridization immediately follows the denaturation step, without further manipulation of the chromosome preparation, after the temperature is lowered to 37°C.

1. Make up 50 µL of hybridization mixture containing 350–500 ng of probe DNA, 7.5 µL of 20× SSC (*see* **Subheading 2.1.**, **step 14**), 2.5 µL of heparin stock solution (10 mg/mL), 21–25 µL of deionized formamide (*see* **Subheading 2.2.**, **step 10** and **Subheading 3.9.2.**); 5 µL of 50% dextran sulfate:distilled water to 50 µL. Fifty microliters is sufficient mix for five slides.
2. Pipet 10 µL of probe mix onto each plastic coverslip (*see* **Subheading 2.2.**, **step 8**), pick up the inverted slide, and seal with rubber cement (*see* **Subheading 3.4.**, **steps 6–8**).
3. Preheat a water bath to 70°C.
4. Select a container, large enough for the slides, that will float on the surface. Line the bottom of the container with filter paper dampened with distilled water. Place the slides above the dampened paper on glass rods.
5. Heat at 70°C for 1 h, then turn the water bath back to 37°C and allow the slides to come back to this temperature overnight (about 15–18 h).
6. Detect the probe/DNA hybrid as described in **Subheading 3.5.** and **3.6.**

3.9.5. In Situ *Hybridization to Polytene Chromosomes from Trichogen Cells of Calliphoridae and Tephritidae*

These notes apply to *in situ* hybridization of trichogen polytene chromosomes of the Australian sheep blowfly *Lucilia cuprina*, the Old World screwworm *Chrysomya bezziana*, and the Mediterranean fruit fly, *Ceratitis capitata*. In these species trichogen cells contain excellent quality polytene chromosomes. Salivary gland chromosomes are unsuited to cytological study in *L. cuprina* and *C. bezziana* because of their poor banding and fragmented morphology. In *C. capitata* the salivary glands yield good polytene chromosomes *(17)*, which have been used for *in situ* studies *(18)*. The method described is for the study of euchromatic portions of the karyotype in polytene chromosomes, but the same *in situ* methodology can be used to analyze heterochromatin in mitotic chromosomes.

When compared to *Drosophila melanogaster* salivary glands, trichogen preparations yield very few usable cells. In *L. cuprina* and *C. bezziana,* where trichogen cells from the posterior scutellar margin are used, a good preparation will have 8–10 large cells with optimal chromosomes. *C. capitata* has only two large trichogen cells beneath the spatulate orbital bristles of male pupae *(19)*. Working with this material therefore requires extra care and patience.

The method outlined uses tritium-labeled, nick-translated DNA or RNA probes, and has been used successfully in the detection of repeated sequences *(20)*, ribosomal genes *(21)* (**Fig. 4**), and single-copy genes *(22)*. Nonradioac-

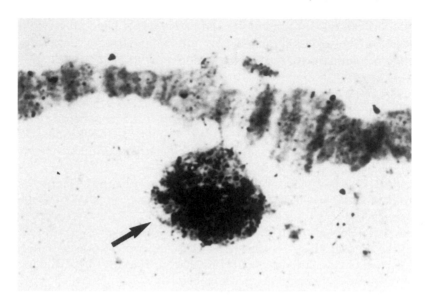

Fig. 4. *In situ* hybridization of a ³H-labeled ribosomal probe to a nucleolar fragment (*arrow*) adjacent to an unlabeled autosome. The nucleolar fragment is connected to the autosome by ectopic pairing (shown as a weakly labeled thread). The chromosome is part of a trichogen cell preparation from *Lucilia cuprina*. The banding on the autosome is quite well preserved and of sufficient clarity to allow the band to be identified when related to the standard polytene chromosome map.

tive probes have proved more difficult to apply to trichogen polytene chromosomes but good results were obtained using salivary gland chromosomes of *C. capitata (18)*. The genome structure of *L. cuprina* makes it essential that probes are checked prior to use for possible contaminating dispersed repeated sequences. This can be carried out using Southern blots of whole genomic DNA (*see* **Note 4**).

3.9.6. Preparation of Chromosome Spreads

Methods for dissecting the posterior scutellar margin of *L. cuprina* and *C. bezziana* pupae and preparing the slides are as previously published *(23)*. At the optimal stage (6 days) pupae are fully formed with orange eyes and unpigmented or faintly pigmented bristles. After removal of the puparium, the pupae are kept in a freshly prepared 3:1 ethanol:acetic acid fixing solution, overnight at 5°C. The fixed pupae are placed in alcohol and the posterior scutellar margin is dissected as a U-shaped piece into which the large bristles insert. Transfer the tissue into a drop of 45% acetic acid on a clean microscope slide and separate the cuticle from the tissue containing the trichogen cells. Discard the cuticle. Macerate the remaining tissue with dissecting needles and, if the polytene cells are visible, remove as much unwanted tissue as possible. Cover the tissue with a siliconized coverslip and apply gentle pressure with a dissecting needle to spread the chromosomes. Monitor chromosome spreading using a phase-contrast microscope. When further spreading cannot be achieved place blotting paper over the slide and apply gentle vertical pressure to flatten the chromosomes. The coverslip must not move laterally during flattening.

The procedure for preparation of orbital bristle cells in *C. capitata* has also been outlined *(18)*. Remove the puparium from 5–6-day male pupae (recognized from the large spatulate orbital bristles) and detach the heads. Fix the heads in freshly prepared 3:1 ethanol:acetic acid solution overnight at 5°C. Dissect a segment of cuticle containing the orbital bristles from the dorsal third of the head including a narrow margin of eye tissue. Remove internal tissues from the segment so that a layer of fat remains next to the cuticle. Transfer the segment to a drop of 45% acetic acid on a clean slide, then separate and discard the cuticle. Apply a siliconized coverslip. Spread and flatten the chromosomes as described previously while monitoring the process with phase optics.

The low number of suitable cells present means that loss of material during preparation of the chromosome spreads is a common cause of experimental failure. To minimize cell loss all slides used must be thoroughly cleaned (**Subheading 2.1., step 1**) and coverslips should be freshly siliconized (old coatings tend to increase the risk of tissue tearing). For best results the polytene cells should be separated from as much surrounding tissue as possible, and cellular debris should be reduced to a minimum. In favorable material the polytene cells can be seen during dissection as translucent spheres. The dissection must be done on a dark background and with bright incident illumination. Individual cells can then be isolated using a finely drawn-out glass needle and transferred to a new slide *(24)*. Chromosomes should be well flattened and no birefringence should be seen when the spreads are examined under phase contrast.

3.9.7. In Situ *Hybridization*

The most important difference between the method outlined in the following steps, which is based on that of Board and Webb *(25)*, and the method described previously for *Drosophila* is the denaturation step. Heat denaturation in the presence of 70% formamide is the preferred method for trichogen polytene chromosomes. This preserves the banding morphology while allowing good levels of hybridization. This protocol can be used with radioactive or nonradioactive probes (*see* **Subheading 3.9.5.**).

1. Pretreat the slides in 2× SSC at 65°C, for 1 h. Wash 3× with tap water and once with distilled water. Dehydrate the slides through a 50%, 75%, 95%, and 100% ethanol series (5 min at each concentration). Air dry the slides.
2. Photograph the chromosomes under phase contrast.
3. Denature the chromosomal DNA. Pour 70% formamide/2× SSC at 73°C into a preheated Coplin jar containing the slides. Incubate for 3 min, shaking the slides intermittently. Immediately transfer the slides to 70% ethanol at 0°C, leave for 3 min with shaking. Dehydrate the slides through a 75%, 95%, and 100% ethanol series and air dry.
4. Prepare the probe in a hybridization mix containing 40% deionized formamide, 10% dextran sulfate, 3× SSC, 0.3 ng/mL of probe DNA, and 3 µg/mL carrier DNA.
5. Denature the probe by incubating at 70°C for 15 min and cool quickly on ice. Apply 12 µL of probe mix per slide if using 22 mm square coverslips, or 24 µL for 22 × 50 mm coverslips. Gently lower a coverslip onto the slide to obtain even spreading of the probe with a minimum of air bubbles. Seal the coverslip to the slide with rubber cement.
6. Incubate the slides overnight at 37°C in a moist chamber (made by placing a tissue moistened with 3× SSC into a box with a tight-fitting lid, or into a large Petri dish).
7. Peel off the rubber cement and rinse off the coverslips in 50% formamide. Wash the slides (with shaking) 3× in 50% formamide/2× SSC at 40°C for 5 min each wash. Wash 5× in 2× SSC at 40°C for 5 min each wash. Hold the slides in 2× SSC at room temperature.
8. Make a solution of 50 µg/mL RNase in 2× SSC. Apply 22 µL per slide and add a large coverslip. Incubate the slides for 2 h at 37°C in a moist chamber. Rinse off the coverslips in 2× SSC and wash 3× in 2× SSC for 5 min each wash. Dehydrate and air dry as described previously.
9. Apply Ilford L4 autoradiographic emulsion, diluted 1:1 with 2% glycerol, to the slides using standard methods. Polytene chromosomes hybridized with repeated sequence probes are well labeled after 11 days of exposure while single-copy probes require about 22 days of exposure.
10. Allow the slides to dry thoroughly before applying stain. Stain the slides with 2% Giemsa stain for 5–15 min. The staining time varies between slides and the best staining time should be determined empirically by monitoring the staining process.

4. Notes

1. If the larvae are at the wrong developmental stage, isolation of good chromosomes is impossible. As larvae approach pupation, resolution of the increasingly fuzzy bands becomes more difficult, disruption of the nuclear membrane may be impossible, and eventually the salivary glands are lost. It is therefore important to ensure that any problems associated with obtaining good preparations are not due to larval age. Establish fly cultures on a number of consecutive days and test larvae from each bottle. A bottle with third instar wandering larvae should yield a good chromosome spread from the majority of larvae sampled.

2. The most common problem among inexperienced workers is failure to obtain flat well-adhered chromosome spreads. This causes a loss of the chromosomes during processing and/or birefringence (resulting in the chromosomes having a beaded appearance, and in poor band resolution). To prevent this it is necessary to take great care in the original cleaning and coating of slides and to monitor all stages of chromosome preparation with the microscope. If the chromosomes are not clearly banded, well spread, and ribbonlike (flattened), be ruthless and discard the slide early in the experiment.

3. Failure to obtain a hybridization signal.

   a. Insufficient denaturation of the chromosomes may result in failure to obtain hybridization. It is important to use freshly prepared NaOH. A range of denaturation times can be tested. Alternatively, try the heat denaturation methods described in **Subheadings 3.9.4.** and **3.9.7.** In cross-species studies, the time required for adequate denaturation is increased and may lead to considerable loss of band resolution. This problem can be overcome by photographing the spreads before denaturation *(4)* and this approach should be considered whenever loss of band resolution is a problem. It is necessary to take clear and exact records of the location of each photographed cell.

   b. Deterioration of reagents. We are aware of experienced workers who have encountered detection problems resulting from inactive alkaline phosphatase. The problem occurs when the enzyme used is not conjugated to the streptavidin. Conjugated streptavidin–alkaline phosphatase is commercially available and in our hands is stable for more than a year when stored at 4°C.

      The BCIP can degrade if improperly stored, with consequent failure of the staining reaction. BCIP can give a positive result on nitrocellulose when tested but be unsuitable for staining chromosome preparations, and probes that hybridize well to nitrocellulose do not necessarily hybridize to chromosomes. As biotinylated probes are stable for long periods, probing slides with previously used probes is a useful positive control for the hybridization and staining processes.

   c. Weak probes. In the method outlined the probe is separated from unincorporated label by ethanol precipitation. Many laboratories use nick-translated probes without any purification. Chromatography should not be used to isolate the DNA from other reactants as it results in substantial loss of probe. As mentioned in (b), a previously used probe can act as a control. If it is known that the control and experimental probes will not hybridize to the same locus both probes can be applied to the same slide. This allows the identification of failure of the nick-translation procedure, or loss of probe DNA.

      If control experiments suggest that failure to *see* a signal is due to a probe fault it is often useful to concentrate the probe further by ethanol precipitation and to rescreen. The sensitivity can also be increased by lowering the hybridization stringency, e.g., by hybridizing at room temperature.

   d. Need for alternative labeling. Biotin-labeled dATP and dCTP are now available. These can be used together with biotin-11-dUTP to increase the amount of label incorporated into a probe. Biotin-labeled dCTP would be particularly useful where short G-C rich sequences are to be used as probes. Where fluorescence microscopy is available the use of fluorescein- and/or rhodamine-labeled antibodies can be used with nonradioactive labels to identify the hybrid. These are particularly useful for double-labeling experiments. Hybridization to a dark chromosomal band with a short probe may be more easily detected with fluorescent labeling.

4. Repeat sequences. Cloned genomic DNA that is to be used for mapping should be hybridized to Southern blots of whole genome DNA prior to use on chromosomes to determine if the clone contains repeat sequence or transposable elements. It may be necessary to subclone unique restriction fragments prior to attempting in situ hybridization.

Acknowledgments

These techniques have been developed during the course of projects supported by grants UME96 from the Wool Research and Development Corporation (A. M. P.), A09231367 (A. M. P.), and D18315889 (J. M.) from the Australian Research Council, and MCB-9204837 from the National Science Foundation to Dr. S. T. Case, University of Mississippi Medical Center, Jackson, MS (J. M.). The assistance of Dr. Case with the PCR-labeling technique is gratefully acknowledged. We thank Quentin Lang for assistance in preparation of the photographs and Cheryl Grant for formatting the manuscript. The *in situ* studies of DGB were undertaken in the Division of Entomology, CSIRO, Canberra.

References

1. Pardue, M. L. (1986) *In situ* hybridization to DNA of chromosomes and nuclei, in *Drosophila: A Practical Approach* (Roberts, D. B., ed.), IRL Press, Washington D.C., pp. 111–137.

2. Engels, W. R., Preston, C. R., Thompson, P., and Eggleston, W. B. (1985) *In-situ* hybridization to *Drosophila* salivary chromosomes with biotinylated DNA probes and alkaline phosphatase. *Focus* **8**, 6–8.

3. Whiting, J. H. Jr., Farmer, J. L., and Jeffrey, D. E. (1987) Improved *in situ* hybridization and detection of biotin-labelled *D. melanogaster* DNA probes hybridized to *D. virilis* salivary gland chromosomes. *Dros. Inform. Serv.* **66**, 170–171.

4. Whiting, J. H. Jr., Pliley, M. D., Farmer, J. L., and Jeffrey, D. E. (1989) *In situ* hybridization analysis of chromosomal homologies in *Drosophila melanogaster* and *Drosophila virilis*. *Genetics* **122**, 99–109.

5. Schmidt, E. R., Keyl, H.-G., and Hankeln, T. (1988) *In situ* localization of two haemoglobin gene clusters in the chromosomes of 13 species of *Chironomus*. *Chromosoma* **96**, 353–359.

6. Martin, J. (1990) *In situ* hybridization to pre-fixed polytene chromosomes. *TIG* **6**, 238.

7. Blackman, R. K. (1996) Streamlined protocol for polytene chromosome *in situ* hybridization. *BioTechniques* **21,** 226–228.
8. Delbridge, M. L. and Kelly, L. E. (1990) Sequence analysis and chromosomal localization of a gene encoding a cystatin-like protein from *Drosophila melanogaster. FEBS Lett.* **247,** 141–145.
9. Phillips, A. M., Bull, A., and Kelly, L. E. (1992) Identification of a *Drosophila* gene encoding a calmodulin-binding protein with homology to the *trp* phototransduction gene. *Neuron* **8,** 631–642.
10. Morgan, M. M. (1991) The cloning and initial characterization of the *Shibire* gene of *Drosophila melanogaster.* Ph.D. Thesis, The University of Melbourne.
11. Davies, A. G. and Batterham, P. (1991) Analysis of an unstable P insertion mutation at the *lozenge* locus in *Drosophila melanogaster. Dros. Inform. Serv.* **70,** 60.
12. Petrovich, T. Z. (1990) Molecular and genetic studies of the *Suppressor of Stoned* locus in *Drosophila melanogaster.* Ph.D. Thesis, The University of Melbourne.
13. Denhardt, D. T. (1966) A membrane-filter technique for the detection of complementary DNA. *Biochem. Biophys. Res. Commun.* **23,** 641–646.
14. Martin, J., Hoffman, R. T., and Case, S. T. (1996) Identification of divergent homologs of *Chironomus tentans* sp185 and its Balbiani Ring 3 gene in Australasian species of *Chironomus* and *Kiefferulus. Insect Biochem. Molec. Biol.* **26,** 465–473.
15. Kato, K. -L., Perkowska, E., and Sirlin, J. L. (1963) Electro- and immunoelectrophoretic patterns in larval salivary secretion of *Chironomus thummi. J. Histochem. Cytochem.* **11,** 485–488.
16. Walter, L. (1973) Syntheseprozesse an den Riesenchromosomen von *Glyptotendipes. Chromosoma* **41,** 327–360.
17. Zacharopoulou, A. (1990) Polytene chromosome maps in the Medfly *Ceratatis capitata. Genome* **33,** 184–197.
18. Zacharopoulou, A., Frisardi, M., Savakis, C., Robinson, A. S., Tolias, P., Konsolaki, M., Komitopoulo, K., and Kafatos, F. C. (1992) The genome of the Mediterranean fruit fly *Ceratatis capitata:* localization of molecular markers by *in situ* hybridization to salivary gland chromosomes. *Chromosoma* **101,** 448–455.
19. Bedo, D. G. (1987) Polytene chromosome mapping in *Ceratatis capitata* (Diptera: Tephritidae). *Genome* **29,** 598–612.
20. Perkins, H. D., Bedo, D. G., and Howells, A. J. (1992) Characterisation and chromosomal distribution of a tandemly repeated DNA sequence from the sheep blowfly, *Lucilia cuprina. Chromosoma* **101,** 358–364.
21. Bedo, D. G. and Webb, G. C. (1990) Localisation of the 5S RNA genes to arm 2R in polytene chromosomes of *Lucilia cuprina* (Diptera: Calliphoridae). *Genome* **33,** 941–943.
22. Bedo, D. G. and Howells, A. J. (1987) Chromosomal localization of the white gene of *Lucilia cuprina* (Diptera: Calliphoridae) by *in situ* hybridization. *Genome* **29,** 72–75.
23. Bedo D. G. (1982) Differential sex chromosome replication and dosage compensation in polytene trichogen cells of *Lucilia cuprina* (Diptera Calliphoridae). *Chromosoma* **87,** 21–32.

24. Bedo, D. G. (1992) Nucleolar fragmentation in polytene trichogen cells of *Lucilia cuprina* and *Chrysomya bezziana* (Diptera: Calliphoridae). *Genome* **35,** 283–293.

25. Board, P. G. and Webb, G. C. (1989) Isolation of a cDNA clone and localization of human glutathione S-transferase-2 genes to chromosome band 6p12. *Proc. Natl. Acad. Sci. USA* **84,** 2377–2381.

26. Hoffman, R. T., Schmidt, E. R., and Case, S. T. (1996) A cell-specific glycosylated silk protein from *Chironomus thummi* salivary glands. *J. Biol. Chem.* **271,** 9809–9815.

7

In Situ Hybridization to Polytene Chromosomes

Robert D. C. Saunders

1. Introduction

Since its development by Pardue et al. *(1)*, the technique of *in situ* hybridization to polytene chromosomes has played a central role in the molecular genetic analysis of *Drosophila melanogaster.* The power of *in situ* hybridization is due largely to the scale of polytene chromosomes and consequently the high degree of resolution they offer the researcher. The use of radiolabeled probes has now been largely superseded by nonradioactive signal detection systems, generally using biotin- or digoxygenin-substituted probes, which offer still greater resolution because there is less scatter of signal with immunochemical and immunofluorescent detection than with silver grains. The utilization of *in situ* hybridization technology is of particular interest to those engaged in chromosome walking or genome mapping projects, in which it is essential to check all clones along a chromosome walk by *in situ* hybridization to identify clones containing repetitive DNA, and to avoid the isolation of clones derived from regions outside that of interest. It is also useful when orienting a chromosome walk, and when determining if a particular clone is derived from DNA uncovered by a deficiency. At least one *Drosophila* genome mapping project *(2)* relied on *in situ* hybridization to accurately map sets of overlapping cosmids (contigs) to the polytene chromosome map, while another *(3)* used *in situ* hybridization as the sole means of ordering yeast artificial chromosome (YAC) clones along the genome.

In this chapter I shall describe the use of biotin-labeled probes for *in situ* hybridization to polytene chromosomes. These protocols are generally applicable to polytene chromosomes from all Diptera, with modifications to the protocol principally concerning the preparation of tissues; a protocol for preparation of Anopheline chromosomes is described.

From: *Methods in Molecular Biology*, Vol. 123: In Situ *Hybridization Protocols*
Edited by: I. A. Darby © Humana Press Inc., Totowa, NJ

2. Materials

1. Clean microscope slides (*see* **Note 1**).
2. Clean siliconized coverslips, 24 mm^2 (*see* **Note 1**).
3. Clean siliconized coverslips, 22 mm × 50 mm (*see* **Note 1**).
4. 0.7% NaCl.
5. 45% acetic acid.
6. 1:2:3 fix: 1 part lactic acid, 2 parts distilled water, 3 parts acetic acid.
7. Carnoy's fixative: 3 parts ethanol to 1 part acetic acid.
8. 50% Propionic acid in distilled water.
9. 2× Saline sodium citrate (SSC).
10. 70 mM Sodium hydroxide, freshly prepared.
11. 70% Ethanol.
12. 96% Ethanol.
13. Coplin jars, or similar, for incubating slides.
14. Oligolabeling buffer: prepared from the following solutions (as described in **ref. 4**):
    Solution O:
    a. 1.25 M Tris-HCl, pH 8.0.
    b. 125 mM MgCl$_2$.
    Solution A: To 0.47 mL of solution O add:
    a. 9 µL of β-mercaptoethanol.
    b. 12.5 µL of 20 mM dATP.
    c. 12.5 µL of 20 mM dCTP.
    d. 12.5 µL of 20 mM dGTP.
    Solution B: 2 M HEPES, pH 6.6.
    Solution C: Random hexanucleotides (Pharmacia) at a concentration of 90 A$_{260}$ U/mL.
    Prepare oligolabeling buffer by mixing solutions A, B, and C in the proportions 2:5:3. Oligolabeling buffer and its constituents should be stored at –20°C.
15. TE: 10 mM Tris-HCl, pH 8.0; 1 mM EDTA.
16. 1 mM Biotin-16-dUTP (Boehringer). Store at –20°C (*see* **Note 2**).
17. Fluorescein-12-dUTP (Boehringer). Store in the dark, at –20°C (*see* **Note 2**).
18. 2× Hybridization solution: 8× SSC, 2× Denhardt's, 20% dextran sulfate, 0.8% sonicated salmon sperm DNA. Store at –20°C.
19. Plastic box with tightly fitting lid, lined with moist tissue paper.
20. Detek-1 streptavidin–horseradish peroxidase detection kit (Enzo), or extravidin–HRP conjugate (Sigma). The Enzo dilution buffer is phosphate-buffered saline (PBS) supplemented with 1% bovine serum albumin (BSA), 5 mM EDTA (*see* **Note 2**).
21. PBS: 8 g of NaCl, 0.2 g of KCl, 1.44 g of Na$_2$HPO$_4$, 0.24 g of KH$_2$PO$_4$ per liter.
22. PBS-TX: PBS containing 0.1% Triton X-100.
23. DAB solution. 5 mg/mL of diaminobenzidine in PBS, supplemented with 0.01% H$_2$O$_2$. DAB is a potent mutagen. Care should be taken at all times when working with solutions containing DAB. Gloves should be worn, and DAB should be dis-

pensed in the fume hood. DAB should be inactivated in 50% bleach before disposal (*see* **Note 2**).
24. 0.89% Giemsa's staining solution in methanol/glycerol (Gurrs/BDH). Used as a 1:20 dilution in 10 m*M* sodium phosphate buffer, pH 6.8.
25. DPX mountant (Fluka).
26. Avidin–fluorescein D (FITC) (Vector labs). Store at 4°C, in the dark (*see* **Note 2**).
27. Glycerol mounting medium: 85% glycerol, 2.5% *n*-propyl gallate, 1 µg/mL of propidium iodide. Store at 4°C, in the dark.
28. Polytene chromosome maps of the species of interest (*see* **Note 3**).

3. Methods
3.1. Preparation of Polytene Chromosomes

Preparation of chromosomes from various species and tissues is broadly the same: tissues are dissected from appropriate staged animals, and generally fixed in an organic acid fixative. The preferred fixative for *Drosophila* salivary gland chromosomes is 45% acetic acid, followed by 1:2:3 lactic acid: water: acetic acid. In the case of *Anopheles* ovarian chromosomes, the fixative of choice is modified Carnoy's followed by 50% propionic acid. The choice of fixative is not immutable, however; e.g., many Drosophilists use merely 45% acetic acid.

Use only slides with good quality chromosomes. These appear flat and gray, with clear banding. Chromosomes that appear bright and reflective under dry phase examination will have poorer morphology, hindering accurate interpretation. The region of the slide where the chromosomes are located should be visible when the slide is dry, and should be marked with a diamond pencil on the reverse side of the slide.

3.1.1. Drosophila *Larval Salivary Gland Chromosomes*

Drosophila stocks should be maintained on a medium suitable for the species in use. The best larvae are collected from well-yeasted, uncrowded cultures. Select large third instar larvae that are still crawling, and have not everted their anterior spiracles.

1. Dissect out the salivary glands in a drop of 0.7% NaCl, and transfer them to a drop of 45% acetic acid. Allow to fix for approx 30 s.
2. Transfer the glands to a drop of 1:2:3 fixative on a clean siliconized coverslip. Fix for 3 min, then pick up the coverslip with a clean slide, by touching it to the drop. The slide does not have to be coated or "subbed" before use.
3. Spread the chromosomes by tapping the coverslip with a pencil in a circular motion. Check the chromosomes using phase contrast. When the chromosomes are suitably spread, fold the slide in blotting paper, and press gently to remove excess fix. Leave the slide at room temperature for 1 h to overnight. This step squashes the chromosomes as the fix evaporates, and the coverslip sinks toward the surface of the slide. Alternatively, the slide can be squashed firmly between blotting pa-

per, and frozen immediately. Take care not to allow the coverslip to slide sideways, or the chromosomes will be overstretched.

4. Freeze the slide in liquid nitrogen. While the slide is still frozen, flip off the coverslip with a scalpel blade and proceed to **step 5**.
5. Place the slide in 70% ethanol for 5 min. Transfer the preparation through two 5-min changes of 96% ethanol, and air dry. The chromosomes can be stored desiccated at room temperature.

3.1.2. Chromosomes from non-Drosophilids

3.1.2.1. ANOPHELES

Good polytene chromosomes are found in the fourth instar larval salivary glands and the ovarian nurse cells of many species of Anopheline mosquitoes, and chromosome maps are available for many of these species. This protocol is used for the preparation of chromosomes from ovaries of *Anopheles gambiae*. A fuller description is given in **ref. 4**.

1. Immobilize a half-gravid mosquito by chilling, or with ether, place it in a drop of 0.7% NaCl, and extract the ovaries by pulling away the terminal two abdominal segments.
2. Place the ovaries in modified Carnoy's fixative, and store at 4°C for at least 12 h. The ovaries can be kept in Carnoy's for several weeks.
3. Transfer one ovary to a droplet of 50% propionic acid, and incubate at room temperature for 10 min.
4. Remove the 50% propionic acid with absorbent tissue, and add three or four drops of fresh 50% propionic acid. Tease apart the ovarioles with a pair of fine needles or forceps. Cover with a siliconized 24 × 24 mm coverslip, fold between layers of blotting paper or 3M paper, and spread the chromosomes by tapping gently with a pencil point, then blot off excess fix.
5. Store the slides at 4°C for a few hours. Do not let the fix dry and let air bubbles form under the coverslip.
6. Freeze the slide in liquid nitrogen, remove the coverslip and dehydrate the specimen through alcohols as described in **Subheading 3.1.1**.

3.1.2.2. OTHER SPECIES

Polytene chromosomes from any source should be amenable to analysis by *in situ* hybridization. In the first instance protocols for preparation of chromosomes for cytological analysis should be consulted for fixation details. Otherwise, procedures similar to those described earlier may be investigated.

3.2. Pretreatment and Denaturation of Chromosomes

To yield a single-stranded target to which the DNA probe may hybridize, the chromosomal DNA must be denatured. I describe two protocols used to denature polytene chromosomes prior to hybridization. The first uses alkali treatment, and the second heat, to denature the chromosomes. Although the latter is a quicker method, some probes can give a rather dispersed signal compared with chromosomes prepared using alkali denaturation.

3.2.1. Alkali Denaturation

1. Incubate the slides in 2× SSC at 65°C for 30 min. This step is intended to help preserve the morphology of the chromosomes.
2. Transfer the slides to 2× SSC at room temperature for 10 min, then denature the chromosomes by incubating the slides in 70 mM NaOH for 2 min. The 70 mM NaOH must be freshly prepared, using NaOH pellets rather than a concentrated stock solution.
3. Rinse the slides in 2× SSC.
4. Dehydrate through ethanol as described in **Subheading 3.1.1.**, **step 5**, and air dry. The slides should be used the same day.

3.2.2. Heat Denaturation

1. Place the slides directly in gently boiling 10 mM Tris-HCl, pH 7.5, for 2 min.
2. Quickly transfer the slides to cold 70% ethanol, and dehydrate through alcohol as described in Section 3.1.1, sub**step 5** above.

3.3. Preparation of Labeled Probe

Probes are most conveniently prepared by the random priming method of Feinberg and Vogelstein *(5,6)*. The DNA can be in the form of intact plasmid, λ, cosmid, or YAC clones, or a restriction fragment isolated by agarose gel electrophoresis. In the latter case, the gel used should be cast using low melting point agarose, and the band excised in a minimum volume of gel. Three volumes of sterile distilled water are then added, and the mixture boiled for 7 min before adding it to the labeling reaction. Alternatively, the DNA can be extracted from the gel, using a variety of methods, typically "freeze-squeezing," electroelution, phenol extraction, or a proprietary method such as Geneclean.

3.3.1. Synthesis of Biotinylated Probes

1. Place 5 μL of oligolabeling buffer and 1 μL of 1 mM biotin-16-dUTP in a microcentrifuge tube.
2. Boil 100–500 ng of DNA in 20 μL of water or TE for 3 min, then add 18 μL to the tube containing oligolabeling buffer and biotin-16-dUTP.
3. Add 1 μL (5 U) of Klenow fragment of DNA polymerase I and incubate the reaction at 37°C for 1 h to overnight.
4. Ethanol precipitate the labeled DNA and resuspend it in 50 μL of sterile distilled water, then add 50 μL of 2× hybridization solution and mix well. This is sufficient probe for five slides. Any unused probe may be stored at –20°C, and may be reused after boiling.

3.3.2. Synthesis of Probes from PCR Amplified DNA

Probes can be made from DNA amplified by the polymerase chain reaction (PCR), using additional PCR cycles in the presence of biotinylated nucleotide.

1. Remove unincorporated nucleotides from the amplified DNA, e.g., by gel electrophoresis.
2. Set up the PCR reaction as used to amplify the DNA. Substitute 10 mM biotin-16-dUTP for the dTTP in the PCR reaction, and include 100–500 ng of DNA as template. Five cycles of synthesis are usually sufficient. Increasing the length of the polymerization step to 10 min is advised, as the concentration of biotin-16-dUTP is low.

3.3.3. Preparation of Fluoresceinated Probes

The use of fluorescein-11-dUTP for direct labeling of DNA probes for in situ hybridization to metaphase chromosomes has been described by Wiegant et al. *(7)*. *In situ* hybridization to polytene chromosomes can also be carried out using such probes, although with much greater sensitivity than can be achieved with metaphase chromosomes. Because this is a single-step detection system, it is very quick, but is less sensitive that the two-step immunofluorescent system.

1. Place 5 µL oligolabeling buffer and 1 µL of 1 mM fluorescein-12-dUTP in a microcentrifuge tube.
2. Boil 100–500 ng of DNA in 20 µL of water or TE for 3 min, then add 18 µL to the tube.
3. Add 1 µL (5 U) of Klenow fragment of DNA polymerase I and incubate the reaction at room temperature for 1 h to overnight.
4. Ethanol precipitate the labeled DNA and resuspend it in 50 µL of sterile distilled water to which 50 µL of 2× hybridization solution is added.

3.4. Hybridization

1. Boil the probe for 3 min by suspending the tube in boiling water, and quench on ice. Check the volume after boiling, and restore to the initial volume with sterile distilled water.
2. Pipette 20 µL of the probe onto the chromosomes, then cover the chromosomes and probe with a clean siliconized coverslip. Avoid trapping air bubbles underneath the coverslip. There is no need to seal the coverslip.
3. Place the slides in a humid box (a plastic box lined with moist tissue) to prevent evaporation from the preparation, seal the lid, and place in a 58°C incubator overnight.
4. Remove the slides from the humid box. Dip them in 2× SSC to allow the coverslip to slide off, then wash them in 2× SSC, 53°C, for 1 h. Three changes of wash solution should be made.

3.5. Signal Detection

Three protocols for signal detection are presented. In each case, the chromosomes are counterstained with a stain contrasting the signal. Giemsa's stain yields bluish-purple chromosomes that contrast well with the brown DAB signal, whereas propidium iodide yields chromosomes that fluoresce in red. In particular, the choice of counterstain with fluorescent detection depends on the microscope filters available, alternatives might include DAPI or Hoechst 33258.

3.5.1. Signal Detection Using Streptavidin-HRP

1. The slides should be taken from the final wash and passed through two 5 min washes in PBS.
2. Wash the slides for 2 min in PBS-TX.
3. Rinse in PBS. Do not allow the slides to dry out during signal detection.
4. While the slides are being washed, make a 1:250 dilution of streptavidin–HRP conjugate in the buffer supplied with the Enzo kit. Alternatively, extravidin–HRP conjugate (Sigma) can be used, at the same dilution. Apply 50 μL to the chromosomes and cover with a 22 × 50 mm coverslip, avoiding air bubbles. Replace the slides in the humid box, and incubate at 37°C for 30 min.
5. Wash off unbound streptavidin conjugate by passing the slides through the PBS and PBS-TX washes, as described in **steps 1–3** above. Drain the slides, but do not allow them to dry.
6. Place 50 μL of DAB solution onto the chromosomes, and cover with a 22 × 50 mm coverslip. This solution should be made up fresh because hydrogen peroxide decays rapidly. Take care when working with diaminobenzidine, as it is a potent carcinogen. Follow the guidelines for use and disposal described in **Subheading 2., step 23**.
7. Incubate at room temperature for 10–15 min in the humid box. Rinse the slides in PBS and examine under phase contrast. The signal appears blackish-brown, sometimes quite refractile in strong cases. If the signal seems weak, add more DAB solution, and incubate for longer (*see* **Note 6**). If the signal is strong enough, rinse the slide well with distilled water.
8. Stain in Giemsa's stain for 1 min, rinsing off excess stain in running tap water for a few seconds. A metallic scum of stain may form on top of the staining solution; avoid coating the slide with it. Allow the slides to air dry. Check that the staining is sufficiently intense. Overstained chromosomes can be destained in 10 mM sodium phosphate buffer, pH 6.8, and understained chromosomes can be restained. The preparation should be mounted under a siliconized coverslip with DPX mounting medium. DPX is a xylene-soluble mountant, which does not affect either the Giemsa's stain or DAB deposit, and the slides should last for many years. It is convenient to seal the edges of the coverslip with nail varnish to prevent immersion oil from seeping under the coverslip.

The preparations should be examined under phase contrast. Some results are shown in **Fig. 1**. Photographic reproduction is best with color film, particularly with weaker signals.

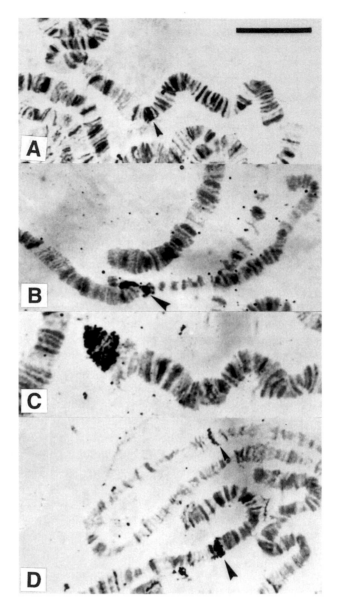

Fig. 1. *In situ* hybridization of biotin labeled probe DNA to *D. melanogaster* polytene chromosomes. The signals have been detected using streptavidin–horseradish
peroxidase conjugate, with DAB as the substrate, and the chromosomes are counterstained with Giemsa's stain. (**A**) Mapping a cosmid clone, using a 2.8-kb restriction
fragment as a probe. The signal is indicated by the arrow, and lies in bands 96A21–25.
(**B**) Mapping a clone relative to a chromosome rearrangement. The probe is a cosmid/

continued on next page

3.5.2. Signal Detection Using Avidin–Fluorescein

Biotin-labeled probes can be detected using avidin–fluorochrome conjugates.

1. Wash the slides in PBS and PBS-TX as described in **Subheading 3.5.1., steps 1–3**.
2. Make a 1:50 dilution of avidin-fluorescein in PBS. Pipet 50 μL onto the chromosomes, cover with a 22 × 50 mm coverslip, and incubate in a dark humid box at room temperature for 30 min.
3. Pass the slide through the PBS and PBS-TX washes as described in **step 1**, drain the slide, and mount in mounting medium (**Subheading 2., step 27**), under a siliconized coverslip. The mounting medium contains propidium iodide to stain the chromosomes.
4. Seal the coverslip with nail varnish. These preparations do not last as long as those described in **Subheading 3.5.1.**, but will keep for several weeks at 4°C in the dark, if sealed as described. They are best examined using a microscope with facility for image merging. A confocal microscope is ideal.

3.5.3. Signal Detection Using Fluoresceinated Probes

1. The hybridization and washing steps are performed as described in **Subheading 3.4., steps 1–4**.
2. Mount the preparation directly in mounting medium as described in **Subheading 3.5.2., steps 3 and 4**.

4. Notes

1. One of the most important factors in successful *in situ* hybridization experiments is the quality of the polytene chromosomes. There are many ways in which polytene chromosomes can be prepared, differing mostly in the manner by which the chromosomes are spread and squashed. Allowing the coverslip to slip sideways when spreading causes the chromosome arms to stretch. Overstretched chromosomes can make analysis of the *in situ* hybridization difficult. Make sure the slides and coverslips are clean, especially of lint from tissue paper used to clean them; thus only lint-free tissue paper should be used. Photographers air brush cans (such as Kenair) are useful for removing fibers and particles from coverslips and slides. Poor chromosome morphology can result from denaturing chromosomes for too long in alkali, and from poorly understood fixation problems. If the chromosome morphology is repeatedly found to be puffy and swollen, try the

---

phage containing an insert derived from bands 96B1–10, hybridized to T(Y;3)B197/+ chromosomes. The signal lies on chromosome 3, in bands 96B1–10, proximal to the breakpoint. (**C**) Hybridization of PCR-amplified DNA derived from microdissection of division 1 (*10*). (**D**) Hybridization of PCR-amplified DNA derived from microdissection of subdivision 25A (*10*). Scale bar: 20 μm.

alternative denaturation method of boiling, which often preserves the morphology better than alkali denaturation.

2. The two methods of probe labeling described earlier are not the only methods available to the researcher; digoxygenin (Boehringer) labeling can be used, and alternatives to random priming for incorporation of label, such as photobiotin labeling can be used. When selecting a detection system for an experiment, several factors must be borne in mind. First, the durability of the specimen is important. Signals visualized with HRP and DAB as described earlier offer the advantage of long-term stability, compared with the fluorescent methods, an important feature when engaged in long-range chromosome walking or genome mapping. Second, one should assess the degree of sensitivity required for a given experiment. For example, the direct fluorescein labeling method, although quicker, does not offer the same degree of sensitivity as do the two-step detection systems. Third, when choosing a method that employs an enzymatic reaction for detection, such as alkaline phosphatase or HRP, care should be taken to select a substrate whose reaction product is both stable and insoluble in the mountant in use, and which contrasts well with the chromosome counterstain.

3. Polytene chromosome maps are available for many *Drosophila* species. Sorsa *(8)* has compiled a list of all published maps of drosophilid polytene chromosomes. For *Drosophila melanogaster*, the 1935 Bridges map *(9)* and the Lefevre *(10)* photomap are indispensable. These are available from Academic Press, in a folder together with the Bridges' revised maps.

   Chromosome maps for many species of Anopheline mosquitoes have also been published, although those for the malaria mosquito *Anopheles gambiae* remain unpublished (although photomaps are available through the World Wide Web, at http://konops.imbb.forth.gr/AnoDB/Cytomap.html).

4. The presence of repetitive DNA within a cloned segment of DNA can prevent easy determination of the chromosomal site of origin of the clone. The use of sibling species can resolve this problem. For example, *Drosophila simulans* and *Drosophila mauritiana* polytene chromosomes have been used *(2)* in mapping cosmids containing cloned segments of *D. melanogaster* DNA. This is possible because the sibling species have different amounts of repetitive DNA, and different populations of transposable elements.

5. High background on preparations is generally associated with poor incorporation of biotinylated nucleotide, and inefficient removal of unincorporated nucleotides prior to hybridization. However, background can be exacerbated by the nature of the salivary gland, e.g., *Rhynchosciara americana* salivary gland polytene chromosomes can show high background as a consequence of the large amount of cocoon protein present in the gland.

6. If no signal is seen when using a biotinylated probe, test it by filter hybridization. A systemic problem where no signals are obtained with a variety of probes may indicate that the DAB solution has decayed. Generally this can be rectified by using fresh hydrogen peroxide. Stocks of hydrogen peroxide should be replaced regularly.

References

1. Pardue, M.-L., Gerbi, S. A., Eckhardt, R. A., and Gall, J. G. (1969) Cytological localization of DNA complementary to ribosomal RNA in polytene chromosomes of *Diptera. Chromosoma* **29,** 268–290.
2. Siden-Kiamos, I., Saunders, R. D. C., Spanos, L., Majerus, T., Trenear, J., Savakis, C., Louis, C., Glover, D. M., Ashburner, M., and Kafatos, F. C. (1990) Towards a physical map of the *Drosophila melanogaster* genome: mapping of cosmid clones within defined genomic divisions. *Nucleic Acids Res.* **18,** 6261–6270.
3. Ajioka, J. W., Smoller, D. A., Jones, R. W., Carulli, J. P., Vellek, A. E. C., Garza, D., Link, A. J., Duncan, I. W., and Hartl, D. L. (1991) *Drosophila* genome project: one-hit coverage in yeast artificial chromosomes. *Chromosoma* **100,** 495–509.
4. della Torre, A. (1997) Polytene chromosome preparation from anopheline mosquitos, in *The Molecular Biology of Insect Disease Vectors* (Crampton, J. M., Beard, C. B., and Louis, C., eds.), Chapman and Hall, London, pp. 329–336.
5. Feinberg, A. P. and Vogelstein, B. (1983) A technique for radiolabeling DNA restriction endonuclease fragments to high specific activity. *Anal. Biochem.,* **132,** 6–13.
6. Feinberg, A. P. and Vogelstein, B. (1984) Addendum to Feinberg and Vogelstein (1983) *Anal. Biochem.* **137,** 266–267.
7. Wiegant, J., Ried, T., Nederlof, P. M., vander Ploeg, M., Tanke, H. J., and Raap, A. K. (1991) *In situ* hybridization with fluoresceinated DNA. *Nucleic Acids Res.* **19,** 3237–3241.
8. Sorsa, V. (1988) Chromosome maps of *Drosophila.* 2 vols., CRC Press, Boca Raton, FL.
9. Bridges, C. B. (1935) Salivary chromosome maps with a key to the banding of the chromosomes of *Drosophila melanogaster. J. Hered.* **26,** 60–64.
10. Lefevre, G. (1976) A photographic representation and interpretation of the polytene chromosomes of *Drosophila melanogaster* salivary glands, in *The Genetics and Biology of* Drosophila, vol. 1a (Ashburner, M. and Novitski, E., eds.), Academic Press, New York, pp. 31–66.

8

Comparative Gene Mapping in Exotic Species Using FISH

Stephen A. Wilcox, Roland Toder, Pino Maccarone, and Jennifer A. Marshall-Graves

1. Introduction

Comparative genome analysis between two distantly related species allows the organization of genes to be traced from a common ancestor. When several genes are mapped in one species and these genes are then localized in the distantly related species, then the genomic content of this region can be inferred in the common ancestor. If two species are closely related in evolutionary terms, then larger blocks of the genome will be conserved. Therefore these segments can be easily traced back to a common ancestor. Such comparisons have made it possible to trace the evolutionary origins of regions of the mammalian X chromosome (1).

Previously, comparative genome analysis used radioactive *in situ* hybridization of heterologous probes to assign genes from one species to a distantly related species (2). The advent of fluorescent *in situ* hybridization and the easily constructed λ genomic libraries allows homologous probes of sufficient size to now be cloned from the species of interest, and be directly hybridized *in situ* to homologous metaphase chromosomes (3).

This chapter provides a detailed account of fluorescent *in situ* hybridization (FISH) localization of homologous λ clones to marsupial chromosomes. Problems encountered and their solutions are also addressed. This procedure can be readily adapted to any species provided a λ genomic library has already been generated. A detailed protocol for probe isolation and characterisation for λ clones has already been described (4).

From: *Methods in Molecular Biology*, Vol. 123: In Situ *Hybridization Protocols*
Edited by: I. A. Darby © Humana Press Inc., Totowa, NJ

2. Materials

2.1. Chromosome Preparations

2.1.1. Fibroblast Cell Lines

1. Phosphate-buffered saline (PBS): 0.14 M NaCl, 3 mM KCl, 1 mM CaCl$_2$, 2 mM MgSO$_4$, 2 mM KH$_2$PO$_4$, 3 mM Na$_2$HPO$_4$, and 0.002% phenol red.
2. Trypsin versene (TV): 0.1% trypsin and 0.1% EDTA (tetra sodium salt), in PBS.
3. Colcemid (Sigma, St. Louis, MO, USA): 10 mg/mL (1%) in PBS. Store at 4°C.
4. Fixative: 3:1 mix of methanol and glacial acetic acid. Make fresh on the day of use.
5. 0.05 M KCl prewarmed to 37°C on day of use.
6. 0.1 M HCl.
7. Ethanol (70%, 90%, and 100%).
8. RNase A (Sigma, St. Louis, MO, USA). Remove contaminating DNase by heating to 100°C for 15 min and allowing to cool prior to storage at –20°C.
9. 2× Saline sodium citrate (SSC): 20× SSC stock comprises 3 M NaCl and 0.3 M sodium citrate, pH 7.0.
10. Slide rack holders.
11. Gauze pads.
12. Diamond tipped pencil.
13. 10-mL (2×) and 50-mL (1×) centrifuge tubes.
14. 37°C incubator.
15. Slides and coverslips.

2.1.2. Whole Blood Cultures

1. 25-mL cell culture flasks.
2. 1 mL of whole blood from species of interest.
3. Phytohemagglutinin A (Murex, Norcross, GA, USA).
4. Complete medium: 10% fetal calf serum in Dulbecco's modified Eagle's medium.
5. 37°C Incubator.
6. 0.075 M KCl prewarmed to 37°C on day of use.
7. Colcemid 10 mg/mL stock, store at 4°C.
8. 10-mL Centrifuge tube.
9. Fixative: 3:1 mix of methanol and glacial acetic acid. Make fresh on the day of use and store at –20°C.

2.2. Removal of Endogenous RNA

1. 100 µg/mL of RNase A.
2. 2× SSC (0.3 M NaCl, 0.03 M sodium citrate), pH 7.0.
3. Ethanol 70%, 90%, and 100%.
4. Plastic resealable box. These are plastic containers with tight-fitting lids that prevent any significant loss of moisture from the atmosphere within the container. Larger containers (250 mm × 200 mm × 100 mm) are more manageable than smaller ones.

5. Paper toweling.
6. 37°C Incubator.

2.3. Preparation of Fluorescent Probes

1. 0.1 M β-Mercaptoethanol.
2. 10× Digoxigenin (DIG) labeling mix (Roche Molecular Biochemicals, Indianapolis, IN, USA).
3. 0.4 mM Biotin-11-UTP (Sigma, St. Louis, MO, USA).
4. 10× Nick-translation buffer: 0.5 M Tris-HCl (pH 8.0), 50 mM MgCl$_2$, and 0.5 mg/mL of bovine serum albumin fraction V (BSA).
5. Nick-translation kit (Gibco-BRL, Grand Island, NY, USA).
6. 15°C Water bath.
7. Agarose.
8. 0.5 × TBE gel: 5× TBE stock comprises 0.45 M Tris borate and 0.01 M EDTA, pH 8.0.
9. 10 mg/mL Ethidium bromide.
10. Birko boiling water bath.
11. φX174 *Hae*III marker DNA (Roche Molecular Biochemicals).
12. Electrophoresis equipment.
13. UV lamp.
14. Stop buffer: 0.1% bromophenol blue; 0.5% dextran blue; 0.1 M NaCl; 20 mM EDTA; and 20 mM Tris-HCl, pH 7.5.
15. 5% Sodium dodecyl sulfate (SDS).

2.4. Determination of Biotin/DIG Incorporation

1. Protran membrane (Schleicher and Schuell, Dassel, Germany).
2. Glass Petri dish with lid.
3. 80°C oven.
4. AP1 buffer, pH 7.5: 0.1 M Tris-HCl, pH 7.5; 0.1 M NaCl; 2 mM MgCl$_2$; and 0.05% Triton X-100.
5. AP2 buffer: 3% BSA in AP1 buffer.
6. AP3 buffer: 0.1 M Tris-HCl, pH 9.5; 0.1 M NaCl; and 50 mM MgCl$_2$.
7. 37°C Oven.
8. 10-mL (2×) centrifuge tubes.
9. Anti-DIG alkaline phosphatase conjugate.
10. Streptavidin–alkaline phosphatase conjugate (Gibco-BRL, Grand Island, NY).
11. DIG DNA labeling and detection kit.

2.5. Probe Precipitation

1. Sonicated genomic DNA or Cot-I DNA (1 mg/mL) from species of interest.
2. 10 mg/mL Sonicated salmon sperm DNA (Sigma, St. Louis, MO, USA).
3. Ice-cold 100% ethanol.
4. 3 M Sodium acetate.
5. –20°C Freezer.

6. Benchtop microfuge.
7. 70% Ethanol.
8. 75°C Water bath.
9. 37°C Oven.

2.6. Chromosome Denaturation

1. Dry ice.
2. Ethanol 70%, 90%, and 100%. Keep at –70°C until ready for use.
3. Denaturing solution: 70% formamide in 2× SSC, pH 7.0. (Use 70 mL of deionized formamide and add 30 mL of 2× SSC, pH 7.0). Deionize ultrapure formamide with a mixed bed ion-exchange resin (5 g/100 mL) in a conical flask completely covered in aluminum foil. Stir at room temperature for a minimum of 1 h. Filter the formamide once through Whatman no. 1 filter paper. The deionized formamide is ready for use or storage at –20°C. (Warning: Formamide is a known carcinogen. Carry out all steps involving formamide in a fume hood.)

2.7. Hybridization

1. Deionized formamide.
2. Probe cocktail mix: 20% dextran sulfate and 4× SSC, pH 7.0. Make in bulk and store indefinitely at –20°C.
3. 37°C Incubator.
4. Heating block.
5. Rubber cement.
6. Resealable plastic boxes.
7. Paper toweling.

2.8. Posthybridization Treatment

1. Slide washing solution: 50% formamide and 2× SSC, pH 7.0.
2. 0.5× SSC, pH 7.0.
3. Antibody washing solution: 4× SSC and 0.2% Tween-20, pH 7.0.
4. 45°C Water bath (*see* **Note 1**).
5. 37°C Oven.
6. Histology (Coplin) jars.
7. Slide rack holders.
8. Resealable plastic boxes (e.g., Decor brand).
9. Blocking reagent: 4× SSC, 0.2% Tween-20, and 5% BSA (fraction V).

2.9. Antibody Detection

1. Antibody washing solution: as described previously.
2. Blocking reagent: as described previously.
3. 37°C oven.
4. a. Monoclonal anti-DIG antibody raised in mouse (Sigma, St. Louis, MO, USA).
   b. Anti-mouse IgG (whole molecule) - TRITC conjugate raised in goat (Sigma, St. Louis, MO, USA).

5. a. Fluorescein avidin DCS (Vector Laboratories, Burlingame, CA, USA).
   b. Biotinylated anti-avidin D antibody raised in goat (Vector Laboratories, Burlingame, MO, USA).
6. Antibody chamber (*see* **Note 2**).
7. Coplin jar.
8. DAPI (4',6-diamidino-2-phenylindole dihydrochloride), 0.2 mg/mL stock solution (Sigma, St. Louis, MO, USA).
9. Vectashield (Vector Laboratories, Burlingame, CA, USA).
10. Coverslips, 50 mm × 22 mm.

3. Methods

3.1. Chromosome Preparations

It is necessary to arrest dividing cells during mitosis to obtain chromosome spreads. The ideal chromosome preparation requires (1) a high mitotic index; (2) no background cytoplasm; and (3) well spaced chromosomes with clear morphology. Our laboratory routinely uses diploid fibroblast cell lines derived from embryos or adult skin, as well as blood cultures used for lymphoblast growth. Marsupial karyotypes are ideal for fluorescent *in situ* hybridization because of their few large, easily identifiable chromosomes.

3.1.1. Fibroblasts

1. Grow cells until flasks are approx 70–80% confluent.
2. Add Colcemid to a final concentration of 0.0002–0.0005% to the cells.
3. Leave at 37°C for a period ranging from 30 min to 16 h depending on rate of cell growth.
4. Concentrate mitotic cells by washing off the monolayer as follows:
   a. After gentle tapping of the flask, remove the media from the cell culture flasks and place it in a 10-mL centrifuge tube.
   b. Wash the cells once with 1 mL of PBS by pipetting the PBS over the cells several times.
   c. Remove the PBS and place it in the centrifuge tube with the media. Cells that are arrested at metaphase will be dislodged from the cell culture flask, while cells that remain attached are still viable and may be maintained in culture by the addition of appropiate media.
   d. Centrifuge for 10 min at 1000g to pellet the cells.
   e. Discard the supernatant, wash the cells twice with PBS, and pellet as described previously.
5. Resuspend the cells in 2 mL of 0.05 M KCl, added dropwise from a Pasteur pipet. The optimal time for hypotonic treatment depends on the cell line used. Generally we have found 12–18 min at 37°C is sufficient.
6. Add 2 mL of fixative to the suspension and centrifuge for 10 min at 1500g.
7. Carefully decant supernatant, and resuspend the cells in 2 mL of fixative and centrifuge. Repeat this step once.

8. Resuspend the cells in 1 mL of fixative and allow one drop to fall from a Pasteur pipet onto a cleaned slide from a height of approx 10 cm.
9. Stain the chromosome preparations with 10% filtered Giemsa stain and examine under microscope (*see* **Note 3**).
10. Store chromosome preparations, either by placing in the fixative at –20°C or dropping them onto microscope slides.
11. Mark the position of the drop on the slide with a diamond tipped pencil to locate the cells during the hybridization procedure.

3.1.2. Whole Blood Cultures

1. To 1 mL of blood add 100 μL of phytohemagglutinin stock and 10 mL of complete medium.
2. Incubate blood culture at 37°C with 5% carbon dioxide for 71.5 h. (Shake blood culture flasks after 24 h: they should appear red and no clots visible. If they are dark brown and/or clotted then discard the cultures).
3. Add 0.1 mL of Colcemid and incubate at 37°C for 2 h.
4. Transfer the blood culture into a 10-mL centrifuge tube and centrifuge for 10 min at 2000g.
5. Discard the supernatant and slowly add 1 mL of prewarmed 0.075 M KCl. Gently add a further 8 mL of KCl. Incubate for 35-40 min at 37°C.
6. Add 1 mL of –20°C fixative and pipet vigorously.
7. Centrifuge for 10 min at 2000g.
8. Discard supernatant and gently resuspend the pellet in 10 mL of fixative.
9. Repeat **steps 7** and **8** two more times.
10. Resuspend the cells in 1 mL of fixative and allow one drop to fall from a Pasteur pipet onto a cleaned slide from a height of approx 10 cm.
11. Stain the chromosome preparations with 10% filtered Giemsa stain and examine under microscope (*see* **Note 3**).
12. Store chromosome preparations either by placing in the fixative at –20°C or dropping them onto microscope slides.
13. Mark the position of the drop on the slide with a diamond tipped pencil to locate the cells during the hybridization procedure.

3.2. Removal of Endogenous RNA

The presence of native RNA on the slides can cause the lessening of signal intensity, due to competing for binding of the labeled probe.

1. Add 200 μL of a 100 μg/mL solution of RNase A to each slide, making sure the entire slide is covered.
2. Place a 50 mm × 22 mm coverslip on top of the solution.
3. Place paper toweling on the bottom of a Decor box, and saturate with distilled water. Drain any excess water.
4. Place the slides in the box, close the box making sure it is air tight, and incubate in a 37°C oven for 1 h.

5. Remove slides from the decor box, and remove coverslips by allowing them to slip off.
6. Rinse slides in 2× SSC, pH 7.0, 4× for 2 min each.
7. Dehydrate the slides through an ethanol series 70%, 90%, and 100% for 2 min each.
8. Allow the slides to air dry.

3.3. Digoxigenin and Biotin Labeling of Probes

The labeling of λ clones by nick-translation generates randomly cleaved biotin or DIG-labeled DNA fragments. When these molecules are denatured, fragments comprising unique but homologous segments can reanneal with homologous sequences present in the chromosome pair on the slides. When several overlapping fragments hybridize to the corresponding chromosomal sequences, then a network of labeled DNA molecules is built up at the site of hybridization. As a result, an intense signal can be visualized down the microscope.

1. Into an Eppendorf tube place 2 μg of λ phage DNA containing the cloned gene of interest in a total volume of 65 μL for DIG reactions and 75 μL for Biotin (BIO) reactions.
2. a. (DIG labeling): Add 10 mL of 0.1 M β-mercaptoethanol, 5 mL of 10× DIG DNA labeling mix, 10 μL of 10× nick-translation buffer, and 10 μL of DNA polymerase/DNase I enzyme solution (Solution C from Gibco Kit).
   b. (BIO labeling): Add 10 mL of buffer and nucleotides (solution A4 from Gibco Kit), 5 μL of 0.4 M biotin-11-UTP, and 10 mL of DNA polymerase/DNase I enzyme solution (Solution C from Gibco Kit).
3. Incubate reaction for 90 min in a 15°C water bath.
4. Using a thin toothed comb, prepare a 1.6% agarose gel in 0.5× TBE and 10 μg/mL of ethidium bromide.
5. After 90 min place the Eppendorf tube on ice.
6. Aliquot 200 ng from the reaction (10 μL) and boil for 5 min. Chill on ice.
7. Load the boiled 200 ng into the 1.6% agarose gel. Also load a marker, in this instance *Hae*III digested φX174 DNA is suitable.
8. Run the gel at 90 mV/cm for approx 30 min. Visualize the DNA fragments over a UV lamp. The size range of the DNA fragments should range between 200 bp and 600 bp.
9. If the DNA fragments are larger, place the Eppendorf tube at 15°C for a further 30 min and repeat **steps 4–8** until the optimum range of DNA fragments has been achieved.
10. Once achieved, stop the reaction by adding the stop buffer and 2.5 μL of 5% SDS.

3.4. Determination of Biotin/Digoxigenin Incorporation

Once the probes have been nick-translated and the fragments are between 200 bp and 600 bp, the amount of incorporation of the biotin and DIG molecules needs to be determined. This will allow the sensitivity of the probe to be assayed.

1. Prepare a serial dilution of the probe (1:10, 1:100, and 1:1000). Make an initial 1:1 dilution of the probe (2 µL of probe and 2 µL of water). Then transfer 1 µL serially.
2. Cut a small piece of Protran membrane, and label in pencil where each sample and corresponding dilution is placed.
3. Spot onto membrane 1 µL from each dilution near the appropiate mark. Place the membrane in a Petri dish.
4. Bake the membrane at 80°C for 30 min.
5. Pour 5 mL of AP1 buffer onto the membrane and carefully pour off.
6. Pour 10 mL of AP2 blocking solution into the Petri dish. Incubate at 37°C for 15 min.
7. Pour off AP2 solution and then briefly rinse with 10 mL of AP1 buffer.
8. a. For DIG detection: add 2 µL of anti-DIG alkaline phosphatase conjugate in 10 mL of AP1 (1:5000 dilution) in a 10-mL centrifuge tube.
   b. For BIO detection: add 10 µL of streptavidin alkaline phosphatase in 10 mL of AP1 (1:1000 dilution) in a 10-mL centrifuge tube.
9. Pour this onto the membrane and incubate in the dark at room temperature for 10 min.
10. Pour off the antibody solution and wash twice in AP1 for 3 min each.
11. Pour off AP1 and wash 3× in AP3 for 3 min each.
12. In a 10-mL centifuge add 7.5 mL of AP3 buffer, 25 µL of X-phosphate, and 33 µL of 4-nitroblue tetrazolium chloride (NBT).
13. After carefully pouring off the last AP3 wash, pour on the NBT mix. Leave in the dark at RT for at least 2 h.
14. The 1:1000 dilution should be visible, and this means that the probe has sufficient incorporation for ensuing hybridization.

3.5. Probe Precipitation

1. Calculate the amount of DNA remaining in the original nick-translation reaction.
2. Approximately 600–700 ng of probe DNA should be precipitated with a 10-fold excess of sonicated genomic DNA, 40 µg of sonicated salmon sperm DNA, 2.5 vol of 100% ethanol, and 1/10th vol of 3 M sodium acetate.
3. Place at –20°C for 2 h.
4. Centrifuge for 10 min in a microfuge. Remove the supernatant.
5. Wash pellet with 100 µL of 70% ethanol, and remove ethanol.
6. Briefly microfuge, and remove any residual ethanol using a micropipet.
7. Resuspend the pellet in 15 µL of hybridization mix (50% formamide/50% probe cocktail mix). Carefully resuspend the pellet until completely dissolved.
8. Denature the probe in a 75°C water bath for 6 min.
9. Briefly microfuge, and incubate the probe at 37°C for a minimum of 30 min. This step allows for preannealing.

3.6. Chromosome Denaturation

To allow hybridization of the probe to complementary sequences in the fixed chromosomes, the chromosomal DNA needs to be denatured. The use of formamide allows the temperature required for denaturation to be reduced, leaving the chromosome structure intact for identification.

1. Place 70 mL of denaturing solution into a Coplin jar. Put the jar in a water bath and allow the solution to reach 70°C.
2. Plunge the slides (up to two at a time) into the formamide solution for 2 min (or optimal determined time).
3. Take the slide through an ethanol series (70%, 90%, and 100%) for 2 min each.
4. Place the slides in a slide rack to air dry.

3.7. Hybridization

The annealing of the probe DNA to the chromosomal DNA depends on salt concentration, temperature, and probe DNA concentration.

1. Carefully pipet the 10 µL of the preannealed probe DNA solution over the marked area of the slide harboring the chromosome preparation.
2. Gently lower the coverslip over this region with care not to create any air bubbles under the coverslip.
3. Seal the coverslips with an overabundance of rubber cement. This will make removing the coverslips after hybridization much easier.
4. Place absorbant toweling on the bottom of the Decor box and saturate with water. Drain off any excess water.
5. Place the slides in the Decor box, close the box, and ensure that it is air tight.
6. Incubate slides at 36°C for 16 h. In some cases 2-day hybridization is also sufficient.
7. The next day set up a water bath with two tanks inside. These tanks rest directly over magnetic stirrers. Place 400 mL of the 50% formamide/2× SSC, pH 7.0, solution in each of these tanks and allow to reach 45°C. Place a small magnetic stirrer bar in the bottom of these tanks.
8. Prewarm 800 mL of 0.5× SSC to 45°C.

3.8. Posthybridization Treatment

This step ensures that only the labeled probe DNA remains hybridized to homologous sequences in the chromosome pair fixed to the slide. By washing the slides in formamide and SSC, any nonhomologous annealing of probe DNA will be removed, thus facilitating later antibody detection.

1. Remove the cement and the coverslips from the slides using a pair of forceps, and discard these into a sharps waste container.
2. Place the two slides in a rack at opposite ends with the chromosomes facing inwards.
3. Briefly rinse the slides in 300 mL of 2× SSC.
4. Place the slide rack in one of the 45°C tanks containing the 50% formamide/2× SSC solution. Make sure the flea lies between the two slides and turn the stirrer on. Wash the slides 3× for 5 min each with continuous stirring.

5. Remove slide rack and place in next tank and proceed as described in **step 4**. While these slides are being washed, remove the formamide wash solution from the first tank (discard the formamide properly) and replace it with 400 mL of prewarmed 0.5× SSC, pH 7.0 (*see* **Note 4**).
6. Place the slide rack in the first 0.5× SSC tank and wash with stirring for 5 min. (Remove the formamide wash solution from tank 2 as described previously). Add the remaining 400 mL of prewarmed 0.5× SSC to the last tank.
7. Place the slides into the last 0.5× SSC wash and leave for 5 min with stirring.
8. After washing the slides place them in a histology jar in antibody washing solution.
9. Prepare a plastic sealable box with absorbant toweling saturated with 2× SSC and place in a 37°C oven.
10. Place the slides upright in the box and flood the slide with 1 mL of blocking solution. Cover box and leave for 30 min.
11. Remove slides and place in slide rack in antibody washing solution.

3.9. Antibody Detection

Although the probe DNA has been hybridized to homologous sequences in the chromosome on the slide, visualization of where the probe anneals needs to be performed. The use of fluorescent antibodies will aid in detection.

1. Centrifuge the antibodies at 10,000*g* for 3 min to pellet any antibody conglomerates that may have formed.
2. Make up antibody mix for each of the antibodies in blocking solution (*see* **Notes 4 and 5**).
3. Place 50 mm × 22 mm coverslips on a flat surface and pipet 200 μL of antibody mix onto the coverslip.
4. Carefully touch the slide onto the antibody mix on the coverslip, allowing surface tension to uptake the coverslip.
5. With extreme care place the inverted slide (coverslip facing down) on top of the four drawing pins in the antibody chamber.
6. Place the chamber in the 37°C oven for a minimum of 30 min.
7. Put 300 mL of antibody washing solution and a magnetic stirrer bar into a histology jar on top of a magnetic stirrer.
8. Remove the coverslips and place the slides into a slide rack.
9. Insert the slide rack into the histology jar with the chromosomes facing inwards, and wash for 5 min at room temperature with a stirring flea.
10. Change the antibody wash solution a further 2× and repeat **step 8**.
11. After washing repeat **steps 3–10** using the next antibody mix.
12. Repeat until all antibodies have been added to the slide.
13. After final antibody wash, place the slides face up on a flat surface and flood with 1 mL of DAPI solution (7 μL of stock in 10 mL of 2× SSC).
14. Incubate in the dark for 7 min.
15. Place slides in a Coplin jar, and rinse under a running tap for 20 s.

16. Remove slides, place in rack, and allow to air dry.
17. Once dry, pipet 15 mL (in a series of drops) of Vectashield onto the slide.
18. Carefully lower a 50 mm × 22 mm coverslip onto the slide, and press firmly to squeeze any excess Vectashield from between the coverslip and slide. Blot gently with tissue.
19. The chromosomes are now ready to be viewed with a microscope (**Fig. 1**).

4. Notes

1. The preferred method for removing unbound labeled probe on the slides is to use constant stirring of the washing solution. We have constructed a large acrylic water bath to accomodate this. The water bath has dimensions of 20 cm deep, 20 cm wide, and 60 cm long using 0.5 cm thick acrylic. We place this water bath on two magnetic stirring units and place a heating element at one end. Distilled water is added within and heated to 45°C. We have also constructed two independent washing chambers that are placed in the water bath directly above the magnetic stirrers. The dimensions for these are: 22 cm deep, 12 cm wide, and 12 cm long. Along the bottom of these chambers are two 0.5-cm strips (10 cm long) fixed 6 cm apart. This allows a magnetic flea to be placed below the slide rack holder that rests on these two strips. It is important to place the slides facing inwards toward the spinning magnetic flea. This allows optimum washing of the chromosomes to remove unbound labeled probe.
2. For the antibodies to bind to their targets on the slide, we have devised a chamber that will maximize binding while reducing the amount of nonspecific background binding to the slide. Antibody conglomerates are removed during centrifugation. However, formation of conglomerates can occur during antibody detection. We have constructed an antibody chamber from a large plastic Petri dish. Into the Petri dish a circular piece of cardboard is placed through which eight drawing pins are pushed upwards. Four pins are placed at a distance to hold a 25 mm × 75 mm microscope, likewise for the other four pins. This allows two slides to be incubated at a time. Through these pins two circular pieces of 3M Whatman filter paper are pushed. Prior to the antibody binding step, the chamber is flooded with 2× SSC, allowed to drain, and then placed in the 37°C oven ready for the antibody detection step.
3. The integrity of the chromosome preparations is paramount. Once chromosome preparations have been obtained that meet the criteria previously mentioned, these preparations can be stored in two ways. We prefer to drop our preparations onto slides immediately. We usually have two drops per slide. Once the preparation has been dropped onto slides, we air dry them and then take them through an ethanol series (70%, 90%, and 100%). These are allowed to air dry overnight. The next day they are placed in slide boxes, along with silica wrapped in gauze at one end, and the box placed in a –70°C freezer. Conversely, if space is a problem then the chromosome preparation can be stored at –70°C in fixative. Prior to use the fixative is changed and the chromosomes are visualized to see if they retained their initial criteria. Rather than staining the chromosomes, they can be visual-

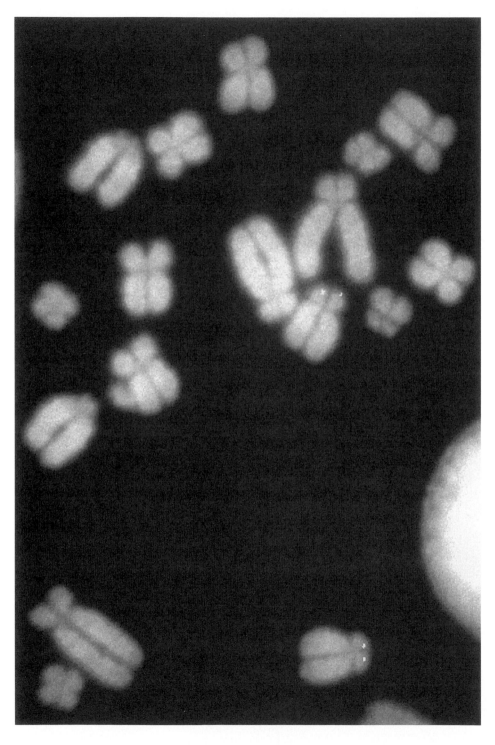

ized through a phase-contrast microscope. This allows the chromosomes on the slide to be viewed before use, after freezing, and even after the chromosome denaturing step to determine if they retain their morphology. We routinely check our chromosome slides after taking them out of –70°C and after the chromosome denaturing step.

4. In this step 0.5× SSC is considered a low-stringency wash. This step can be modified according to stringency constraints. For instance, a high-stringency wash could replace 0.5× SSC with 0.1× SSC, whereas a moderate wash could use 0.2× SSC. The first SSC wash is always 0.5×, and the stringency washes follow this.

5. Recommended dilutions in blocking reagent for the antibody solutions are as follows:
   a. Anti-DIG (raised in mouse) — 1:500,
   b. Anti-mouse-TRITC (raised in goat) — 1:160,
   c. Avidin–FITC — 1:250, and
   d. Biotinylated anti-avidin (raised in goat) — 1:100.

   Stock solutions of these reagents can be aliquotted and stored at –20°C. An individual vial from each can then be thawed and stored at 4°C in the dark. They remain viable for up to 2 mo. These tubes are then used for the above dilutions. These dilutions once made can be stored at room temperature in the dark. If two probes are to be detected on the same slide then antibody dilution mixtures can be made. We use the following antibody solutions for two-color detection:
   a. First round — avidin-FITC (4:1000).
   b. Second round — anti-avidin (10:1000)/anti-DIG (2:1000).
   c. Third round — avidin-FITC (4:1000)/anti-mouse TRITC (6:1000).

References

1. Andersson, L., Archibald, A., Ashburner, M., Audun, S., Barendse, W., Bitgood, J., Bottema, C., Broad, T., Brown, S., Burt, D., Charlier, C., Copeland, N., Davis, S., Davisson, M., Edwards, J., Eggen, A., Elgar, G., Eppig, J. T., Franklin, I., Grewe, P., Gill, T., Graves, J. A., Hawken, R., Hetzel, J., and Womack, J. (1996) Comparative genome organization of vertebrates. The First International Workshop on Comparative Genome Organization. *Mamm. Genome* **7**, 717–734.

2. Maccarone, P., Wilcox S. A., Watson, J. M., Marshall-Graves, J. A. (1994) Gene mapping using ^{3}H-labeled heterologous probes, in *Methods in Molecular Biology* (Choo, K. H. A., ed.), Humana Press, Totowa, NJ, pp. 159–172.

3. Toder, R., Wilcox, S. A., Smithwick, M., and Graves, J. A. M. (1996) The human/mouse imprinted genes IGF2, H19, SNRPN and ZNF127 map to two conserved autosomal clusters in a marsupial. *Chromosome Res.* **4**, 295–300.

4. Wilcox, S. A., Toder, R., and Foster, J. W. (1996) Rapid isolation of recombinant lambda phage DNA for use in fluorescence *in situ* hybridization. *Chromosome Res.* **4**, 397,398.

Fig. 1 *(opposite page)*. FISH analysis of a marsupial clone for the human X-linked sex reversing gene, DAX-1 (isolated from a *Macropus eugenii* λ phage genomic library), shows that this clone localises to the short arm of chromosome 5 in *M. eugenii*.

PART 2

TISSUE TECHNIQUES

9

In Situ Hybridization with Nonradioactive Probes

Gwen V. Childs

1. Introduction

In situ hybridization takes advantage of the reaction between two complementary single-stranded nucleic acid molecules. These molecules bind by means of hydrogen bonding of complementary base pairs. Hybridization performed *in situ* includes techniques that detect these hybrids in a cell or tissue section by cytochemistry.

When it was first introduced, *in situ* hybridization was applied to cell nuclei to detect DNA targets or amplified RNA genes *(1–3)*. Another early use for this approach was termed fluorescence *in situ* hybridization (FISH) in which genes were detected on specific sites on chromosomes or in nuclei *(4,5)*. *In situ* hybridization for cytoplasmic RNA was first used to detect viral nucleic acid sequences within infected tissues *(6,7)*.

Radiolabeled probes were used by early groups of investigators. However, during the past decade more efficient detection systems have been developed. They may involve the attachment of a signaling hapten molecule such as biotin or digoxigenin to the cDNA or cRNA probe. The detection systems for these molecules include fluorophores, enzyme reactants, or heavy metals (colloidal gold). They may be enhanced by sandwich techniques that apply different layers of reactants, or enzyme reactions, or both.

The purpose of this chapter is to describe a nonradioactive method for the detection of mRNAs *(8–15)*. Biotinylated complementary oligonucleotide or cRNA probes hybridize with the cytoplasmic mRNAs and are detected by antibiotin. The reaction is amplified by a sandwich technique that provides layers of biotin and streptavidin–peroxidase *(15,16)*. The protocol can be combined with a classical immunolabeling protocol that also detects antigens with a contrasting-color system *(8–15)*.

From: *Methods in Molecular Biology*, Vol. 123: In Situ *Hybridization Protocols*
Edited by: I. A. Darby © Humana Press Inc., Totowa, NJ

Table 1
Vendors for *In Situ* Hybridization Histochemistry, Including Uniform Resource Locator (URL, Internet link)

1. DAKO Corporation, 6392 Via Real, Carpenteria, CA 93013, USA; 800-235-5763. Web page: http://www2.multinet.net/dli/dako.htm
2. Fisher Scientific, 10700 Rockley Road, Houston, TX 77099, USA; 800-876-1900 or 800-766-7000; 800-395-5442 (instrument service). Web page: http://www.fisher1.com
   JRH Biosciences, PO Box 14848 Lenexa, KS 66325, USA; 800-255-6032.
3. Sigma Chemical, PO Box 14508; St. Louis, MO 63178, USA; 800-325-3010. Web page: http://www.sigma.sial.com/sigma/sigma.html
4. Thomas Scientific, 99 High Hill Road, PO Box 99, Suresdesboro, NJ 08085-0099, USA; 800-345-2100.
5. Vector Laboratories, 30 Ingold Road, Burlingame, CA 94010, USA; 800-227-6666. Web page: http://www2.multinet.net/dli/vector.htm

The protocol was first described by McQuaid and Allan *(16)*. It was used for tissue sections. We have modified it to detect mRNA and antigens in whole pituitary cells in culture. The rationale is that these cells will present more mRNA and antigen binding sites than cells in tissue sections. Therefore, all experiments are begun with freshly dispersed, whole pituitary cells. However, the protocol can also be used on frozen sections or paraffin sections. This protocol is described in the following paragraphs. After dissociation, the pituitary cells are plated for 1–36 h on 13-mm glass coverslips (Thomas Scientific, cat. no. 6672A75) in 24-well trays (Fisher Scientific, cat. no. 08-757-156). The protocol can also be adapted for use with dispersed cells at the electron microscopic level, if they are fixed and centrifuged between each step.

2. Materials

Sterile conditions are used for all fixation, washing, and handling methods to prevent RNase contamination. We make the buffer or diluent solutions ahead and Millipore filter them. Then, they are stored frozen in small aliquots. The following list includes the solutions needed for *in situ* hybridization. **Table 1** provides information about some of the major vendors that supply these reagents.

2.1. Solutions

1. Phosphate-buffered saline (PBS): 0.1 M phosphate buffer + 0.9% NaCl, pH 7.2.
2. Triton X-100 (0.3%): 300 μL of Triton X-100 (Sigma Chemical) and bring to 100 mL with 0.1 M PBS.

3. EDTA (50 mM): add 1.46 g of EDTA to 100 mL of 0.1 M Tris buffer (Sigma Chemical, cat. no. T-5030), (20 mL of 0.5 M Tris + 80 mL of Millipore-filtered water), pH 8.0.

4. Paraformaldehyde (4%): 4.0 g of paraformaldehyde in 100 mL of 0.1 M PBS, pH 7.2.

5. Acetic anhydride (0.25%) + 0.1 M triethanolamine: 250 μL of acetic anhydride and bring to 100 mL using Millipore-filtered water, then add 1.856 g of triethanolamine, pH 8.0 (Sigma Chemical).

6. Deionized formamide (50%) in 2X saline sodium citrate (SSC): make a 1:10 dilution from 20× SSC with Millipore-filtered water and add 1:1 (v/v) formamide (Sigma Chemical, cat. no. F-7503).

7. SSC stock: make 20× SSC stock by adding 3 M sodium chloride + 0.3 M sodium citrate to Millipore-filtered water. Make 4× solution with a 1:5 dilution from stock 20× SSC with Millipore-filtered water

8. The *in situ* hybridization buffer: The components are added in sequence in the following instructions. Preparation can be done in advance and usually takes about a day. Most of these components come from Sigma Chemical, unless otherwise noted (*see* **Table 1**).

   a. Deionize formamide: add 1.5 g of REXYN I-300 (Fisher Scientific, cat. no. R-208) to 100 mL of formamide, stir for 45-60 min and filter through Whatman filter paper. Combine 100 mL of deionized formamide + 100 mL of 4× SSC (20 mL of 20× SSC +80 mL Millipore-filtered water). The final concentration will be 50% formamide.

   b. Add 7.88 g of Tris salt to a small volume of distilled water, and warm to about 37°C to dissolve Tris. Add this solution to the buffer.

   c. With a sterile syringe and needle, add the following to the warm formamide–Tris solution: 20 g of dextran sulfate; 0.5 g of bovine serum albumin (RIA grade) (Sigma Chemical, cat. no. A-7638); 0.5 g of Ficoll-400; 0.5 g of polyvinyl pyrrolidone-360; 1.0 g of sodium pyrophosphate; 1.0 g of lauryl sulfate, sodium salt.

   d. Add 200:1 of dissolved salmon sperm DNA (ssDNA) to 200 mL of hybridization buffer.

   e. Aliquot into 10-mL units and store frozen (–20°C) in 15-mL centrifuge tubes.

3. Methods

3.1. Preparation of Pituitary cells

The pituitary cells are grown in Dulbecco's modified Eagle's medium (DMEM, JRH Biosciences, cat. no. 56499-10L). Just before hybridization, they are fixed in 2.5% glutaraldehyde (in 0.1 M phosphate buffer) for 30 min at room temperature. Then cells are washed for 1 h in four changes of phosphate buffer + 4.5% sucrose. We work under sterile conditions even during the fixation process and store the cells no longer than 1 wk at 2–4°C. These steps prevent RNase contamination, which would obviously eliminate RNA from the tissue or cells.

3.2. Preparation of the Complementary Probes

For the hybridization, we biotinylate and hybridize either cRNA probes *(8–10,15)* or complementary oligonucleotide probes that are at least 30 mer long *(12–14,17–20)*. The probes are either biotinylated by the Vector Photobiotin kit (Vector Laboratories, Burlingame, CA, USA) following kit instructions *(8,19,20)*. Or, the oligonucleotide probes are produced commercially with biotin attached. This does not add significantly to the cost of the oligonucleotide probe. A third option is to add biotin-UTP to the cRNA probes produced during in vitro transcription *(16)* (*see* **Note 1**). Controls include substitution of labeled sense sequences, or omission of the labeled probe (*see* **Note 2**).

3.3. Prehybridization Protocol

The prehybridization protocol prepares the cells or tissues for the entry of the probes and the hybridization itself. The steps may be categorized by the following functions. The first group aids penetration of the probes. The second group prevents nonspecific reactions of the probe or detection system.

On the day of the hybridization, first rinse the cells (which are either fixed on coverslips or in suspension) with fresh, sterile 0.1 *M* PBS for 5 min at room temperature with agitation. Expose the cells to the following sequence of solutions at room temperature. (All of the following chemicals but the paraformaldehyde are purchased from Sigma Chemical Co.).

1. To aid penetration of reagents, the cells are first treated with 0.3% Triton X-100 for 15 min at room temperature. Washing twice with 0.1 *M* PBS, 3 min each, follows, at which point the cells are gently shaken on a side-to-side shaker.
2. To further improve penetration and remove associated proteins, cells are then treated with proteinase K (1 µg/mL) for 15 min at room temperature.
3. The next step involves stabilizing the cells by exposing them to 4% paraformaldehyde/0.1 *M* PBS for 5 min at room temperature. This is followed by washing twice with 0.1 *M* PBS, 3 min each, while gently shaking.
4. To cover nonspecific reactive sites, the cells are treated with 0.25% acetic anhydride for 10 min while shaking.
5. The final prehybridization step prepares the cells for the hybridization buffer and helps break weak hydrogen bonds in the nonspecific linkages between the complementary probe and the surrounding tissue. This involves treatment of the cells with 50% deionized formamide in 2× SSC. The incubation is initially done while shaking the tray at room temperature for 10 min and it then continues at 37°C for an additional 10 min. Note: adjustments in this formamide step may be needed to prevent loss of hybrids. If there is a 2–3% mismatch in the complementary probe, the concentration of this solution should be lowered to 40–45%, because the higher concentration may break bonds in the mismatched regions of the hybrid. This same lower concentration should also be used to make the hybridization buffer (*see* earlier solutions).

3.4. In Situ *Hybridization*

The cells are then placed in the hybridization solution. This is hybridization buffer containing 10–500 ng/mL of biotinylated oligonucleotide or cRNA-probe (*see* **Note 3**). The cells are then incubated for 12–15 h at 37°C. The temperature of hybridization may be varied according to the melting temperature of the probe. A range from 36 to 42°C works for most of our probes against gonadotropin (8, 12, 17, 18) or proopiomelanotropin mRNAs *(19,20)*.

3.5. Posthybridization Protocol

Posthybridization steps have two functions. First, the posthybridization washes are designed to remove unbound probes while preserving the hybrids. If a cRNA probe is used, an RNase treatment step is added *(8,15)*. Second, the post-hybridization steps may prepare the cells or tissues for the detection system.

Initial washes are done at "low stringency" to prevent removal of the specific probe that has been hybridized to the mRNA. This involves washing in a high salt solution at room temperature. In our protocol, the cells are first washed 3× with 4× SSC for 15 min each wash. This is done at room temperature (with gentle shaking during the last 5 min). The stringency of these washes is low enough to remove the excess unreacted probe without removing the probe that has specifically hybridized to the mRNA (*see* **Note 4**).

The above washes may be sufficient to remove background stickiness as well. However, if there is background with the reaction, higher stringency washes may be added. This involves further washing in 2×, 1×, or even 0.5× SSC. Also, one can raise the temperature of the washes. We have found this high stringency washing to be unnecessary in our protocol.

After the washes, the cells are prepared for the detection protocol. The first step is a blocking step in a solution containing proteins. These proteins cover nonspecific sites and prevent background reactions, but they do not react with any of the components of the detection protocol. In our protocol, we block with 0.05 M Tris-buffered saline containing 1% bovine serum albumin + 10% normal horse serum (Vector Laboratories) for 15 min at room temperature (pH 7.6).

3.6. Detection of the Biotin

The biotin on the hybrids is detected with the following sequence of solutions:

1. Monoclonal anti-biotin (1:30) (Dako Corp) for 30 min at 37°C. They are then washed twice with 0.05 M Tris-buffered saline for 3 min each (pH 7.6).
2. Biotinylated horse anti-mouse IgG (rat absorbed, Vector Laboratories) (1:100) for 10 min at room temperature. This is followed by two washes with 0.05 M Tris-buffered saline, 3 min each. Note: one must use rat-absorbed anti-mouse IgG because most anti-mouse antisera will react nonspecifically with rat tissues.

3. A second layer of monoclonal anti-biotin (1:100) should be added by incubation for 30 min at 37°C. Coverslips are then washed twice with 0.05 *M* Tris-buffered saline, for 3 min each.

4. Apply a second layer of biotinylated horse anti-mouse IgG (rat absorbed) 1:100 for 10 min room temperature and wash twice with 0.05 *M* Tris-buffered saline, 3 min each.

5. Add a 1:10 dilution of peroxidase-conjugated streptavidin (Dako Corp, cat. no. P-397) (*see* **Note 5**) for 5 min at room temperature. The cells are then washed twice with 0.05 *M* Tris-buffered saline.

During the incubation in the streptavidin solution, the nickel-intensified diaminobenzidine (DAB) is prepared by dissolving 0.45 g of nickel ammonium sulfate and one DAB tablet (Sigma Chemical, cat. no. D-5905) in 30 mL of 0.05 *M* acetate buffer. Then add 20:1 of 30% hydrogen peroxide. The solution is filtered on a Whatman filter paper and added immediately to the cells for 6 min. The reaction product should be blue-black (*see* **Note 6**). This reaction cannot be improved by longer times. There is a limited life and time span of this solution. Therefore, to increase signal, the concentrations of the probe or other components of the detection protocol should be varied (*see* discussion).

After the DAB step, wash the coverslips twice with 0.05 *M* acetate buffer, dehydrate, dry, and mount onto glass slides, cell side up with Permount. A second square coverslip should be mounted over the cells. It is important to note that the DAB reaction product is dissolved in water-soluble solutions and therefore cannot be used with water-soluble mounting media or glycerol. Therefore, use only organic solvent soluble mounting media. One can view the slides for a brief period mounted in glycerol; however, eventually this will remove the reaction product.

After the mRNA is detected with the blue-black peroxidase substrate, the cell can be further labeled by immunocytochemistry for its protein or antigen content. For this we use the dual-labeling protocol with contrasting color peroxidase substrates *(9–15)*.

3.7. Results and Interpretation

This protocol produces finely distributed labeling for mRNA in dark patches or lines in the cell. **Figures 1** and **2** illustrate labeling for mRNA transcripts encoding the alpha subunit of the L calcium channel in pituitary corticotropes. The cell in **Fig. 1** shows a fine distribution of label. The cell in **Fig. 2** is from a population of corticotropes treated with corticotropin-releasing hormone (CRH) for 4 h. **Fig. 2** illustrates the fact that many of these cells contain label that is more dense and spreads out over a wider area of the cell. **Figure 3** shows that, at the electron microscopic level, the label is associated with dilated profiles of rough endoplasmic reticulum. This field shows labeling for luteinizing

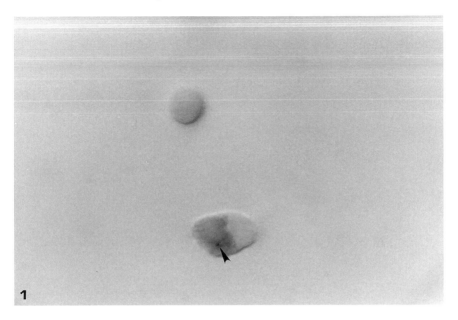

Fig. 1. Pituitary corticotrope from an enriched population labeled for mRNA for the α-subunit of the L calcium channel. Label is in lines or patches throughout the cell. Original magnification: ×1490.

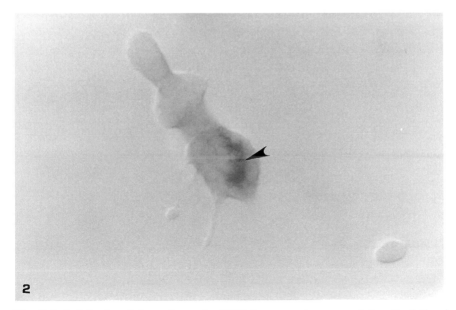

Fig. 2. Label for L calcium channel mRNA increases in area and density following stimulation for 4 h with corticotropin stimulating hormone (CRH). Original magnification: ×1490.

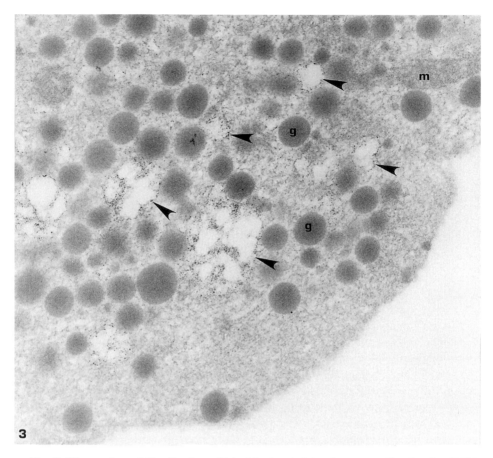

Fig. 3. Illustration of distribution of label for luteinizing hormone–β-subunit mRNA at the electron microscopic level. Shows label on profiles of rough endoplasmic reticulum. No label is associated with mitochondria or granules. Original magnification: ×60,000.

hormone β-subunit mRNA on dilated rough endoplasmic reticulum (RER) in a pituitary gonadotrope.

4. Notes

1. Some low-abundance mRNAs may require cRNA probes for adequate detection. We have found that cRNA probes (illustrated by **Figs. 1** and **2**) usually are more sensitive because they carry more biotin molecules/probe and they are longer. Yet, they are sticky and difficult to work with. They also require expertise in molecular biology techniques for their production. Oligonucleotide probes can be engineered to carry one or more biotins. Their sensitivity may also vary with

the size and number of biotin molecules. Nevertheless, for most of our studies, biotinylated oligonucleotide probes label expected populations of cells at concentrations of 1–100 ng/mL One could also engineer several biotinylated probes directed against different parts of the mRNA transcript. This may increase sensitivity. All probes must react with unique sequences, however, or specificity will be compromised.

2. We have controlled for interference by eliminating the complementary RNA probe from the first sequence or substituting sense probes for the antisense probes. In each case, no labeling results or it is reduced to 1–3% of the cell population. Tests of nonspecific labeling, if it occurs, can be made by substituting different antisera in the detection protocol. For example, as stated in the Methods section, the anti-mouse IgG should be absorbed with rat tissues (liver powder). If it is not, it will react nonspecifically with rat tissues and cause labeling similar to that seen with the *in situ* hybridization protocol.

   If labeling is seen in the control preparations, one can remove the mouse anti-biotin. This will test the ability of the anti-mouse antibodies to react on their own (with no mouse antibodies in the sequence).

3. The labeling density for the mRNA (sensitivity) can be improved by adding another layer of anti-biotin and streptavidin. However, this may be at the expense of higher background *(16)*. Tests of different temperatures of hybridization may show an optimal temperature that promotes labeling. Finally, one can vary the concentration of probe from 0.1–100 ng/mL and the reaction should increase with increasing concentration (up to a plateau point).

4. There are several ways to reduce background. First, one can remove one layer of the detection system for the mRNA (an anti-biotin and an anti-mouse IgG step) or dilute the components further. This should be done only if the signal is strong, however. Second, one can increase SSC washes, reducing the concentration to 0.1×. This low salt provides high-stringency conditions needed to wash out the nonspecifically bound probes that may cause the background. This is especially a problem with the sticky cRNA probes. Caution must be given, however, to the possibility that the washes will wash out the probe attached in the specific hybrids. Finally, the use of different concentrations of the detecting reagents for the biotin (anti-biotin and streptavidin–peroxidase) will reduce background. If the signal is strong, one can dilute both and achieve optimal results.

5. This protocol has been used successfully for the detection of mRNAs in tissue sections *(16)* and cell cultures *(15)*. It is preferred over the direct avidin detection systems because of its sensitivity. We have also found that some avidin-containing kits produce high background. The anti-biotin sandwich method provides flexibility and amplification potential along with a streptavidin solution that works at neutral pH and gives low background reactions.

6. The label is easy to detect in cells because of its density and linear or patchy pattern. The label should be blue-black or a purplish blue. Ideally, the background should be crystal clear so that the unlabeled cells are difficult to see. A steel-gray background over all the cells indicates background problems.

References

1. Gall, J. G. and Pardue, M. (1969) Formation and detection of RNA–DNA hybrid molecules in cytological preparations. *Proc. Natl. Acad. Sci. USA* **63,** 378–383.

2. John, H. A., Birnstiel, M. L., and Jones, K. W. (1969) RNA–DNA hybrids at the cytological level. *Nature* **223**, 582–587.

3. Buongiorno-Nardelli, S. and Amaldi, F. (1970) Autoradiographic detection of molecular hybrids between rRNA and DNA in tissue sections. *Nature* **225**, 946–948.

4. Pardue, M. L. and Dawid, I. B. (1981) Chromosomal locations of two DNA segments that flank ribosomal insertion-like sequences in *Drosophila*: flanking sequences are mobile elements. *Chromosoma* **83**, 29–43.

5. Fostel, J., Narayanswami, S., Hamkalo, B., Clarkson, S. G., and Pardue, M. L. (1984) Chromosomal location of a major tRNA gene cluster of *Xenopus laevis*. *Chromosoma* **90**, 254–260.

6. Brahic, M. and Haase, A. T. (1978) Detection of viral sequences of low reiteration frequency by *in situ* hybridization. *Proc. Natl. Acad. Sci. USA* **75**, 6125–6127.

7. Haase, A. T., Venture, P., Gibbs, C., and Touretellotte,W. (1981) Measles virus nucleotide sequences: detection by hybridization *in situ*. *Science* **212**, 672–673.

8. Childs, G. V., Lloyd, J., Unabia, G., Gharib, S. D., Weirman, M. E., and Chin, W. W. (1987) Detection of LH mRNA in individual gonadotropes after castration: use of a new *in situ* hybridization method with a photobiotinylated cRNA probe. *Mol. Endocrinol.* **1**, 926–932.

9. Childs, G. V., Unabia, G., Weirman, M. E., Gharib, S. D., and Chin, W. W. (1990) Castration induces time-dependent changes in the FSH-mRNA-containing gonadotrope cell population. *Endocrinology* **126**, 2205–2213.

10. Childs, G. V., Patterson, J., Unabia, G., Rougeau, D., and Wu, P. (1991a) Epidermal growth factor enhances ACTH secretion and expression of POMC mRNA by corticotropes in mixed and enriched cultures. *Mol. Cell. Neurosci.* **2**, 235–243.

11. Childs, G. V., Taub, K., Jones, K. E., and Chin, W. W. (1991b) Tri-iodothyronine receptor β-2 mRNA expression by somatotropes and thyrotropes: effect of propylthiouracil-induced hypothyroidism in rats. *Endocrinology* **129**, 2767–2773.

12. Childs, G. V., Unabia, G., and Rougeau, D. (1994) Cells that express luteinizing hormone (LH) and follicle stimulating hormone (FSH) beta (β) subunit mRNAs during the estrous cycle: the major contributors contain LH, FSH and/or growth hormone. *Endocrinology* **134**, 990–997.

13. Kaiser, U., Lee, B. L., Unabia, G., Chin, W., and Childs, G. V. (1992) Follistatin gene expression in gonadotropes and folliculostellate cells of diestrous rats. *Endocrinology* **130**, 3048–3056.

14. Lee, B. L., Unabia, G., and Childs, G. (1993) Expression of follistatin mRNA in somatotropes and mammotropes early in the estrous cycle. *J. Histochem. Cytochem.* **41**, 955–960.

15. Fan, X. and Childs, G. V. (1995) EGF and TGF mRNA and their receptors in the rat anterior pituitary: localization and regulation. *Endocrinology* **136**, 2284–2324.

16. McQuaid, S. and Allan, G. M. (1992) Detection protocols for biotinylated probes. Optimization using multistep techniques. *J. Histochem. Cytochem.* **40,** 569–574.

17. Childs, G. V., Unabia, G., and Lloyd, J. (1992a) Recruitment and maturation of small subsets of luteinizing hormone (LH) gonadotropes during the estrous cycle. *Endocrinology* **130**, 335–345.

18. Childs, G. V., Unabia, G., and Lloyd, J. M. (1992b) Maturation of FSH gonadotropes during the rat estrous cycle. *Endocrinology* **131,** 29–36.

19. Wu, P. A. and Childs, G. V. (1990) Cold and novel environment stress affects AVP mRNA in the paraventricular nucleus, but not the supraoptic nucleus: an *in situ* hybridization study. *Mol. Cell. Neurosci.* **1**, 233–249.

20. Wu, P. A. and Childs, G. V. (1991) Changes in rat pituitary POMC mRNA after exposure to cold or a novel environment detected by *in situ* hybridization. *J. Histochem. Cytochem.* **39**, 843–852.

10

Quantitative *In Situ* Hybridization Using Radioactive Probes to Study Gene Expression in Heterocellular Systems

Catherine Le Moine

1. Introduction

For several years, quantitative *in situ* hybridization has emerged as a particularly powerful technique to study gene expression within complex heterocellular systems, including the central nervous system, by measuring mRNA levels and their variations under experimental or physiological conditions. In comparison with other quantitative techniques such as Northern blot or dot blot, this approach makes possible the analysis of gene expression, not only at the regional level but also at the cellular level, which appears particularly useful for the study of heterogeneous cellular populations *(1–3)*. Quantitative *in situ* hybridization has been greatly improved during the past 10 years, especially by the development of various image analyzer systems. Moreover, the technique has reached a very high level of sensitivity by using appropriate standardization and methods of measurement *(1–9)*. Two levels of analysis yield different and complementary information: (1) macroscopic study using densitometry on X-ray films allows the measurement of variations in mRNA at the anatomical level and (2) microscopic study allows quantitation of mRNA at the cellular level (*see* **Note 1**).

Potential changes in the expression of genes coding for a wide range of molecules in heterocellular systems can be investigated by quantitative *in situ* hybridization. Our group has particularly investigated dopaminergic control in the nigro-striatal complex, and especially the interactions between dopamine neurons and striatal target neurons (e.g., **refs.** *10–13*). Using pharmacological treatments, we have attempted to elucidate how changes in the dopaminergic

From: *Methods in Molecular Biology*, Vol. 123: In Situ *Hybridization Protocols*
Edited by: I. A. Darby © Humana Press Inc., Totowa, NJ

environment modify the expression of genes coding for neuropeptides, biosynthetic enzymes, neuroreceptors, and transcription factors in the striatum *(10–13)*. Given the anatomical complexity and heterogeneity of this structure, the analysis of alterations in gene expression, as well as the study of cellular subpopulations and/or the number of cells expressing a given gene, required us to develop a procedure of quantitative *in situ* hybridization that would allow the measurement of mRNA levels not only at the anatomical level, but also at the single cell level. This chapter provides a detailed description of such a procedure that, with minimal modifications, should be readily applicable to study gene expression in other heterocellular systems.

2. Materials

1. Cryostat sections of 1% paraformaldehyde-perfused tissue (e.g., rat brain) or fresh-frozen tissue.
2. ^{35}S- or ^{33}P-labeled nucleotides (NEN Life Science Products).
3. Probes: synthetic oligonucleotides, cDNA clones, or cRNA probes.
4. Materials needed for standard *in situ* hybridization protocols using radioactive probes include kits for probe labeling (tailing for oligonucleotides, nick-translation for cDNA, in vitro transcription for cRNA), RNase-free glassware, DEPC water-treated buffers, humidified incubator and water bath, 8× SSC (1.2 *M* sodium chloride, 0.12 *M* trisodium citrate), absolute ethanol.
5. Hybridization buffer for oligonucleotide and cDNA probes:
   a. 2× SSC, 50% formamide.
   b. 10% Dextran sulfate.
   c. 500 µg/mL Salmon sperm DNA.
   d. 1% Denhardt's solution.
   e. 5% Sarcosyl.
   f. 250 µg/mL Yeast tRNA.
   g. 200 m*M* dithiothreitol (DTT).
   h. 20 m*M* NaH$_2$PO$_4$.
6. Hybridization buffer for cRNA probes:
   a. 20 m*M* Tris-HCl.
   b. 1 m*M* EDTA.
   c. 300 m*M* NaCl.
   d. 50% Formamide.
   e. 10% Dextran sulfate.
   f. 1× Denhardt's solution.
   g. 250 µg/mL Yeast tRNA.
   h. 100 µg/mL Salmon sperm DNA.
   i. 100 m*M* DTT.
   j. 0.1% SDS.
   k. 0.1% Sodium thiosulfate.
7. RNase A (10 mg/mL aliquots) for the posthybridization with cRNA probes.

8. X-ray films (Kodak BIOMAX or X-OMAT) and photographic emulsion (Ilford K5 or Kodak NTB2).
9. Developer and fixative for autoradiography, and material for staining (toluidine blue or Mayer's hematoxylin and eosin).
10. Optical table and microscope fitted out with epipolarization under fluorescence.
11. Image analyzer system allowing densitometric analysis and silver grain counting.
12. Statistical analysis program (e.g., Crunch, Statview, Statgraphics).

3. Methods

Before starting the experiment and especially the *in situ* hybridization procedure, one should clearly define the various parameters to be measured, and choose an appropriate number of samples and animal groups for quantitative analysis. This part is critical to ensure that results can be analyzed statistically, especially for small variations in gene expression. In standard pharmacological experiments as we have performed, six to eight animals per group appears to be a suitable sample size. It is also important to adhere to a precise standardized protocol, from tissue preparation to statistical analysis of the results, as we will describe in the present section. *In situ* hybridization on brain sections can be performed using oligonucleotide, cDNA or cRNA probes, radioactively labeled with ^{35}S- or ^{33}P-nucleotides (*see* **Note 2**). Relative variations in mRNA levels are measured against radioactive standards. Quantitation is performed with an image analyzer system (BIOCOM, Les Ulis, France) at the macroscopic and microscopic levels (*see* **Note 5**).

3.1. Tissue Preparation

1. Maintain adult male rats (Sprague-Dawley, 150–250 g) in standard housing conditions, and perform the pharmacological treatments on different animal groups according to the requirement of the experiments.
2. The rats are killed at various times following treatment. After anesthesia with chloral hydrate, the brains are either perfused with 1% paraformaldehyde, pH 7.2, in 0.1 *M* phosphate buffer and cryoprotected in 15% sucrose/phosphate buffer, pH 7.2, for 16 h, or directly dissected out, before being frozen in liquid nitrogen. The brains are then cut into 12–15 μm thick frontal cryostat sections, collected on gelatin-coated slides, and stored at -80°C until use (*see* **Note 4**).

3.2. Generation of the Radioactive Standards

Radioactive standards are generated using brain paste and are necessary to convert the optical densities (O.D.) (on X-ray film), or the silver grain densities (on tissue sections), into radioactivity concentrations that are a function of the mRNA levels. These brain paste standards are prepared as follows:

1. Homogenize four to eight rat brains, previously perfused with saline buffer, in 0.1 *M* phosphate buffer, pH 7.2.

2. Concentrate the brain paste by centrifugation for 20 min at 10,000g.
3. Freeze in 10-mL aliquots and store at –80°C until use. These aliquots may be stored for several years.
4. Prepare 16 dilutions of ^{35}S-labeled nucleotide (*see* **Notes 2** and **3**) in distilled water (from approx 3500 to 1,500,000 cpm/μL).
5. Thaw an aliquot of brain paste.
6. Incorporate and homogenize carefully 40 μL of each radioactive dilution into 500 μL of brain paste.
7. Transfer the paste into a straw and flash-freeze in liquid nitrogen.
8. When frozen, cut the radioactive brain paste cylinders into 12 μm thick cryostat sections and collect them on gelatin-coated slides.
9. Take 10 sections from each serial radioactive dilution, collect them in scintillant cocktail, and count them to get the radioactive value for each "spot."

The concentration of radioactivity is expressed in cpm/spot or nCi/mg of tissue. For macroscopic analysis, a slide bearing all 16 dilutions is exposed on X-ray film with the hybridized brain sections (as illustrated in **Fig. 1**). For microscopic analysis, four sections of each dilution are collected on slides, and dipped in photographic emulsion (*see* **Fig. 2B**). These standards are then exposed and developed in parallel with the hybridized tissue sections. The slides bearing the radioactive standards are kept in a box at room temperature until used (up to several months).

3.3. In Situ *Hybridization Procedure*

3.3.1. Labeling of the Probes

All the probes are labeled using standard protocols as previously described (*11,12,14*). Oligonucleotide probes are labeled with [^{35}S]- or [^{33}P]dATP by "tailing," i.e., addition of nucleotides at the 3'-end. Nick-translation is used to label cDNA probes by incorporation of the same radiolabeled deoxynucleotides. cRNA probes are synthesized by in vitro transcription from cDNA clones using [^{35}S]- or [^{33}P]UTP (*see* **Note 4**).

3.3.2. In Situ *Hybridization*

In situ hybridization is performed with the three usual steps of section pretreatment, hybridization of the probe and posthybridization washes. The standardization of all these steps is necessary to provide homogeneous and reliable hybridization signals that are suitable for quantitation. Within the same experiment, all the slides must be treated identically, especially with the same buffers and dilutions of the probe.

3.3.2.1. PRETREATMENT.

1. Thaw the slides 10 min at room temperature.
2. Wash twice for 30 min in 4× SSC, 0.1% Denhardt's (only for oligonucleotides and cDNAs).

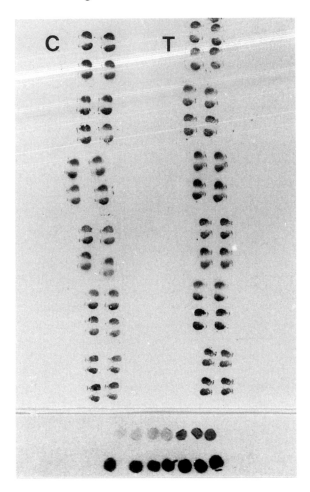

Fig. 1. X-ray film showing autoradiographs of the hybridized sections from control (C) and treated (T) animals and radioactive standards (*bottom two rows*). Measurement of optical densities is done for each standard and a calibration curve is generated (*see also* **Fig. 2A**). The areas of interest are then delineated on each section, and the optical densities are measured and converted into concentration of radioactivity using the calibration curve.

3. Wash the slides twice for 10 min in 4× SSC, then incubate 10 min in 4× SSC, 1.33% triethanolamine, 0.25% acetic anhydride, pH 8.0. This step is useful for decreasing the background by acetylation of free-radicals.
4. Rinse 2× for 5 min each in 4× SSC.
5. Dehydrate in graded ethanols (70%, 80%, 90%, 100%) and air dry.

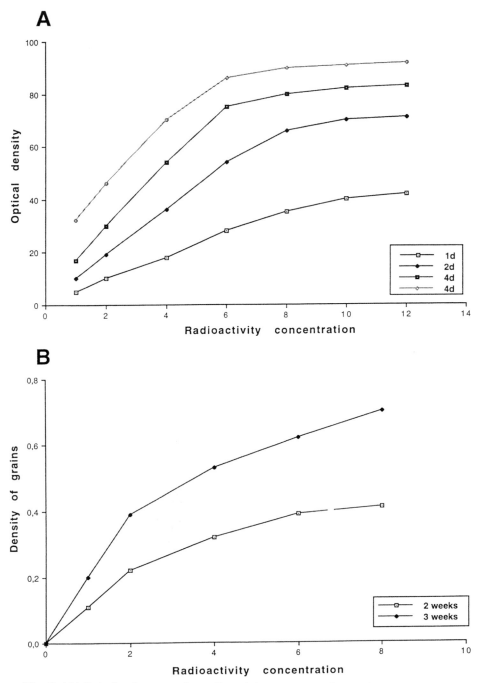

Fig. 2. (**A**) Relation between the radioactivity concentration and the optical density (O.D.) measured on X-ray film. Exposure times for the same film are 1, 2, 4, and 6 d. (**B**) Relation between the radioactivity concentration and the density of silver grains on radioactive standards after micro-autoradiography. Exposure times for the same standards are 2 and 3 wk.

3.3.2.2. HYBRIDIZATION

1. Centrifuge the labeled probe (usualy stored in ethanol/acetate) for 1 h at 10,000*g*.
2. Remove and check the supernatant for radioactivity, and dry the pellet under vacuum.
3. Resuspend the pellet in 1× TE for oligonucleotides and cDNAs and 1× TE/50 m*M* DTT for cRNA probes.
4. For oligonucleotide and cDNA probes, dilute the probes into 35–50 μL of the appropriate hybridization buffer (*see* **Materials**) to a concentration of 30 pg/μL (for cDNA probes, approx 150,000 cpm/slide) or 3 pg/μL (for oligonucleotide probes, 100,000–3,000,000 cpm/slide), and hybridize overnight at 40°C in a humidified chamber.
5. For cRNA probes, dilute the probe into 50 μL of the appropriate hybridization buffer (*see* **Materials**) to a concentration of 1–5 ng/slide (0.5×10^6 to 2×10^6 cpm/slide), and hybridize overnight at 55°C in a humidified chamber.
6. For better homogeneity of the hybridization signal, liquid-phase hybridization in glassware, such as a Coplin jar, gives excellent results but this approach may be constrained by the use of much larger quantities of probe (10–20 mL of hybridization buffer for approx 16 slides) (*see* **Note 4**).

3.3.2.3. POSTHYBRIDIZATION WASHES FOR CDNA AND OLIGONUCLEOTIDE PROBES

1. Wash the slides briefly in cold 4× SSC.
2. Wash twice for 30 min in 1× SSC at room temperature with agitation.
3. Wash twice for 30 min in 1× SSC at 40°C with agitation.
4. For oligonucleotide probes, add an additional washing step of 30 min (twice) in 0.1× SSC at 40°C with agitation.
5. Dehydrate twice in 100% ethanol and air dry.

3.3.2.4. POSTHYBRIDIZATION WASHES FOR CRNA PROBES

1. Wash the slides twice for 5 min in 4× SSC with agitation
2. Wash for 15 min at 37°C in RNase A (20 μg/mL in 0.5 *M* NaCl/10 m*M* Tris-HCl/0.25 m*M* EDTA).
3. Wash the slides again in 2× SSC (5 min, twice), 1× SSC (5 min), 0.5× SSC (5 min) at room temperature, and in 0.1× SSC at 65°C (30 min, twice) with agitation.
4. Place the slides at room temperature in 0.1× SSC to cool.
5. Dehydrate in graded ethanol (70%, 80%, 90%, 100%) and air dry.

The hybridized sections are then exposed on the same X-ray film (Kodak X-OMAT or BIOMAX) together with the radioactive standards for macroscopic analysis (**Fig. 1**). Note that the radioactive standards need to be counted in parallel for each experiment to take into account the radioactive decay. For cellular analysis after exposure on film, the hybridized sections and the standards for microautoradiography (four "spots" of each dilution) are dipped in emulsion (Ilford K5), exposed for the appropriate time at 4°C in the dark (usually 3–12 wk depending on the abundance of the target mRNA). The autoradiographic signal is then developed, and the sections are stained in toluidine blue (or Mayer's hematoxylin and eosin) and mounted (*see* **Note 4**).

3.4. Procedure of Measurement and mRNA Quantitation

Quantitative measurements of changes in gene expression can be performed in two different ways:

1. Macroscopic analysis using densitometry on X-ray film provides information at the regional level.
2. Microscopic analysis by silver grain counting provides information at the cellular level. This step is necessary to study cellular subpopulations and/or cells whose density does not allow a macroscopic analysis.

All the quantitative analysis is performed using an image analyzer system (in our procedure, BIOCOM, Les Ulis, France) fitted with a densitometric image system and a microscope under fluorescent epi-illumination (*see* **Note 5**). A charged-couple device (CCD) camera provides a computerized image on a video monitor on which the regions of interest are delineated and analyzed.

3.4.1. Macroscopic Analysis

1. Preheat the optical table for at least 1 h.
2. Place the X-ray film (**Fig. 1**) on the optical table, and use the camera to record the image on the video monitor.
3. Measure the optical densities of the standards within a constant square area, and make a calibration curve representing optical densities as a function of the radioactivity concentration (**Fig. 2A**). It is important that the probe concentration and the exposure time be determined before the experiment in order to stay within the linear part of the calibration curve.
4. Delineate the areas of interest on the monitor.
5. For each area, measure the optical densities, directly converted into radioactivity concentration by the calibration curve. These values are representative of the quantity of probe hybridized and therefore correspond to the relative mRNA level in the area.

3.4.2. Microscopic Analysis

The microscopic study is performed with a ×50 objective lens under both fluorescent epi-illumination and normal illumination. Silver grain counting is based on the procedure described by Bisconte et al. *(15)* with the number of grains expressed as a function of the quantity of reflected light under epi-fluorescence only. After calibration, the image analysis system provides an estimation of the grain number over identified single cells. Measurements are performed as follows:

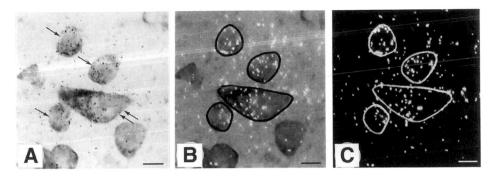

Fig. 3. Protocol for measurement of silver grain densities at the cellular level after hybridization with a [35]S- or ^{33}P-labeled probe. **(A)** Bright-field image showing two types of cells containing the same mRNA: medium-sized neurons (*single arrows*) and large-sized neuron (*double arrow*). **(B)** The same image after combined illumination (normal bright-field plus epi-illumination), where each labeled cell is delineated on the video monitor. **(C)** Observation under epi-illumination only. Once the cells are delineated, the bright field is turned off so that only silver grains are visible. The number and density of these grains are then measured for each cell directly by the image analysis system.

1. Construct an internal calibration curve for each section to be analyzed under epi-illumination only (as in **Fig. 3C**), and measure the mean quantity of light reflected by a known number of grains (i.e., 0, 1, or 2 grains) within a constant square area.
2. Check and store a calibration curve (number of grains as a function of the amount of reflected light).
3. Identify the labeled cells under transmitted light (**Fig. 3A**) and delineate them on the video monitor (**Fig. 3B**).
4. The quantity of light reflected in the area is then measured under epi-illumination *only* (**Fig. 3C**). The optical density will be directly converted into number of grains, then into grain density (grain number per mm^2) using the calibration curve.
5. Measure the grain density of the background over randomly-chosen areas other than over the labeled cells.
6. Subtract the mean background value from the grain density measured on each cell.
7. Convert the final data for each cell into radioactive concentration using the calibration curve generated by the measurement of silver grain densities over the brain paste standards (**Fig. 2B**). This value gives an index of the cellular mRNA level.

3.4.3. Statistical Analysis of the Data

Depending on the groups that have been defined in the experiment, the variations of mRNA expression, at the macroscopic and microscopic levels, need to be analyzed using a multifactorial analysis of variance (one-way or two-way ANOVA), followed by the appropriate posthoc comparisons (e.g., Student's *t*-

test, Newman–Keuls, Mann–Whitney). For the microscopic studies, the frequency distribution of the grain densities over the single cells is plotted for the experimental and the control groups, and compared using a two-tailed Kolmogorov–Smirnov sample test.

4. Notes

1. Quantitative *in situ* hybridization procedure has several advantages and disadvantages. In comparison with other quantitative approaches such as Northern blot or dot blot, this procedure provides additional information at the regional and cellular levels, and allows the study of gene expression in heterocellular systems. Also macroscopic analysis on X-ray films is relatively simple and convenient, and allows rapid testing of several experimental conditions, such as different pharmacological treatments (within a time frame of several days). However, the density of labeled cells has to be high enough to give a signal detectable on X-ray films. Such analysis does not provide information at the cellular level, and does not allow the detection of cellular subpopulations. In this case, the procedure for microscopic analysis, although requiring more effort to develop, is absolutely necessary for the study of dispersed single cells. It is important to note that quantitative *in situ* hybridization also has some disadvantages. The entire procedure is quite complicated (from cryostat sections to quantitation) and needs careful standardization and analysis which requires experienced users.

2. ^{35}S and ^{33}P are useful radioisotopes for quantitation experiments for the following reasons:
   a. They can be handled without major shielding.
   b. They provide good cellular resolution (especially when compared with ^{32}P) and reasonably short exposure times (especially as compared with ^{3}H).
   c. For the radioactive standards, the half-life of ^{35}S (87 d) allows the use of the same material for several months if these are made with a wide range of dilutions as described here (as compared with 25 d for ^{33}P and 14 d for ^{32}P).
   d. There are no quenching problems (e.g., compared with ^{3}H).

3. The use of brain paste standards instead of commercial standards, ^{35}S- or ^{14}C-converted standards (*see also* **ref. 8**) has been developed so that consistency and thickness are similar to that of the tissue sections used, thereby optimizing comparison between hybridized sections and standards. Note that it takes 3–4 d to make the radioactive standards using brain paste, but the same standards can be used for up to 12–18 mo with good results. Despite the absence of absolute quantitation of mRNA copies, calibration of the radioactivity using standards is still necessary to ensure that the *in situ* hybridization signals do not fall within the saturation part of the curve (as shown in **Fig. 2**). In our own experiments, we measure relative variations of radioactivity (i.e., mRNA levels) between experimental and control groups, to examine variations in gene expression. Note that other groups have compared hybridized mRNA to total mRNA level using a poly-U labeled probe (*1*) or using internal radioactive standards with competitive hybridization between labeled and nonlabeled probes (*2,5*). However, given the lack

of accuracy inherent to these procedures and the multiple steps that need to be standardized, the method described here appears to offer a more reasonable index of changes in gene expression.

4. Although the protocols described here are routinely used in our laboratory, they can be readily modified for use with other types of histological materials. Nevertheless, to ensure accurate measurement, it is necessary to get a homogeneous signal at both the macro- and microscopic levels. An important parameter for this is the thickness of the tissue sections. Indeed, 12–15 µm gives a much better homogeneity compared with 8–10 µm, and more intense hybridization signals. As previously mentioned, *in situ* hybridization in "liquid phase" gives good homogeneity both for macroscopic and microscopic analyses, as compared with the classical procedure using coverslips, but this requires significantly much larger quantities of radioisotopes and probes (especially cDNA or cRNA probes) because of the larger volumes of hybridization buffer involved. Nevertheless, oligonucleotide probes may allow such quantities to be used at a reasonable cost. Note that when using ^{35}S, the same dilution of probe can be reused several times for about 1 month. Homogeneity of signals for microscopic analysis is less easy to obtain because the sections are dipped vertically, resulting in the emulsion thickness being different from one section to another. This problem can be minimized by always comparing measurements made at the same location on the slides.

5. To perform macroscopic analysis, the density of cells expressing a given gene has to be high enough inside the area of interest to give a homogeneous signal on X-ray film. If the labeled cells are widely dispersed over the area, cellular analysis should be used. However, following densitometric analysis at the regional level, a detailed study at the cellular level is often necessary to allow a precise analysis of the observed changes in gene expression. For example: (1) variations of the number of cells expressing a given mRNA, (2) variation of mRNA level within individual cells, and (3) variation between distinct cell populations expressing the same gene. For macroscopic study, an even and constant illumination of the optical table is required to get reproducible and accurate measurements. Thus depending on the analyzer systems, the optical table may need to be turned on 1 or 2 h before the measurements are made. For microscopic analysis, it is necessary to perform a good calibration of the silver grains and to check that the counting is accurate on each analyzed section. It should also be verified that the staining used does not result in autofluorescence (as may happen with cresyl violet, for example).

6. **Figure 4** shows an example of application of our procedure to study the expression of preproenkephalin A (PPA) and D_2 dopamine receptor genes in the rat striatum following haloperidol treatment, an antagonist of D_1/D_2 dopamine receptors *(10–12)*. Quantitative analysis at the macroscopic level has demonstrated that haloperidol treatment provokes an increase of both PPA and D_2 receptor mRNA levels in the rat striatum. For PPA, microscopic analysis shows that this variation is not due to an increase in the number of labeled cells but is due to an increase of mRNA level in all the neurons expressing the PPA gene. For the D_2

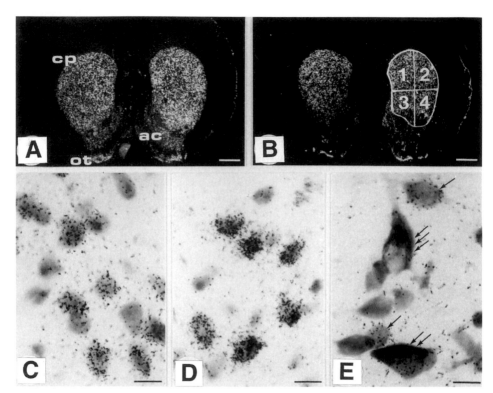

Fig. 4. Detection of PPA and D_2 mRNA in neurons of the rat striatum (*see* **Note 6**). Microautoradiographic analysis. (**A,B**) Darkfield views of rat brain sections after *in situ* hybridization with PPA (**A**) and D_2 receptor (**B**) probes showing intense labeling in the caudate-putamen (cp), accumbens nucleus (ac), and olfactory tubercle (ot). **B** shows an example of the different regions that may be used to count silver grains and neuronal densities (1, dorsomedial; 2, dorsolateral; 3, ventromedial; 4, ventrolateral). (**C–E**) Brightfield views of neurons after hybridization with a PPA probe on control (**C**) and haloperidol-treated (**D**) rats, and with a D_2 receptor probe (**E**). From **C** to **D**, the increase in labeling is obvious after haloperidol treatment. With the D_2 probe (**E**), labeling is observed in two types of neurons: medium-sized neurons (*single arrow*) and large-sized neurons (*double arrow*). Some large-sized neurons are not labeled (*triple arrow*). Scale bars = 1 mm (**A,B**) and 10 μm (**C–E**). Note that with powerful image analysis systems, all these populations of neurons can be identified, and mRNA quantitation may be done separately in each subpopulation and in relation to their distribution within each quadrant.

receptor, as shown in **Figs. 3** and **4E**, two types of neurons containing the D_2 mRNA can be analyzed in the striatum: medium-sized neurons (i.e., enkephalin neurons) and large-sized neurons (i.e., acetylcholine neurons). In this case, quantitative *in situ* hybridization allows the demonstration that the increase in D_2 mRNA occurs in both these neurons, but with a different amplitude (+54% vs

+119%). This result shows that these two neuronal populations may respond differently to modifications of the dopamine environment. In addition, microscopic analysis allows the study of these populations inside the same structure in relation to their subregional localization, as illustrated in **Fig. 4B**. Similar cellular analysis of gene expression has been described, on the basis of another quantitative procedure, by Gerfen et al. *(5,6)*.

Acknowledgments

The author is especially indebted to V. Bernard, E. Normand, and B. Bloch, as well as to BIOCOM (Les Ulis, France) for their help in the development and the improvement of the quantitation procedure with the image analyzer system. The author also thanks M. Jaber, F. Georges and V. Fauchey for helpful discussion, and C. Vidauporte for photographic artwork.

References

1. Griffin, W. S. T. (1987) Methods for hybridization and quantification of mRNA in individual brain cells, in *In Situ Hybridization, Applications to Neurobiology* (Valentino, K. L., Eberwise, J. H., and Barchas, J. D. eds.), Oxford University Press, New York, pp. 97–110.
2. Gerfen, C. R. (1989) Quantification of *in situ* hybridization histochemistry for analysis of brain function. *Methods Neurosci.* **1**, 79–97.
3. Young, W. S. III (1990) *In situ* hybridization histochemistry, in *Handbook of Chemical Neuroanatomy. Analysis of Neuronal Microcircuits and Synaptic Interactions* (Björklund, A., Hökfelt, T., Wouterlood, F. G., and van den Pol, A. N., eds.), Elsevier, Amsterdam, pp. 481–512.
4. Salin, P., Mercugliano, M., and Chesselet, M. F. (1990) Differential effects of chronic treatment with haloperidol and clozapine on the level of preprosomatostatin mRNA in the striatum, nucleus accumbens, and frontal cortex of the rat. *Cell. Mol. Neurobiol.* **10**, 127–143.
5. Gerfen, C. R., McGinty, J. F., and Young, W. S. III (1991) Dopamine differentially regulates dynorphin, substance P, and enkephalin expression in striatal neurons: *in situ* hybridization histochemical analysis. *J. Neurosci.* **11**, 1016–1031.
6. Gerfen, C. R., Keefe, K. A., and Gauda, E. B. (1995) D1 and D2 dopamine receptor function in the striatum: coactivation of D1 and D2–dopamine receptors on separate populations of neurons results in potentiated immediate early gene response in D1-containing neurons. *J. Neurosci.* **15**, 8167–8176.
7. Baskin, D. G., Filuk, P. E., and Stahl, W. L. (1989) Standardization of tritium-sensitive film for quantitative autoradiography. *J. Histochem. Cytochem.* **37**, 1337–1344.
8. Miller, J. A. (1991) The calibration of ^{35}S or ^{32}P with ^{14}C-labeled brain paste or ^{14}C-plastic standards for quantitative autoradiography using LKB Ultrofilm or Amersham Hyperfilm. *Neurosci. Lett.* **121**, 211–214.
9. Smolen, A. J. and Beaston-Wimmer, P. (1990) Quantitative analysis of *in situ* hybridization using image analysis, in *In Situ Hybridization Histochemistry* (Chesselet, M. F., ed.), CRC Press, Boca Raton, FL, pp. 175–188.

10. Normand, E, Popovici, T., Fellmann, D., and Bloch, B. (1987) Anatomical study of enkephalin gene expression in the rat forebrain following haloperidol treatment. *Neurosci. Lett.* **83**, 232–236.

11. Le Moine, C., Normand, E., Guitteny, A. F., Fouqué, B., Teoule, R., and Bloch, B. (1990) Dopamine receptor gene expression by enkephalin neurons in rat forebrain. *Proc. Natl. Acad. Sci. USA* **87**, 230–234.

12. Bernard, V., Le Moine, C., and Bloch, B. (1991) Striatal neurons express increased level of dopamine D2 receptor mRNA in response to haloperidol treatment: a quantitative *in situ* hybridization study. *Neuroscience.* **45**, 117–126.

13. Le Moine, C., Svenningsson, P., Fredholm, B. B., and Bloch, B. (1997) Dopamine-adenosine interactions in the striatum and the globus pallidus: inhibition of striatopallidal neurons through either D2 or A2A receptors enhances D1 receptor-mediated effects on c-fos expression. *J. Neurosci.* **17**, 8038–8048.

14. Le Moine, C. and Bloch, B. (1995) D1 and D2 dopamine receptor gene expression in the rat striatum: sensitive cRNA probes demonstrate prominent segregation of D1 and D2 mRNAs in distinct neuronal populations of the dorsal and ventral striatum. *J. Comp. Neurol.,* **355**, 481–426

15. Bisconte, J. C., Fulcrand, J., and Marty, R. (1968) Analyse autoradiographique dans le système nerveux central par photométrie et cartographie combinées. *C. R. Scéanc. Soc. Biol.* **161**, 2178–2182.

11

In Situ End-Labeling of Fragmented DNA and the Localization of Apoptosis

Tim D. Hewitson, Teresa Bisucci, and Ian A. Darby

1. Introduction

Apoptosis is a morphologically distinct form of programmed cell death. It has a role in such processes as embryogenesis, immune regulation, and defense against viruses, and can also be induced by a variety of physical and chemical stimuli *(1)*. Importantly apoptosis leads to the safe removal of cells by phagocytosis, whereas in contrast, necrosis provokes tissue injury and inflammation.

Apoptotic cell death is associated with a number of biochemical and morphological changes including *de novo* gene expression, condensation of chromatin, and DNA degradation. Many attempts have been made to utilize these changes in the identification of apoptosis, with varying degrees of success.

Up-regulated expression of a number of gene products, including *clusterin*, *myc*, and *p53* is known to occur during programmed cell death *(2)*. However, although of biological interest, to date none of these has been shown to be a specific marker of apoptosis.

Apoptosis is also characterized by the condensation of nuclear chromatin and the formation of membrane-bound cell fragments termed apoptotic bodies. These are then engulfed by professional and recruited phagocytes. Although the morphological criteria of apoptosis are easily recognized in vivo by electron microscopy (**Fig. 1**), the sparsity of its occurrence makes its quantitation by this means problematic.

Activation of endonuclease activity during the process of cell death *(3)* causes DNA fragmentation in apoptotic cells. This produces a characteristic ladder of oligonucleosome-sized DNA fragments on agarose gel electrophoresis. Although applicable to pure cell populations, gel electrophoresis is difficult to apply to *in situ* studies with mixed cell populations.

From: *Methods in Molecular Biology*, Vol. 123: In Situ *Hybridization Protocols*
Edited by: I. A. Darby © Humana Press Inc., Totowa, NJ

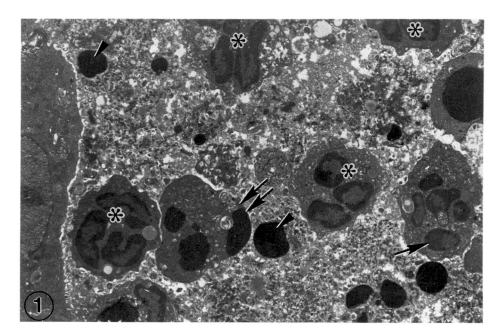

Fig. 1. Electron micrograph of polymorphonuclear granulocyte apoptosis in a rat model of experimental renal infection. Resolution of tubulointerstitial nephritis occurs through apoptosis of inflammatory cells and drainage of apoptotic cells through the tubular lumen. Various stages of polymorph apoptosis can be seen. Several normal polymorphs are visible with irregular cell outlines and granular appearance of nuclear chromatin (*asterisk*). Early stages of apoptosis are recognized by the margination of nuclear chromatin in a characteristic crescent shape (*arrow*). At later stages, the increased electron density of chromatin can be seen with the early stages of budding recognizable (*double arrow*). Finally, numerous apoptotic bodies (examples shown with *arrowheads*) are distributed throughout the lumen.

More recently, the presence of DNA fragmentation has been adapted to the in vivo identification of apoptosis. Apoptotic cells may be localized by the *in situ* labeling of this fragmented DNA using terminal transferase-mediated UTP nick end-labeling (TUNEL) *(4)*. In this technique, labeled dUTP is attached to the 3'-end of these breaks by terminal transferase (TdT) and is then detected using immunohistochemical techniques. TUNEL has been used successfully to study apoptosis in a diverse range of biological systems, including among others, developmental biology *(5)*, immunoselection *(6)*, wound healing *(7)*, and glomerulonephritis *(8)*.

It is however important to remember that DNA fragmentation is not confined to apoptosis exclusively. Fragmentation of DNA is also found in the late

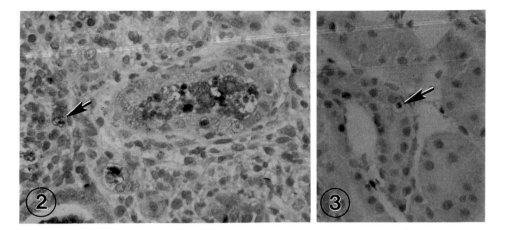

Fig. 2. (*See* Color Plate 2 following page 176.) Double labeling for monocytes/ macrophages and TUNEL in a rat model of experimental renal infection. Alkaline phosphatase immunohistochemical staining of monocytes/macrophages was performed with the clone ED-1. The alkaline phosphatase reaction product was detected with a fast red substrate. End-labeling of fragmented DNA was colocalized with DAB substrate (brown reaction product). Phagocytosis of an apoptotic body by a macrophage can be seen (*arrow*). (Reproduced from Hewitson et al. (1996) *Nephrology* **2**, 127– 132, with the permission of Blackwell Science.)

Fig. 3. (*See* Color Plate 2 following page 176.) End-labeling of fragmented DNA in a rat tubular epithelial cell. Condensation of nuclear chromatin (*arrow*) is easily seen. The rapid kinetics of apoptosis frequently result in the formation of "halos" around the apoptotic cells.

stages of necrosis, although the nuclear floculation and diffuse labeling pattern results in differences in histological appearance. Several in vitro studies have, however, confirmed that end-labeling of fragmented DNA correlates with the incidence of apoptosis, as measured by other parameters *(2)*. Furthermore, there are some data to suggest that single-strand DNA breaks, as occur in necrosis, are labeled less easily than the double-strand fragmentation in apoptosis *(9)*.

In summary, despite the limitations of the TUNEL method, the judicious use of this technique may aid considerably in the identification and quantitation of apoptosis *in situ.*

2. Materials

2.1. TUNEL on Paraffin Embedded Tissue

1. Paraffin-embedded sections of 10% neutral buffered formalin or 4% paraformaldehyde fixed tissue collected onto 3'-aminopropyltriethoxysilane (Sigma, St. Louis, MO, USA) coated slides (*see* **Notes 1** and **2**). A positive control tissue

section should be included to verify that the TUNEL reaction has worked (*see* **Note 3**).

2. Proteinase K (Sigma, St. Louis, MO, USA): stock solution of 20 µg/mL made up in distilled water and stored at –20°C.

3. Biotin-labeled dUTP (Boehringer-Mannheim, Mannheim, Germany).

4. Terminal deoxynucleotidyl transferase (TdT) (25 U/µL) (Boehringer-Mannheim, Mannheim, Germany).

5. Buffer 1: sodium chloride (30 mL of 5 M stock), sodium citrate (15 mL of 1 M stock), distilled water added to a total volume of 500 mL (final composition 0.3 M sodium chloride, 0.03 M sodium citrate).

6. Tris/EDTA (TE) buffer: Tris-HCl, pH 7.4 (0.5 mL of 2 M stock), EDTA, pH 8.0 (20 µL of 0.5 M stock), distilled water to 100 mL (final composition 0.01 M Tris, 0.0001 M EDTA).

7. TdT buffer: 15.0 g of sodium cacodylate, 0.119 g of cobalt chloride, 15 mL of 1 M stock Tris-HCl, distilled water added to a total volume of 500 mL (final composition: 140 mM cacodylate, 1 mM cobalt chloride, 30 mM Tris).

8. Phosphate-buffered saline (PBS), pH 7.2.

9. Vector ABC Elite staining kit (Vector laboratories, Burlingame, CA, USA).

10. Chromogen substrate: 4 mL of 3.3'-diaminobenzidine tetrahydrochloride (DAB) (DAKO, Glostrup, Denmark) mixed with 3 µL of 30% H_2O_2 (BDH, Poole, UK). An equivalent commercial preparation (e.g., metal-enhanced diaminobenzidine [DAB], Pierce, Rockford, IL, USA) may be substituted.

11. Humidified staining tray.

12. Microscope.

13. Staining racks and baths.

14. Wax pen (DAKO or equivalent).

15. Harris' hematoxylin.

16. Scott's tap water: 20 g of $MgSO_4 \cdot 7H_2O$, 3.5 g of $NaHCO_3$, made up to 1 L with distilled water (final concentrations: 0.081 M of $MgSO_4$, 0.041 M of $NaHCO_3$).

17. DNase I (Sigma, St. Louis, MO, USA).

18. Absolute ethanol (BDH, Poole, UK).

19. Xylene (BDH, Poole, UK).

20. 0.45 µM Syringe filter.

21. Depex (BDH, Poole, UK) or other nonaqueous mounting media.

2.2. Additional Materials for Combined Immunohistochemistry and TUNEL

1. Tris-buffered saline (TBS), pH 7.4 (25 mM Tris base, 0.9% NaCl, adjust pH to 7.4 with HCl).

2. Primary cell specific antisera (e.g., leukocyte marker).

3. Alkaline phosphatase anti-alkaline phosphatase complex (Silenus, Victoria, Australia or Dako, Alostrup, Denmark).

4. Alkaline phosphatase conjugated anti-IgG serum (species specific for primary antiserum) (Silenus, Victoria, Australia or Dako).

5. Fast red substrate (Sigma, St. Louis, MO, USA).
6. Aquamount (BDH, Poole, UK) or other aqueous mounting media.

3. Methods

3.1. TUNEL Protocol

1. Heat-paraffin-embedded tissue sections in a 60°C oven for 10 min.
2. Dewax tissue sections by transferring slides immediately to a xylene bath for 5 min.
3. Rehydrate tissue in graded (100%, 75%, 50%) alcohols and wash in distilled water for 5 min.
4. Digest sections by treatment with proteinase K (1 μL of 20 μg/mL stock in 1000 μL of TE buffer) for 15 min at room temperature (*see* **Note 4**).
5. Wash tissue sections in TdT buffer for 5 min.
6. Outline the section with a wax pen to create a hydrophobic barrier around the tissue section, thereby reducing the volume of reagents required.
7. Incubate sections at 37°C with 1 μL of biotin-labeled dUTP and 1 μL of TdT in 100 μL of TdT buffer for 30 min in a humid atmosphere.
8. Wash in buffer 1 to terminate the reaction.
9. Detect *in situ* incorporation of biotinylated dUTP by incubating sections with biotin horseradish peroxidase complex (ABC Elite kit) for 15 min or according to the manufacturer's instructions.
10. Wash in PBS 2×5 min.
11. Develop reaction product by staining with DAB/H_2O_2 substrate for 3–7 min at room temperature. Most DAB solutions made from DAB in powder or tablet form should be filtered through a 0.45 μm syringe filter prior to use. Monitor DAB/H_2O_2 on wet slides using a microscope with a low-power ($\times$10) objective.
12. Terminate the staining by rinsing in distilled water when the ratio of positive staining of apoptotic nuclei to background staining is maximal.
13. Wash in distilled water (5 min).
14. Counterstain sections with Harris' hematoxylin (1–2 min depending on strength of hematoxylin), rinse in tap water, dip briefly in Scott's tap water (3$\times$) until the sections are visibly "blued," and wash in tap water for 5 min.
15. Finally, dehydrate in graded (50%, 75%, 100%) alcohols, rinse in xylene and mount with Depex.

3.2. Combined Immunohistochemistry and TUNEL Protocol

The TUNEL technique can be combined with immunohistochemistry for cytoplasmic or surface markers to identify apoptotic cell phenotype *(6)* and/or phagocytosis of apoptotic bodies *(10)* (**Fig. 2**).

1. Follow the TUNEL protocol in **Subheading 3.1.** for **steps 1–11** inclusive.
2. Wash in TBS (*see* **Note 5**) for 3 min $\times$ 2 (*see* **Note 8**).
3. Incubate with primary antiserum for 1 h at room temperature in a humidified container. (Determine appropriate dilution empirically.)

4. Wash sections in TBS for 3 min × 2.
5. Incubate for 20 min with alkaline phosphatase conjugated anti-IgG serum (species specific for primary antiserum) (1:50 dilution in TBS).
6. Wash in TBS for 3 min × 2.
7. Incubate for 20 min with alkaline phosphatase anti-alkaline phosphatase complex (1:50 dilution in TBS).
8. Wash in TBS (3 min × 2).
9. Detect alkaline phosphatase reaction product by incubation with fast red substrate for 10–20 min at room temperature. Monitor substrate reaction microscopically, terminating the reaction by washing in distilled water for 5 min.
10. Counterstain sections as described previously.
11. Mount sections in Aquamount. (N.B.: dehydration will remove fast red reaction product.)

4. Notes

1. Fixatives: 4% paraformaldehyde or paraformaldehyde–lysine–periodate (PLP) may be substituted for neutral buffered formalin. Fixatives that denature DNA (e.g., mercuric formalin) may result in labeling of all nuclei and are therefore unsuitable for TUNEL.

   False-positive labeling is an artefact associated with any delay between retrieval of tissue and fixation.

2. 3'-Aminopropyltriethoxysilane (APES) coating of slides: Pretreatment of microscope slides with APES prevents sections falling off during protease digestion. Importantly it also avoids having to "bake" sections in a hot oven, as is often routinely used in histology laboratories to ensure adherence of the tissue.

   APES-coated slides can be prepared by sequentially washing microscope slides in (a) dilute laboratory detergent overnight, (b) running tap water for 3 h, (c) distilled water for 2 × 5 min, and (d) 95% alcohol for 2 × 5 min. Slides are then air dried before being dipped in a freshly prepared 2% solution of APES (Sigma, St. Louis, MO, USA) in 100% acetone (BDH, Poole, UK) for 10 s. Slides are then washed in (a) acetone for 2 × 5 min and (b) distilled water for 2 × 5 min and air dried at 40°C in a hot air oven for 12 h. Slides may be stored at room temperature in a dust-free container until use.

3. Controls: Positive controls consist of TUNEL labeling of rat ovary sections and of tissue sections pretreated with DNase I at a concentration of 1 μg/mL TBS for 5 min. TUNEL labeling of tissue sections in the absence of TdT is used as a negative control. DNase digestion results in positive labeling of all nuclei *(4)*, while the TUNEL reaction in the absence of TdT leaves all nuclei unlabeled.

   Sections of small intestine may be not be a suitable positive control. Despite Gavrieli et al. *(4)* describing labeling of the intestinal villus in their original description of the TUNEL method, doubts have been raised about the specificity of this reaction *(2)*.

4. Digestion: Optimal concentration and time of protease digestion may have to be established empirically for individual tissues. Microwave treatment has been sub-

stituted for protease digestion by some investigators *(11)*. Overdigestion using proteinase K may result in false-positive results; therefore digestion times need to be established for each tissue.

5. Tris-buffered saline is used in all washes in the double labeling procedure as phosphate-buffered saline is incompatible with subsequent alkaline phosphatase detection.

6. Interpretation: The definition of apoptosis is still essentially based on morphological criteria. The presence of apoptosis may therefore need to be substantiated by electron microscopy (**Fig. 1**) or the identification of morphological features in TUNEL-positive cells. Apoptotic cells are often recognizable on light microscopy as condensed TUNEL positive chromatin lying within a "halo" of what was the cell cytoplasm (**Fig. 3**). Such a phenomenon is presumably due to the contraction of cell cytoplasm in combination with the rapid kinetics of apoptosis.

7. Quantitation: Apoptotic cells are usually expressed as a proportion (%) of total cell number or as the number of apoptotic cells per unit area (e.g., 10×0.25 μm fields).

   Kinetic studies have suggested that apoptotic cells are histologically recognizable for 0.5–2.0 h before clearance *(8)*. Identification of a low rate of apoptosis therefore still indicates the removal of a significant proportion of the cell population. For example, if apoptotic bodies are visible for 2 h, a 1% incidence of visible apoptosis may represent clearance of up to 12% of cells within 24 h. This study highlights that for statistical validity it is necessary to examine sufficient nuclei (at least 1000) when enumerating the incidence of apoptosis.

8. *In situ* end-labeling (ISEL) as described by Wijsman et al. *(12)* is a variant of the above technique where the enzyme Klenow DNA polymerase I is substituted for terminal deoxynucleotidyl transferase (TdT). There is some evidence that TdT is preferable to DNA polymerase because more favorable kinetics are reflected in shorter incubation times *(13)*.

References

1. Ueda, N. and Shah, N. (1994) Apoptosis. *J. Lab. Clin. Med.* **124,**169–177.
2. Ansari, B., Coates, P. J., Greenstein, B. D., and Hall, P. A. (1993) *In situ* end labeling detects DNA strand breaks in apoptosis and other physiological and pathological states. *J. Pathol.* **170,** 1–8.
3. Arends, M. J., Morris, R. G., and Wyllie, A. H. (1990) Apoptosis; the role of endonuclease. *Am. J. Pathol.* **136,** 593–608.
4. Gavrieli, Y., Sherman, Y., and Ben-Sasson, S. (1992) Identification of programmed cell death *in situ* via specific labeling of nuclear DNA fragmentation. *J. Cell Biol.* **119,** 493–501.
5. Fekete, D. M., Homburger, S. A., Waring, M. T., Riedl, A. E., and Garcia, L. F. (1997) Involvement of programmed cell death in morphogenesis of the vertebrate inner ear. *Development* **124,** 2451–2461.
6. Smith, K. G. C., Hewitson, T. D., Nossal, G. J. V., and Tarlinton, D. M. (1996) The phenotype and fate of the antibody-forming cells of the splenic foci. *Eur. J. Immunol.* **26,** 444–448.

7. Darby, I. A., Bisucci, T., Hewitson, T. D., and MacLellan, D. G. (1997) Apoptosis is increased in a model of diabetes-impaired wound healing in genetically diabetic mice. *Int. J. Biochem. Cell Biol.* **29,** 191–200.
8. Baker, A. J., Mooney, A., Hughes, J., Lombardi, D., Johnson, R. J., and Savill, J. (1994) Mesangial cell apoptosis: the major mechanism for resolution of glomerular hypercellularity in experimental mesangial proliferative nephritis. *J. Clin. Invest.* **94,** 2105–2116.
9. Hockenbery, D. (1995) Defining apoptosis. *Am. J. Pathol.* **146,** 16–19.
10. Hewitson, T. D., Smith, K. G. C., and Becker, G. J. (1996) Apoptosis and resolution of experimental renal infective tubulointerstitial nephritis. *Nephrology* **2,** 127–132.
11. Strater, J., Gunthert, A. R., Bruderlein, S., and Moller, P. (1995) Microwave irradiation of paraffin-embedded tissue sensitizes the TUNEL method for *in situ* detection of apoptotic cells. *Histochemistry* **103,** 157–160.
12. Wijsman, J. H., Jonker, R. R., Keijzer, R., Van de Velde, C. J. H., Cornelisse, C. J., and Van Dierendonck, J. H. (1993) A new method to detect apoptosis in paraffin sections: *in situ* end-labeling of fragmented DNA. *J. Histochem. Cytochem.* **41,** 7–12.
13. Gorczyca, W., Gong, J., and Darzynkiewicz, Z. (1993) Detection of DNA strand breaks in individual apoptotic cells by the *in situ* terminal deoxynucleotidyl transferase and nick translation assays. *Cancer Res.* **53,** 1945–1951.

12

Combined Immunohistochemical Labeling and *In Situ* Hybridization to Colocalize mRNA and Protein in Tissue Sections

Malcolm D. Smith, Angela Parker, Riyani Wikaningrum, and Mark Coleman

1. Introduction

Sensitive techniques developed to detect biologically relevant proteins at the mRNA and protein levels have been major research tools in basic and applied biomedical research *(1–6)*. The combination of these two techniques has been particularly valuable when the biological protein being studied is a secreted cell product that has the ability to bind to components of the extracellular matrix or to receptors on target cells within the same tissue *(4,5,7–9)*. The best example of this is the study of cytokine production in human tissue sections, where standard immunohistochemical labeling techniques will detect cytokine production, secreted cytokine, and cytokine bound to receptors on target cells *(4,5,9)*. If it is important to determine which of the cellular components of the tissue under study is producing the biological protein, the combination of *in situ* hybridization to detect mRNA and immunohistochemical labeling to detect the protein offers the ability to distinguish between the cells responsible for production of the protein and its target cells.

We describe here a method that combines *in situ* hybridization and immunohistochemical labeling of tissue sections *(8)*. The technique utilizes nonradioactive methods, which avoid the use of radioactivity, allow a permanent recording of mRNA detection with colorimetric techniques, more precise localization of mRNA within cells than radioactive methods, and a greater speed and efficiency compared with radioactive methods *(10,11)*. The use of riboprobes with the higher affinity and thermal stability of RNA–RNA hybrids compared with RNA–DNA hybrids avoids the problems of competitive hy-

From: *Methods in Molecular Biology*, Vol. 123: In Situ *Hybridization Protocols*
Edited by: I. A. Darby © Humana Press Inc., Totowa, NJ

bridization to the complementary strand of nick-translated cDNA probes and permits higher stringency washing conditions, resulting in reduced nonspecific background hybridization signals *(8)*. This technique therefore offers several advantages over previous techniques including the ability to use archival, paraffin-embedded sections, the use of a nonradioactive label for *in situ* hybridization, the inclusion of an amplification step to enhance the sensitivity of *in situ* hybridization, and the ability to vary the color end products in the immunohistochemical detection steps as determined by the particular requirements of the tissue sections and the research plan.

2. Materials

2.1. Solutions for Immunoperoxidase Technique

1. 1 M Tris-HCl buffer (BDH Chemicals, Poole, UK), pH to 8.0 and sterilize before use.
2. 5× Phosphate-buffered saline (PBS) buffer: 45 g of NaCl, 15.2 g of Na_2HPO_4 (BDH Chemicals, Poole, UK), 3.93 g of NaH_2PO_4 (BDH). Dissolve in 1 L of distilled water.
3. 0.05 M Tris-1× PBS buffer, pH 7.5: 50 mL of 1 M Tris-HCl buffer, pH 8.0; 200 mL of 5× PBS buffer. Make up to 800 mL with distilled water and adjust pH to 7.5 using HCl. Add distilled water to 1 L. Autoclave before use.
4. Endogenous peroxidase block: 3% (v/v) hydrogen peroxide (Univar, Auburn, Sydney, NSW, Australia) in methanol.
5. Proteinase K solution: dilute proteinase K (Boehringer Mannheim) stock (5 mg/mL) to the desired concentration using proteinase K buffer (*see* **Note 1**).
6. Proteinase K buffer: dissolve 0.185 g of EDTA (BDH chemical, Poole, UK) in 100 mL of 1× PBS. Autoclave before use.
7. Normal serum: 20% normal donkey serum (Jackson Immuno Research, Avondale, PA, USA) in 0.05 M Tris-1× PBS buffer, pH 7.5.
8. Primary antibody: dilute primary antibody in 0.05 M Tris-1× PBS buffer, pH 7.5. The dilution must include 10% normal serum and 1:200 of 1× PBS–0.1% Thimerosal (Sigma).
9. Secondary antibody (linking antibody) — biotinylated donkey anti-mouse antibody (Jackson Immuno Research, Avondale, PA, USA): for use, a 1:200 dilution in 0.05 M Tris-1× PBS buffer, pH 7.5, plus 5% normal serum in solution, is required.
10. ABC Complex (Standard Elite ABC Kit — Vector Laboratories, Burlingame, CA, USA): this must be made 30 min prior to use. 10 μL of Bottle A (Avidin); 10 μL of Bottle B (biotinylated horseradish peroxidase); 1000 μL of 0.05 M Tris-1× PBS buffer, pH 7.5.
11. DAB (3, 3'-diaminobenzidine tetrachloride, Sigma) solution: Add 500 μL of DAB (10 mg/mL) to 4.5 mL of 0.05 M Tris-1× PBS buffer, pH 7.5, and add 2 μL of hydrogen peroxide just before use (*see* **Note 2**).

2.2. Solutions for Alkaline Phosphatase Anti-Alkaline Phosphatase (APAAP) Technique

1. Tris-buffered saline (TBS): 0.02 M Tris-HCl and 0.15 M NaCl, pH 7.6.
2. 20× TBS: 90 g of NaCl; 200 mL of 1 M Tris-HCl, pH 8.0; add distilled water to 500 mL and adjust pH to 7.6. Autoclave before use.
3. 1 M Tris-HCl buffer, pH 8.0: 60.57 g of Tris in 500 mL of distilled water. Adjust pH to 8.0. Autoclave before use.
4. Normal serum: 10% (v/v) normal sheep serum (Jackson Immuno Research) in TBS.
5. Linking antibody: 1:200 dilution sheep anti-mouse in TBS plus 10% normal sheep serum. The linking antibody can be stored in 4°C in 1:10 dilution with 1× PBS + 0.05% Thimerosal.
6. APAAP complex (Boehringer Mannheim): use 1:50 dilution in TBS.
7. Fast red substrate: dissolve 5 mg of naphthol AS–MX phosphate (Sigma) in 200 µL of dimethylformamide (BDH). Add 10 mL of 0.2 M Tris-HCl buffer, pH 8.0, and 2 mg of levamisole (Sigma), and finally add 10 mg of fast red TR salt (Sigma). Filter substrate for immunohistochemistry (*see* **Note 3**).

2.3. Materials for In Situ Hybridization

1. 0.1 M Tris-HCl – 0.2 M glycine solution, pH 7.2: 1.21 g of Tris, 1.50 g of glycine. Dissolve in 100 mL of distilled water, adjust the pH to 7.2 with concentrated HCl. Autoclave and store at 4°C.
2. 20× Standard saline citrate (SSC) solution: 175 g of NaCl, 88 g of Na citrate dihydrate. Make up to 1000 mL in distilled water and adjust pH to 7.0.
3. Prehybridization solution: (a) 10% dextran sulphate (500,000), (b) 0.05% polyvinylpyrrolidone (PVP), (c) 20× SSC, (d) 50% deionized formamide, (e) 0.05% Triton X-100, (f) 500 µg/mL of herring sperm DNA (Boehringer Mannheim). Dissolve the dry ingredients in 20× SSC, with manual stirring, then add formamide and Triton X-100. Boil the herring sperm DNA for 10 min and chill on ice before adding to the solution. Store the solution at –20°C
4. Nail polish with enough acetone added to allow free flow from a Pasteur pipet.
5. Acid-cleaned slides and coverslips: Place slides or coverslips in Petri dish with 2.0 N HCl for 20–30 min. Pick up slides or coverslips with clean forceps and place in 70% alcohol for a few minutes before drying on paper towel.

2.4. Materials for Digoxigenin Detection

1. Hybridization buffer I solution: 0.1 M Tris-HCl, 0.15 M NaCl, pH 7.5.
2. Protein block: 10% normal sheep serum in hybridization buffer I.
3. Anti-DIG-alkaline phosphatase (anti-DIG-AP, Boehringer Mannheim); sheep antibody, used in 1:500 dilution with 10% normal sheep serum and hybridization buffer I as diluent.
4. Hybridization buffer III: 0.1 M Tris-HCl, 0.1 M NaCl, 0.05 M MgCl$_2$, pH 9.5.
5. Nitroblue tetrazolium (NBT, Boehringer Mannheim) substrate: 80 µL of dimethylformamide, 135 µL of NBT solution (7.5 mg of NBT in 300 µL of dis-

tilled water), 105 µL of X-P (5-bromo-4-chloro-3-indolylphosphate, toluidine salt, Boehringer Mannheim) solution (5.0 mg of X-P in 300 µL of distilled water), 2.4 mg of levamisole in 10 mL of hybridization buffer III (*see* **Note 4**).

3. Methods

3.1. Preparation of Tissue Sections

3.1.1. Frozen Sections

1. Cut 4–6 µm thick sections on a cryostat.
2. Place sections on acid-cleaned slides coated with 2% 3-aminopropyltriethoxysilane (APES, Sigma–Aldrich, Castle Hill, New South Wales, Australia) (*see* **Note 5**).
3. Fix in acetone for 10 min at 4°C, air dry, and use immediately or store at –20°C, individually wrapped in aluminum foil.
4. Wash sections in Tris-PBS before use (*see* **Note 6**).

3.1.2. Formalin Fixed, Paraffin-Embedded Sections

1. Cut 4 µm thick sections on a microtome and place onto acid-cleaned APES-coated slides.
2. Dewax sections in xylene and rehydrate in 100%, 90%, 80%, and 70% alcohol.
3. Place sections in Tris-PBS before use (*see* **Note 6**).

3.2. Generation of Riboprobe

1. Restrict cDNAs to a size of equal to or less than 2000 bp within the coding sequence, using appropriate restriction endonucleases (*see* **Note 7**).
2. Insert cDNA into an appropriate vector, e.g., SP72, pGEM 3, pGEM 7z- (Promega, Madison, WI, USA).
3. Transform an appropriate strain of *E. coli* using the calcium chloride procedure.
4. Select colonies, grow bacteria, and make plasmid DNA mini-preparations.
5. Check the suitability of selected clones by restriction endonuclease digests of plasmid DNA to confirm the presence of the correct sized insert.
6. Make large scale preparation of plasmid DNA using a Maxiprep kit (Qiagen, Chatsworth, CA, USA)
7. Recheck cDNA by restriction endonuclease digests to confirm the presence of the correct insert.
8. cDNA fragments cloned into the plasmid vector were linearized using appropriate restriction endonucleases in such a manner that the cDNA insert could be transcribed into an RNA copy from either the SP6 or T7 RNA polymerase promoters without including any vector sequence.
9. Generate riboprobes using a commercial riboprobe generation kit containing either biotin-labeled UTP (Gibco-BRL) or digoxigenin-labeled UTP (Boehringer Mannheim) as part of the nucleotide mix. This results in biotin- or digoxigenin-labeled riboprobes that are a copy of either the sense or antisense strand of the cDNA (*see* **Note 8**).

10. Remove template DNA with an RNase-free DNase (Boehringer-Mannheim) and unincorporated nucleotides by passage through affinity columns (Qiagen, Chatsworth, CA, USA).
11. Precipitate riboprobes in 2 M LiCl$_2$ in 100% ethanol and resuspend in diethylpyrocarbonate (DEPC)-treated water.
12. Confirm labeling with biotin or digoxigenin by dot-blot hybridization and appropriate size by Northern blot analysis (*see* **Note 9**).

3.3. In Situ *Hybridization*

1. Place slides in 0.2 N HCl for 20 min at room temperature. Rinse sections in DEPC-treated water for 5 min in two changes (*see* **Note 6**).
2. Digest slides with 50 µg/mL of proteinase K (Boehringer Mannheim) for 30 min at 37°C. Rinse sections in DEPC-treated water for 5 min in two changes (*see* **Note 1**).
3. Incubate sections with 0.1 M Tris-HCl, 0.2 M glycine solution, pH 7.2, for 10 min at room temperature followed by rinsing in DEPC-treated water.
4. Rinse sections in 2× SSC solution for 5 min (two changes).
5. Cover sections in prehybridization buffer and incubate for 90–120 min at 50°C.
6. Remove the prehybridization solution and replace with the desired concentration of riboprobe in prehybridization buffer.
7. Cover each section with an acid-cleaned coverslip, seal with nail polish, and incubate overnight at 50°C (*see* **Note 10**).
8. Remove coverslips with a scalpel blade. Wash slides twice in 2× SSC solution for 10 min, 1× SSC solution for 10 min, 0.1× SSC solution for 15 min at 45°C, in 0.1× SSC for 5 min at room temperature, and finally in water for 5 min at room temperature. The sections are then ready for biotin or digoxigenin detection.

3.4. *Detection of* In Situ *Hybridization*

3.4.1. Biotin Detection System

1. Place sections in endogenous peroxidase blocking solution for 10 min at room temperature, wash in water followed by Tris-PBS buffer wash for 5 min.
2. Remove sections from Tris-PBS, drain, and incubate with 20% normal serum (30 min, room temperature) to block nonspecific protein binding.
3. Incubate sections with anti-biotin antibody (Boehringer Mannheim) at a 1:100 dilution containing 10% normal donkey serum (Jackson Immuno Research) for 45 min at room temperature (*see* **Note 11**).
4. Wash sections in Tris-PBS, then incubate with a biotin-labeled antimouse immunoglobulin (Boehringer Mannheim) 1:200 dilution containing 10% normal donkey serum for 30 min.
5. Wash sections in Tris-PBS, then incubate with ABC complex for 30–60 min.
6. Wash sections in Tris-PBS, then incubate with a DAB solution for 5 min.
7. Wash sections in Tris-PBS (*see* **Note 12**).

3.4.2. Digoxigenin Detection System

1. Wash slides in hybridization buffer I solution for 5 min.
2. Block nonspecific binding by incubating the slides in 10% normal sheep serum (Jackson Immuno Research) diluted in hybridization buffer I for 45 min.
3. Incubate sections with anti-digoxigenin alkaline phosphatase-labeled antibody (anti-DIG AP, Boehringer Mannheim) in 1:500 dilution in hybridization buffer containing 10% normal sheep serum.
4. Wash sections with hybridization buffer I and equilibrate in hybridization buffer solution III for 2 min.
5. Incubate sections with the color substrate, NBT (Boehringer Mannheim) in the dark for 1–4 h. Stop color development by washing slides in water (*see* **Note 13**).

3.5. Immunohistochemical Labeling for Protein

1. Cover with normal sheep serum (Jackson Immuno Research) for 2 h to block nonspecific binding, then wash in TBS buffer.
2. Incubate sections overnight at room temperature with the primary antibody (mouse IgG).
3. Wash sections in TBS buffer the incubate with secondary sheep anti-mouse antibody (Boehringer Mannheim, Castle Hill, NSW, Australia) for 60 min.
4. Wash sections in TBS buffer, then incubate with alkaline phosphatase anti-alkaline phosphatase complex (APAAP-complex) (Boehringer Mannheim, Castle Hill, NSW, Australia) for 60 min at room temperature.
5. Wash sections in TBS buffer, then apply a chromogenic substrate, e.g., fast red (Sigma) to the sections for 15 min.
6. Stop the color development by washing slides in water.
7. Counterstain the sections with dilute hematoxylin and mount in aqeous mountant (e.g., Aquamount, BDH, Poole, UK) (*see* **Note 14**).
8. Quantitation can be performed on hybridized sections (*see* **Note 15**).

4. Notes

1. Proteinase K is necessary to expose epitopes in formalin-fixed tissues and to permeabilize tissue to allow access to the riboprobes with the *in situ* hybridization step. Alternatives include pepsin digestion and microwave retrieval steps which, in our hands, are less reliable than the use of proteinase K digestion.
2. Although silver enhancement of the DAB reagent can be used to produce a darker stain, it adds little to the procedure.
3. Other color substrates can be used to produce a blue, green, or yellow color product but the combination of a brown *in situ* hybridization product with a red immunohistochemical reaction product gave the best color combination when both procedures were combined on the same tissue section.
4. It is possible to vary the color end product of the digoxigenin labeling system including using an anti-digoxigenin antibody conjugated to horseradish peroxidase to convert to a DAB end product.

5. Coating of slides with APES *(12)*, rather than poly-L-lysine, is necessary to avoid loss of part or all of the tissue section, especially when a combination of *in situ* hybridization and an immunohistochemical reaction is performed on the same tissue section.

6. Although DEPC treatment is not necessary in most steps, owing to the inactivation of RNase by the fixation step as well as other steps in the procedure, it may give added security when first establishing this procedure. Autoclaving all solutions is a necessary procedure to prevent bacterial contamination of solutions and the use of DEPC-treated solutions and the wearing of gloves is essential for the *in situ* hybridization steps in **Subheading 3.3**.

7. Where possible, restriction endonucleases were chosen so that the cDNA fragments generated had different restriction enzyme sites at the 5'- and 3'-ends of the cDNA, allowing insertion in a known orientation within the polycloning site of an appropriate plasmid (SP72 , pGEM 3, pGEM 7z-, Promega, Madison, WI, USA) in such a way that the cDNA fragment was flanked by the SP6 and T7 RNA polymerase promoters.

8. The commercial riboprobe generation kits will generate a labeled copy of the antisense (positive probe) and the sense strand (negative probe), which gives an inbuilt control for each probe. The use of riboprobes also increases the strength of hybridization of an RNA probe to a mRNA, allowing greater stringency of washing conditions, which will reduce nonspecific hybridization.

9. Aliquots were stored at –20°C for up to 12 mo. Aliquots of riboprobes were thawed as needed and diluted 100- to 500-fold in hybridization buffer, in which they were stable at 4°C for up to 1 month. Probe concentrations were measured using QuantaGene DNA/RNA Calculator (The Australian Chromatography Company, Pharmacia, Cambridge, UK) and the same concentration of sense and antisense probes were used for *in situ* hybridization.

10. Ensure that there are no air bubbles under the coverslip; otherwise the labeled riboprobe will not be in contact with all parts of the tissue section. Alternatives such as parafilm on top of the sections are rarely adequate to prevent drying of the sections in the overnight hybridization step.

11. This amplification step with anti-biotin antibody greatly enhances the sensitivity of the technique.

12. We have attempted several combinations of *in situ* hybridization with immunohistochemistry including performing either the *in situ* hybridization or immunohistochemical labeling step first and either developing the color reaction immediately or at the end of the combined procedure. Best results were achieved by performing the *in situ* hybridization step first including developing the color reaction, followed by the immunohistochemical reaction *(8)*. **Figures 1** and **2** demonstrate examples of combining *in situ* hybridization for cytokine mRNA with immunohistochemical detection of either cell lineage markers (**Fig. 1**) or cytokine (**Fig. 2**).

13. If excess background is a problem with a biotin-based system because of endogenous biotin in the tissue sections that is not adequately blocked, a digoxigenin

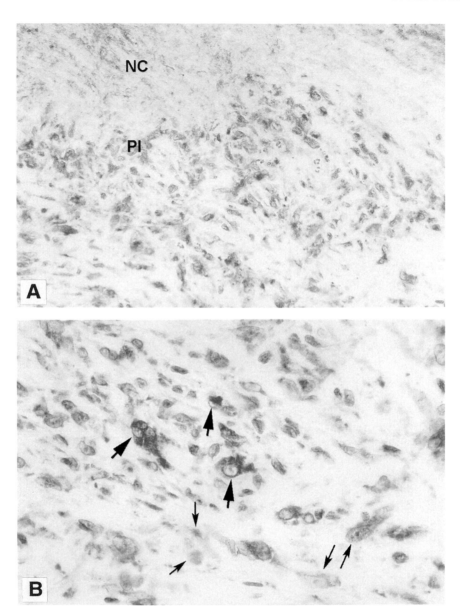

Fig. 1. (*See* Color Plate 3 following p. 176.) Low- (**A**, ×66) and high-power (**B**, ×132) views of combined *in situ* hybridization for IL-1β mRNA (brown) and macrophage lineage marker CD68 (red) in a tissue section from a rheumatoid nodule. *Large arrows* show macrophages containing IL-1β mRNA (red and brown) while small arrows show fibroblasts containing IL-1β mRNA (brown only). NC, necrotic center; Pl, palisade layer; St, stromal region.

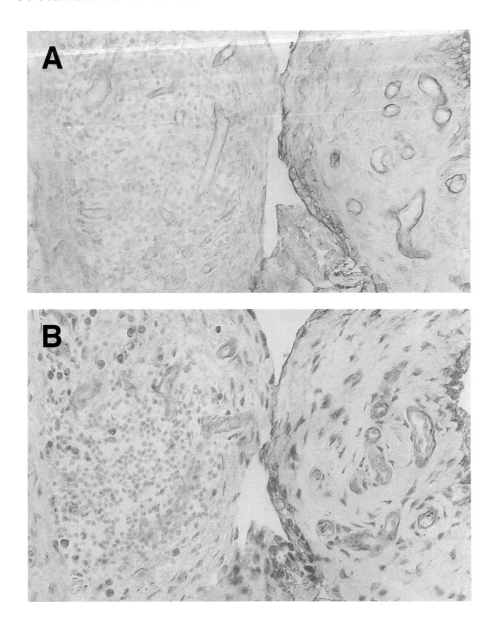

Fig. 2. (*See* Color Plate 4 following p. 176.) Low-power (×66) view of combined *in situ* hybridization for tumor necrosis factor-α (TNFα) mRNA (brown) and immunohistochemistry for TNFα (red) on a synovial membrane biopsy from a patient with rheumatoid arthritis. (**A**) Sense riboprobe and anti-TNFα antibody. (**B**) Antisense riboprobe and anti-TNFα antibody.

system will overcome this and can be converted to a DAB color end product, using an anti-digoxigenin antibody linked to horseradish peroxidase.

14. The color end product from an immuno-alkaline phosphatase reaction is not stable in xylene or alcohol, so an aqueous mounting medium must be used.

15. Quantitation of *in situ* hybridization and immunohistochemical labeling can be performed using computer-assisted image analysis techniques, as previously described *(13–16)*. The reproducibility of this method of quantitation is excellent, with an intra- and inter-observer variability of <10%, mainly owing to operator variability in field selection *(16)*.

References

1. Firestein, G. S., Berger, A. E., Tracey, D. E., Chosay, J. G., Chapman, D. L., Paine, M. M., Yu, C., and Zvaiffler, N. J. (1992) IL-1 receptor antagonist protein production and gene expression in rheumatoid arthritis and osteoarthritis synovium. *J. Immunol.* **149**, 1054–1062.

2. Hoeffler, H., Childers, H., Montminy, M. R., Lechan, R. M., Goodman, R. H., and Wolfe, H. J. (1986) *In situ* hybridization methods for detection of somatostatin mRNA in tissue sections using antisense RNA probes. *Histochem. J.* **18**, 597–604.

3. Denijn, M., de Weger, R. A., Berends, M. J. H., Campier-Spies, P. L., Jansz, H., Van Unnik, J. A., and Lips, C. J. (1990) Detection of calcitonin-encoding mRNA by radioactive and non-radioactive *in situ* hybridization: improved colorimetric detection and cellular localization of mRNA in thyroid sections. *J. Histochem. Cytochem.* **38**, 351–358.

4. Matsuki, Y., Yamamato, T., and Hara, K. (1992) Detection of inflammatory cytokine messenger RNA (mRNA)-expressing cells in human inflamed gingiva by combined *in situ* hybridization and immunohistochemistry. *Immunology* **76**, 42–47.

5. Piguet, P. F., Ribaux, C., Karpuz, V., Grau, G. E., and Kapanci, Y. (1993) Expression and localization of tumor necrosis factor-α and its mRNA in idiopathic pulmonary fibrosis. *Am. J. Pathol.* **143**, 651–655.

6. Smith, M. D., Triantafillou, S., Parker, A., Youssef, P. P., and Coleman, M. (1997) Synovial membrane inflammation and cytokine production in patients with early osteoarthritis. *J. Rheumatol.* **24**, 365–371.

7. Kriegsmann, J., Keyszer, G., Geiler, T., Gay, R. E., and Gay, S. (1994) A new double labeling technique for combined *in situ* hybridization and immunohistochemistry. *Lab. Invest.* **71**, 911–917.

8. Smith, M. D., Triantafillou, S., Parker, A., and Coleman, M. (1997) A nonradioactive method of *in situ* hybridization which utilises riboprobes and paraffin-embedded tissue. *Diagnostic Mol. Pathol.* **6**, 34–41.

9. Wikaningrum, R., Highton, J., Parker, A., Coleman, M., Hessian, P. A., Roberts-Thomson, P. J., Ahern, M. J., and Smith, M. D. (1998) Pathogenic mechanisms in the rheumatoid nodule: comparison of pro-inflammatory cytokine production and cell adhesion molecule expression in rheumatoid nodules and synovial membranes from the same patient. *Arthritis Rheum.* **41**, 1783–1797.

10. Bloch, B. (1993) Biotinylated probes for *in situ* hybridization histochemistry: use for mRNA detection. *J. Histochem. Cytochem.* **41**, 1751–1754.

11. Fleming, K. A., Evans, M., Ryley, C., Franklin, D., Lovell-Badge, H., and Morey, A. L. (1992) Optimization of non-isotopic *in situ* hybridization on formalin-fixed, paraffin-embedded material using digoxigenin-labelled probes, and transgenic tissues. *J. Pathol.* **167**, 9–17.

12. Burns, J., Graham, A. K., Frank, C., Fleming, K. A., Evans, M. F., and McGee, J. O. (1987) Detection of low copy human papilloma virus DNA and mRNA in routine paraffin sections of cervix by non-isoptopic *in situ* hybridization. *J. Clin. Pathol.* **40**, 858–864.

13. Youssef, P. P., Triantafillou, S., Parker, A., Gamble, J., Haynes, D., Roberts-Thomson, P. J., Ahern, M. J., and Smith, M. D. (1997) Effects of pulse methylprednisolone on inflammatory mediators in peripheral blood, synovial fluid and the synovial membrane in rheumatoid arthritis. *Arthritis Rheum.* **40**, 1400–1408.

14. Jarvis, L. R. (1992) The microcomputer and image analysis in diagnostic pathology. *Microsc. Res. Technique* **21**, 292–299.

15. Skinner, J. M., Zhao, Y., Coventry, B., and Bradley, J. (1993) Videoimage analysis in pathology. *Dis. Markers* **11**, 53–70.

16. Youssef, P. P., Triantafillou, S., Parker, A., Coleman, M., Roberts-Thomson, P. J., Ahern, M. J., and Smith, M. D. (1997) Variability in cytokine and cell adhesion molecule staining in arthroscopic synovial biopsies: quantification using colour video image analysis. *J. Rheumatol.* **24**, 2291–2298.

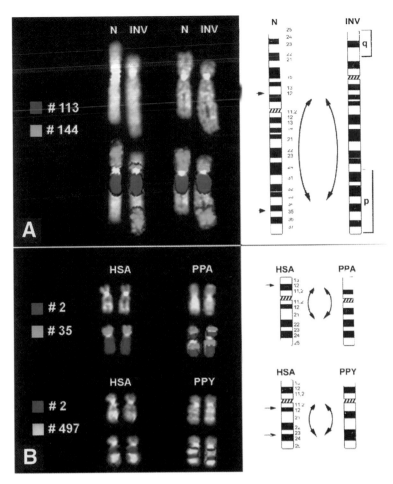

Plate 1, Fig. 4. (*See* full caption and discussion on p. 10 in Chapter 1).

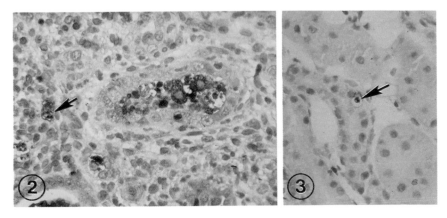

Plate 2, Figs. 2 and 3. (*See* full caption and discussion on p. 159 in Chapter 11).

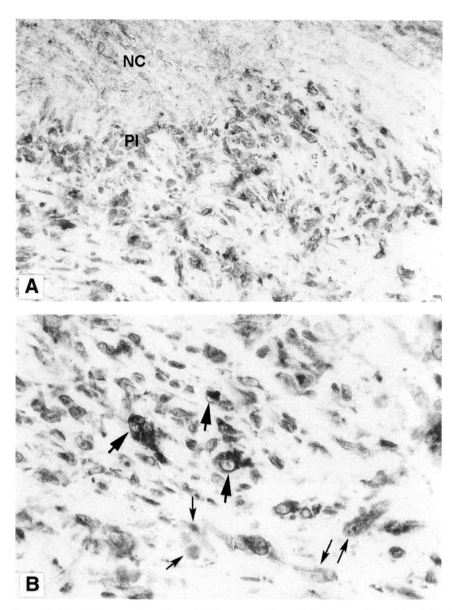

Plate 3, Fig. 1. Low- (**A**, ×66) and high-power (**B**, ×132) views of combined *in situ* hydridization for IL-1β mRNA (brown) and macrophage lineage marker CD68 (red) in a tissue section from a rheumatoid nodule. *Large arrows* show macrophages containing IL-1β mRNA (red and brown) while small arrows show fibroblasts containing IL-1β mRNA (brown only). NC, necrotic center; Pl, palisade layer; St, stromal region.

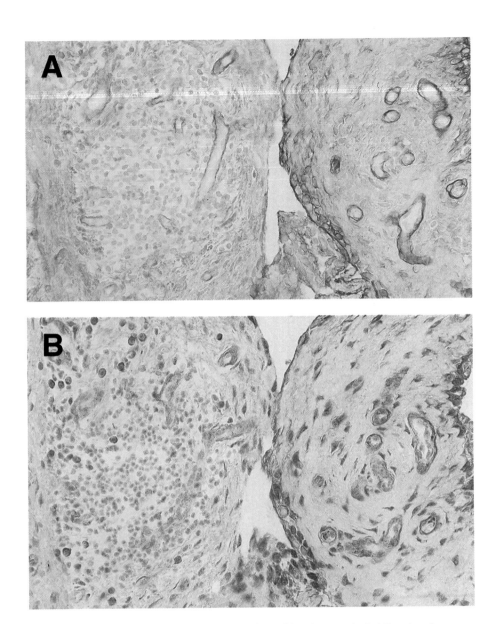

Plate 4, Fig. 2. Low-power (×66) view of combined *in situ* hybridization for tumor necrosis factor-α (TNFα) mRNA (brown) and immunohistochemistry for TNFα (red) on a synovial membrane biopsy from a patient with rheumatoid arthritis. **(A)** Sense riboprobe and anti-TNFα antibody. **(B)** Antisense riboprobe and anti-TNFα antibody.

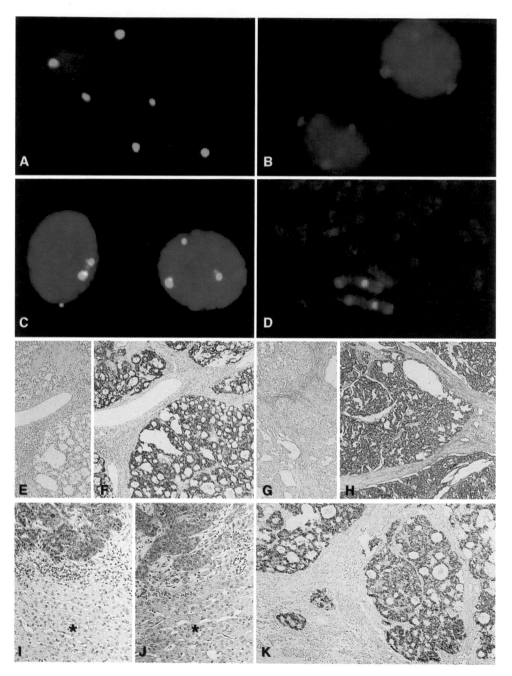

Plate 5, Fig. 3. (*See* full caption on pg. 200 and discussion in Chapter 14).

13

In Situ Hybridization Protocols for Detection of Viral DNA Using Radioactive and Nonradioactive DNA Probes

Allison R. Jilbert

1. Introduction

In situ hybridization can provide accurate intracellular localization of specific viral nucleic acids in infected tissues and cells. The technique, although conceptually simple, is affected by many variables including: stability and accessibility of target sequences; methods of tissue and cell fixation; prehybridization and hybridization conditions; choice of indicator molecules and detection system; and size, specific activity, and nature of the probe used. Each new tissue and target sequence may have different optimal conditions for each of these variables and should be carefully balanced to yield the maximum amount of information. The technique has been used extensively in studies of viral infection of cells. Particular advantages include: (1) high sensitivity when examining tissues where a small percentage of cells contain a high copy of target molecules; (2) ability to correlate target sequences with cell type and distribution, histological features, and intracellular localization (nuclear vs cytoplasmic).

The chapter contains general principles for *in situ* hybridization (with selected references to key articles) and detailed methods for detection of human, duck, and woodchuck hepatitis B virus DNA using radioactive and nonradioactive probes. Relevant publications include those that demonstrate use of tritium-labeled probes *(1–3,30)* iodine-labeled probes *(4–7,30)*, and digoxigenin-labeled probes *(8,9)*.

The notes section contains information for planning in situ hybridization experiments and analysis of the resultant signals. Topics covered include 1) comparison of the advantages and disadvantages of radioactively vs nonradioactively labeled probes;

From: *Methods in Molecular Biology*, Vol. 123: In Situ *Hybridization Protocols*
Edited by: I. A. Darby © Humana Press Inc., Totowa, NJ

2) choice and method of preparation of tissues and cells; 3) control tissues and cells; 4) prehybridization treatment of tissues and cells; 5) calculation of melting temperatures (T_m) and optimal hybridization and washing conditions; 6) examples of hybridization and washing conditions; 7) analysis of *in situ* hybridization signals; and 8) localization of viral DNA and antigen.

2. Materials

2.1. Tissue Preparation

1. Tissue embedding facility.
2. Wax block sectioning microtome.
3. Cryostat for frozen sectioning.
4. Cultured cells grown on slides.
5. Cytospun cells or cell smears.
6. Gelatin- or silane (AAS)-coated microscope slides.

2.2. Prehybridization

1. Staining dishes.
2. Magnetic stirrer.
3. Incubator (37–50°C).
4. Glass/metal staining dishes.
5. Forceps.
6. Shaking water bath.

2.3. Radioactive probe labeling

1. Radiation monitor, e.g., Series 900 Mini monitor with ^{125}I type 444 scintillation probe (Mini-Instruments Ltd, Essex, UK).
2. ^{125}I-lead shields and pots.
3. Drying system with charcoal traps.
4. Radioactive waste disposal system.
5. Scintillation counter to monitor efficiency of probe labeling.

2.4. Hybridization and Washing

1. Acid-washed siliconized or plastic coverslips.
2. Trays of paraffin oil or heating block.
3. Incubator.
4. Shaking water bath (37–70°C).

2.5. Radioactive Detection Systems

1. Dark room with orange/red light source (e.g., Ilford safelight 904F).
2. Lightproof slide exposure boxes and packets of silica gel crystals.
3. Dipping chamber.
4. Chilled glass or metal plate.

2.6. Buffers

2.6.1. 20× Saline Sodium Citrate (SSC)

1. 20× 0.15 M NaCl, 0.015 M trisodium citrate: 175.3 g NaCl, 88 g of trisodium citrate ($Na_3C_6H_5O_7.2H_2O$; mol wt 294.10) per liter of distilled water (DW).
2. Adjust to pH 7.0 with citric acid. Autoclave and store at room temperature.

2.6.2. 5× Hybridization Buffer

1. For a total volume of 50 mL, add 25 mL of 20× SSC, 0.5 mL of 1 M KI, 2.5 mL each of 2% PVP and Ficoll, 5 mL of 500 mM EDTA, 2.5 mL of 1 M PO_4, 10 mL of 1 M Tris-HCl, pH 7.6, and 2 mL of DW.
2. Use autoclaved stocks of 20× SSC, 500 mM EDTA, 1 M PO_4, and 1 M Tris-HCl.
3. Filter through 0.45 μm filters and store at –20°C.

2.6.3. Proteinase K Buffer

1. To make 100 mL of 20× proteinase K buffer, mix 40 mL of 1 M Tris-HCl, pH 7.4, 4 mL of 1 M $CaCl_2$, 50 mL of 400 mM EDTA, and 6 mL of DW. Sterilize by autoclaving and store at room temperature.

2.6.4. Rapid Hematoxylin

Rapid hematoxylin is also available as a solution-hematoxylin Gill No. 2, Sigma-GHS-2-32.

4 g Hematoxylin (Sigma H9627 or H3136), 1 L DW, 50 g ammonium or potassium alum (aluminum ammonium sulfate), 5 g citric acid, 75 g chloral hydrate, 0.3 g sodium iodate (critical-weigh carefully).

1. Dissolve the alum in the DW (without heating).
2. Add the hematoxylin, sodium iodate, citric acid, and chloral hydrate in that order.
3. Allow each to dissolve fully before adding the next compound. The amount of sodium iodate is critical and should be weighed carefully. The solution is stable and may be kept for up to a year at room temperature.

2.6.5. Alcoholic Eosin

Alcoholic eosin is also available as a solution (Sigma HT110-3-32).

1. 137 mL of DW, 2730 mL of 95% alcohol, 350 mL of 1% aqueous eosin Y (eosin Y powder from Koch-Light Laboratories 6940-00 or Sigma E6003 or E4382), 35 mL of 1% aqueous phloxine (phloxine powder from Koch-Light Laboratories 6995-00 or Sigma P2759 or P4030), 14 mL of glacial acetic acid. Mix well and store at 4°C.

2.6.6. DIG Buffer 1

1. 100 mM Tris-HCl, pH 7.5, 150 mM NaCl.

2.6.7. DIG Blocking Reagent (1%)

1. 1 g Blocking reagent (Boehringer Mannheim cat. no. 1096-176), 100 mL of buffer 1.
2. Heat on stirrer to dissolve; autoclave, store room temperature. (This solution is used at 0.5% in buffer 1 when used as diluent for antibody).

2.6.8. DIG Buffer 3

1. 100 mM Tris-HCl, pH 9.5, 100 mM NaCl, 50 mM MgCl$_2$. (Note that the MgCl$_2$ should be dissolved separately and added as a solution to the dissolved Tris/NaCl. A precipitate may form in this solution. If this happens do not use it as it dramatically increases background.)

2.6.9. DIG Developing Solution

1. 337.5 µg/mL NBT (Boehringer cat. no. 1383 213; 4-nitroblue tetrazolium chloride), 175 µg/mL of X-phosphate (Boehringer cat. no. 1383 221; 5-bromo-4-chloro-3 indolyl phosphate) in buffer 3 (*see* **Subheading 2.6.8.**).

3. Methods

3.1. Preparation of Slides and Coverslips

3.1.1. Acid Washing

Slides and coverslips should be acid washed prior to coating or siliconization.

1. Soak 76 × 26 mm glass microscope slides or 13 mm diameter coverslips (Mediglass, Australia) overnight in a solution of 100 g of potassium bichromate, 850 mL of DW, 100 mL of concentrated H$_2$SO$_4$ (added slowly while stirring).
2. Wash the slides or coverslips for several hours in tap water.
3. During this time transfer the slides to metal racks for more complete washing.
4. Wash slides in racks 2× in DW or wash coverslips in a beaker with two changes of DW.
5. Store slides and coverslips in DW, or dry coverslips and store at room temperature.

3.2. Slide Coating and Coverslip Preparation

3.2.1. Gelatin Coating (0.5%) (28)

1. Dissolve 5 g of gelatin and 0.5 g of chromic potassium sulfate in 1 L of DW. Dissolve by heating on a magnetic stirrer until almost boiling. Filter before use.
2. Gelatin solution can be stored at 4°C for 1–2 d before use.
3. Dip acid-washed and dried slides (in metal racks) into the 0.5% gelatin solution for 1–2 s.
4. Drain and dry the slides in a dust-free environment by heating in front of a fan for 1–2 h or O/N at room temperature or 37°C.

3.2.2. Aminoalkylsilane Coating

Glass microscope slides can also be coated with 3-aminopropyl-triethoxy silane (AAS) for improved tissue adhesion of paraffin and frozen sections *(29)*.

1. To make a working solution, dilute 10 mL of AAS (Sigma cat. no. A-3648) in 500 mL of AR ethanol.
2. Place acid washed and dried microscope slides in metal racks.
3. Rinse the dried slides in AR ethanol, dip into the AAS working solution for 3 s, rinse again in AR ethanol, rinse in DW, dry, and store at room temperature. The AAS working solution should be used within 24 h for up to 10 racks of slides and then discarded.

3.2.3. Siliconization of Coverslips

1. Dip clean acid washed and dried coverslips in a siliconization solution of Prosil-28 (PCR Inc.) diluted 1:100 in DW.
2. Rinse 2× in DW and spread singly on trays covered with aluminum foil.
3. Dry the coverslips at 80°C for 2 h, transfer the dried coverslips to a glass Petri dish or other container, and sterilize by hot air.

3.2.4. Flexible Plastic Coverslips

Flexible plastic coverslips can be purchased (Integrated Sciences) or can be made from overhead transparency (acetate) sheets.

1. Transparency sheets are cut to the required size (~25 × 25 mm) and attached to the slides with double-sided 25 mm wide adhesive tape.
2. The area for hybridization is cut from the tape with an 8 mm diameter hole punch.
3. The coverslips can be used when the probe and the target DNA are denatured simultaneously using a heating block or a PCR *in situ* hybridization machine (Omnislide, Hybaid, Integrated Sciences).

3.3. Probe Labeling for Detection of Duck or Woodchuck Hepatitis B Virus (DHBV or WHV) DNA in Ethanol: Acetic Acid Fixed Tissues Using Either ^{125}I- or Digoxigenin (DIG)-Labeled DNA Probes

See **Note 1** for a comparison of radioactive and nonradioactive probe labeling.

1. For labeling DHBV or WHV and control DNA probes with ^{125}I-dCTP by random priming. ^{125}I should be stored at all times in lead pots and labeling reactions should be performed using lead perspex screens.
2. Obtain 250 µCi of ^{125}I-dCTP in 50 µL of 50% ethanol (NEN cat. no. NEX-074). Pierce the vial using a 20-gauge needle with a charcoal filter (NEN cat. no. NEX 033T) attached. Dry the ^{125}I-dCTP under vacuum for 1–2 h to evaporate the ethanol (preferably using a lead-based Perspex block in an ice bath).
3. Label the DNA using an Amersham Megaprime DNA labeling system (cat. no. RPN 1605). In separate tubes mix 3 µL of gel-purified specific viral and control

plasmid DNA (150 ng each). Add 5 μL of primer and 3 μL of DW to each tube. Heat for 3 min at 100°C. Chill on ice for 3 min.

4. In a separate tube mix 2×4 μL each of dATP, dGTP, dTTP, and 2×5 μL bovine serum albumin (BSA) reaction buffer. Remove this mixture and use to dissolve the dried ^{125}I-dCTP, then aliquot into the two Eppendorf tubes containing the denatured probe DNA. Wash out the vial with 2×20 μL of DW and aliquot into the two Eppendorf tubes.

5. Add 2 μL of Klenow enzyme per tube and incubate for 30 min at 37°C. Stop the reaction by adding 2.5 μL of 400 m*M* EDTA, 50 μL of DW, and 12 μL of sodium acetate, pH 4.8, to each tube. Mix. Add 250 μL of AR ethanol, mix, and hold at –80°C for 3 h to precipitate the DNA.

6. Centrifuge and wash the precipitated DNA 4× with 70% AR ethanol. Monitor the supernatant from each wash with a gamma counter to check that the number of counts being removed with each wash is decreasing. Dry the precipitated DNA and redissolve in 25 μL of Tris-HCl:EDTA (T:E; 10 m*M*:1 m*M*).

7. The probe may be heated to help redissolve the precipitated DNA. Count an aliquot of the labeled DNA in a gamma counter. Total expected counts should be approx 1×10^7 cpm. Store the labeled DNA at –20°C. Use approx $0.2–1 \times 10^5$ cpm/μL of hybridization mix.

3.4. Labeling DNA Probes with Digoxigenin DIG-dUTP by Random Priming

1. In separate tubes mix 100 ng each of gel-purified plasmid or control DNA. To each tube add DW to 10 μL, and 5 μL of primer (Amersham Megaprime DNA labeling system cat. no. RPN1605). Heat for 3 min at 100°C. Chill on ice for 3 min.

2. To each tube add 3 μL of BSA reaction buffer, 3 μL each of dATP, dCTP, dGTP, 1.2 μL of 1 m*M* dTTP, 1.0 μL of DIG-11-dUTP (Boehringer cat. no. 1093 088). Final concentration of dTTP = 40 μ*M*; DIG-dUTP = 33 μ*M*.

3. Add 2.0 μL of Klenow enzyme to each tube and incubate at 37°C for 16 h. Stop the reaction by adding 2.5 μL of 400 m*M* EDTA, 60 μL of DW, and 10 μL of 3 *M* sodium acetate, pH 4.8. Mix, then add 220 μL of AR ethanol, mix, and incubate at –80°C > 2 h.

4. Centrifuge for 10 min at 4°C, and wash the precipitated DNA 3× with 70% AR ethanol, dry, redissolve in 25 mL TE (10:1). Assume that the yield of each reaction is 250 ng (according to the manufacturer's advice) and the final probe concentration is 10 ng/μL. Store at –20°C.

3.5. Tissue Fixation

1. Fix fresh $3 \times 3 \times 3$ mm pieces of tissue in fresh ethanol:acetic acid (EAA, 3:1) for 30 min at RT. Wash the tissue 3× for 10 min with 70% AR ethanol at 4°C. Store O/N at 4°C in 70% AR ethanol and process into paraffin wax blocks on a short cycle. (N.B.: Tissues should enter the cycle at the ethanol stage.)

3.6. Section Preparation

1. Section wax-embedded tissues at 6 μm onto coated microscope slides. Transfer the slides to a metal slide rack. Heat the slides to melt the wax and to help the sections adhere to the slides in front of a fan or dry at 37°C overnight.

2. In a fume hood, dewax the slides for 2×10 min in AR xylene. Alternatively dewax slides (2×15 min) in a nontoxic solvent, e.g., Histoclear (National Diagnostics) or Safsolvent (Ajax Chemicals). Rehydrate the sections by incubating in an ethanol series of 100%, 95%, 90%, and 70% for 2 min in each solution. Briefly rinse in phosphate-buffered saline (PBS).

3. Fix in 0.1% glutaraldehyde in PBS (2 mL of 25% electron microscope grade glutaraldehyde (TAAB) in 500 mL of PBS) at 4°C for 30 min. Wash in PBS 2×5 min.

4. Digest EAA fixed sections with 1 µg/mL of proteinase K (Boehringer Mannheim) in 20 mM Tris-HCl, pH 7.4, 2 mM CaCl$_2$, 10 mM EDTA. In each new system the concentration of proteinase K should be titrated from 0.5 to 10 µg/mL on separate sections. For 1 µg/mL, add 50 µL proteinase K stock to 500 mL of prewarmed 1× proteinase K buffer and incubate at 37°C for 15 min, then wash in PBS 2×5 min.

5. Denature sections in 0.1× SSC in a boiling water bath at 100°C for 5 min. Immediately transfer to 0.1% glutaraldehyde for 15 min 4°C. Wash in PBS at room temperature for 2×5 min.

6. Acetylate the sections in 0.25% acetic anhydride in 500 mL of 0.1 M triethanolamine, pH 8.0. Incubate 10 min at room temperature with stirring, wash $2 \times$ SSC, 2×5 min. To make 500 mL of working solution add 6.65 mL of 7.5 M stock triethanolamine (BDH) to 500 mL of DW. Adjust to pH 8.0 with HCl (~1.5 mL). Add 1.25 mL of 100% acetic anhydride (BDH), i.e., 0.25%, and mix thoroughly by shaking.

7. Dehydrate the sections in a graded AR ethanol series 70%, 90%, 95%, 100% 2 min each.

3.7. Hybridization

1. Prepare a stock of 5× hybridization mix containing 10× SSC, 10 mM KI (for ^{125}I probes only), 0.1% PVP, 0.1% Ficoll 400, 50 mM EDTA, 50 mM phosphate buffer, pH 6.5, 200 mM Tris-HCl, pH 7.6.

2. Allow ~3.5 µL of complete probe mix for each section with either 13 mm diameter glass or 25 mm plastic coverslips, e.g., for 25 sections of tissue allow 90 µL of final probe mix. Prepare positive (plasmid DNA containing DHBV or WHV DNA) and negative (PUC19 DNA) probes.

3. The final probe mixture should contain 50% deionized formamide, 500 µg/mL of carrier DNA (stock of 10 mg/mL), 500 µg/mL of tRNA (stock of 10 mg/mL) and labeled DNA probe. For ^{125}I-labeled probes use $\sim 1 \times 10^5$ cpm/section. For DIG-labeled probes use ~2.5 ng of labeled DNA/µL of final probe mix.

4. Heat at 100°C for 3 min to denature the probe and chill on ice for 5 min. Add 1/5 vol of 5 × hybridization mix, 1 mg/mL of BSA (stock of 50 mg/mL) and DW to the final probe volume.

5. Glass coverslips. Apply ~ 3.5 µL of probe to each section. Cover with a heat-sterilized 13-mm siliconized coverslip. To adhere the coverslip and to remove any trapped bubbles, tap the coverslip gently with fine forceps. Hybridize at 37°C for 16 h submerged in trays of paraffin oil preheated to 37°C, i.e., T_m-21°C (*see* **Note 5**). After hybridization, wash the slides for 3×10 min in AR chloroform to

remove the oil. Air dry. Wash the slides in 500 mL of $2 \times$ SSC until the coverslips fall to bottom of the dish.

6. Alternatively, use flexible plastic coverslips adhered with double-sided adhesive tape. Attach the tape to the microscope slide with the tissue section within the 8 mm circle, apply the probe to the section, overlay with the coverslip, and seal. Heat the slides to ~90°C using a heating block or an Omnislide system (Hybaid, Integrated Sciences) to denature the target DNA and the probe on the section. Incubate in a moist chamber at 37°C for 16 h. Remove the coverslips with forceps.

3.8. Posthybridization Washing

For alternate washing protocols, *see* **Note 6**.

1. For ^{125}I probes add 100 μM KI to all wash solutions.
2. Wash in $2 \times$ SSC at room temperature for 2×1 h.
3. Wash in $0.1 \times$ SSC at 55°C for 3×10 min.
4. Wash in $0.1 \times$ SSC at room temperature for 2×1 h.
5. Dehydrate the slides in AR Grade ethanol solutions of 70%, 90%, 95%, 100% for 2 min each. Air dry.

3.9. Development of Signal for Detection Of Bound Probe

1. Autoradiography for sections hybridized with ^{125}I-labeled probes (*see* **Subheading 3.10.**).
2. For DIG-labeled probes *see* DIG color development (*see* **Subheading 3.13.**).

3.10. Autoradiography

Duplicate or triplicate sets of slides (labeled exposure 1, 2, or 1, 2, 3) should be prepared for development at different times as a control of signal intensity. Slides must be coated with emulsion and these steps as well as development of the autoradiographs must be performed in a darkroom. Following development the sections are counterstained with a range of dyes. These steps must be performed with care so that the emulsion layer is not lost from the slides.

1. Sort the slides into their exposure sets (exposure 1, 2, 3, etc.) (*see* **Note 6**) and pack into the exposure boxes for transfer to the darkroom.
2. Use either Ilford K2, K5, or L4 emulsion. Allow 10 mL of diluted emulsion for the first 10 slides, then an extra 2 mL for each additional 10 slides. Add emulsion to an equal volume of DW with 2% glycerol in screw-capped plastic tube.
3. Amersham LM-1 is supplied prediluted and can be remelted and reused.
4. Melt the emulsion at 43–45°C in a water bath for 20–25 min. Gently mix the emulsion by inverting the tube end over end approx 10×. Avoid creating air bubbles in the emulsion. Pour the melted emulsion into a clean, dipping chamber (Hypercoat dipping vessel; cat. no. RPN 39, Amersham). Dip the slides one at a time into the emulsion, drain each slide on the side of the chamber, wipe the back of the slide with a damp cloth, and place on a flat chilled metal or glass plate.

5. After 10–15 min on the chilled plate, transfer the slides to the bench, and dry at room temperature for 1–2 h or until visibly dry and the surface has flattened.
6. Repack the slides into light tight boxes with small packets of silica gel at 4°C for the required exposure time. Wrap the boxes in two layers of aluminum foil for added protection from light. The wrapped boxes may be removed from the darkroom and stored in the refrigerator at 4°C.

3.11. Development of Autoradiographs

1. All buffers and solutions that contact the emulsion should be at 15–20°C to prevent melting of the emulsion layer (especially important in hot weather).
2. Remove the slides stored in light tight boxes from the refrigerator at 4°C and bring to room temperature (approx 20 min in box). In the darkroom, transfer the coated slides to a metal slide rack.
3. Develop in undiluted Kodak D19 developer for 4–6 min at 15–20°C.
4. Fix in Ilford hypam fixer diluted 1:4 in DW for 6 min at 15–20°C. Emulsion should "clear" in 1–2 min. (If necessary, slides can be removed from the darkroom after twice the "clearing" time as they are no longer sensitive to light).
5. Transfer slides to 5% Na_2SO_4 solution and remove from the darkroom.
6. Gradually dilute the Na_2SO_4 with DW (5–10 min) until all salt is removed. Wipe the back of the slides with cloth or paper to remove the excess emulsion.
7. Stain the sections by hematoxylin and eosin.
8. Examples of sections of DHBV-infected duck liver hybridized with an I^{125}-labeled DNA probe are shown in **Fig. 1**.

3.12. Hematoxylin and Eosin Staining of Autoradiographs

1. Incubate slides in Rapid Hematoxylin for 0.5–1 min at 15–20°C. The timing of this step is a critical step as the emulsion layer is sensitive to the acidity of the hematoxylin. Quickly wash and differentiate the stain (red/purple to blue) in three or four changes of low-salt solution, e.g., PBS.
2. Incubate the slides in alcoholic eosin, 1 min at 15–20°C. Rinse with 100% alcohol in a wash bottle. Wash 2 × 2 min in AR ethanol.
3. Transfer to fresh AR xylene 2 × 10 min (use xylene under a fume hood).
4. Coverslip with DePex mounting medium (Gurr, BDH) or nonaqueous mountant.

3.13. DIG Color Development for DNA Probes

1. Section treatments: Wash slides for 5 min in buffer 1, shaking gently. Incubate in blocking reagent for 30 min, shaking gently. Wipe around section to remove excess blocking reagent. Incubate each section with 50 µL of 1:750 dilution of anti-DIG alkaline phosphatase conjugate (dilute antibody in 0.5% blocking reagent), coverslip. Note that the dilution used may need to be optimized for other systems.
2. Incubate in a humidified box for 30 min at 37°C.
3. Carefully remove coverslips, rinse sections several times in buffer 1, and then wash in buffer 1 for 2 × 15 min, with gentle shaking.
4. Rinse briefly in DIG buffer 3.

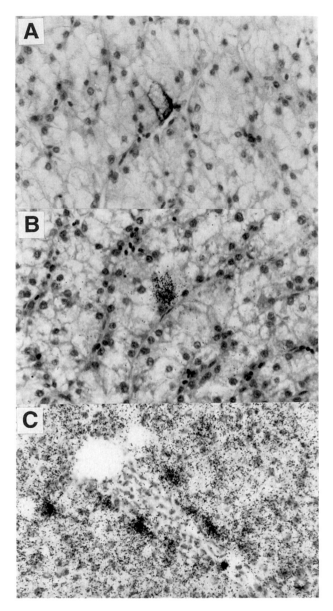

Fig. 1. Detection of cytoplasmic duck hepatitis B virus (DHBV) surface antigen
(**A**) and DNA (**B,C**) and in EAA-fixed liver tissue. (**A**) DHBV surface antigen detected in the cytoplasm of a single hepatocyte by immunoperoxidase staining of duck
liver *(5)*. (**B**) A consecutive section of liver from the same duck was hybridized to
detect DHBV DNA by *in situ* hybridization with an ^{125}I-labeled DNA probe. (**C**)
DHBV DNA was also detected in the cytoplasm of > 95% of hepatocytes in liver from

(continued)

5. Place the slides in a clean Coplin jar and add the developing solution (*see* **Subheading 2.6.9.**).
6. Incubate the slides in the dark at room temperature and check color development at regular intervals. Note that the developing solution can tolerate some exposure to light but avoid leaving developer-coated slides exposed to light for extended periods. Development is generally exhausted after 20 h.
7. Rinse developed slides in tap water to stop the reaction, counterstain the tissue with hematoxylin, and mount in an aqueous mounting medium, e.g., Hydro mount (National Diagnostics) or 25% glycerol in PBS.
8. Although the manufacturer states that the resulting precipitate is "blue" it is not. Generally precipitate is a reddish-purple-brown that can easily be distinguished in tissues counterstained with hematoxylin.
9. Sections may be dehydrated in ethanol and transferred to xylene and mounted in DePex but beware, as this procedure may dissolve the NBT/X-phosphate product and will not be a permanent mount.
10. Examples of sections of WHV-infected woodchuck liver hybridized with a digoxigenin-dUTP-labeled DNA probe are shown in **Fig. 2**.

4. Notes

1. Radioactively vs nonradioactively labeled probes: A comparison of the advantages and disadvantages of radioactive and nonradioactive systems for detection of viral nucleic acids is shown in **Table 1**. The improved resolution and sensitivity of radioactive probes must be balanced against the cost and hazards posed by radioactive probe labeling and detection systems. The choice of radioactive or nonradioactive systems remains a personal one, with nonradioactive systems taking over as the sensitivity of these detection systems is improved.

    Radioactively labeled nucleotides that are commonly used to label probes for *in situ* hybridization include: ^{125}I-dCTP (half-life 59.6 d), ^{33}P-UTP (half-life 25.4 d), ^{35}S-dCTP (half-life 87.4 d). The stability of these reagents is dictated by their radioactive half-life. These reagents are expensive to purchase and must be handled carefully but provide excellent localization of labeled target that is visualized by autoradiography.

    Nonradioactive systems for probe labeling most commonly include: digoxigenin *(17)* and biotinylated cRNA probes *(18)*. Other nonradioactive systems that have also been used include 2-acetyl-aminofluorene *(14)*, chemical labeling *(19)*, and alkaline phosphatase. Tyramide Signal Amplification (TSA) detection systems (NEN Life Science Products) have also been developed that result in enhancement of DIG, biotin, or fluorescence signals and can be visualized by chromogenic or fluorescent detection *(20–23)*.

---

a duck with persistent DHBV infection by *in situ* hybridization with an ^{125}I-labeled DNA probe. Following hybridization and washing the bound probe was detected by autoradiography. Tissue sections were counterstained with hematoxylin (**A**) or hematoxylin and eosin (**B,C**). Original magnification **A, B** × 132; **C** × 66.

Table 1
Advantages and Disadvantages of Radioactive and Nonradioactive
In Situ **Hybridization Systems**

Radioactive		Nonradioactive	
Advantages	Disadvantages	Advantages	Disadvantages
Superior sensitivity and resolution of grains	Radionucleotide half-life, stability, and expense	Extended half-life of labeled probes	Difficult to distinguish low-level labeling from background
Quantitation of signal by grain counting	Breakdown of labeled probes by radioactive damage	Generally less expensive	Quantitation is difficult
Control signal intensity by adjusting autoradiographic exposure time	Radiation hazard	Low hazard	

DNA probes can be labeled to high specific activity by random primer labeling. The preferred template for labeling is plasmid DNA containing viral gene sequences. To ensure labeling of viral DNA sequences only, viral DNA is separated from vector plasmid by restriction enzyme digestion, gel electrophoresis, and purification from agarose.

2. Choice and method of preparation of tissues and cells: (a) frozen tissue sections; (b) tissues fixed in formalin or other fixative, wax embedded and sectioned onto glass slides; and (c) cultured cells grown on slides, cell smears, or cytospins *(10,11)*. Tissue fixation will generally result in decreased sensitivity of detection but results in improved morphology and allows retrospective analysis of tissues *(12–16)*.

3. Control tissues or cells. It is essential to include in every experiment negative control tissues or cells that have been fixed and pretreated in the same way as the test samples. These might include preinfection samples from the same animal (**Fig. 2A**), tissues or cells collected from an uninfected control animal or uninfected cultured cels. Positive control tissues, i.e., samples that are known to contain high levels of viral nucleic acid, are also useful but are not always available. Negative control probes should also be included. These allow identification of nonspecific signals and should be tested using consecutive sections from the same tissue block or sample. Consecutive sections can be mounted on a separate area of the same slide, ensuring identical pretreatment of the sections. This provides a good control and is an economical strategy as it cuts in half the number of microscope slides being handled and the amount of reagents needed.

4. Prehybridization treatment of tissues and cells includes: (a) removal of wax from wax-embedded sections; (b) fixation of sections after dewaxing to maintain tis-

sue morphology and retain target nucleic acids; (c) protease digestion to increase probe access to target nucleic acids (The concentration of protease needs to be optimized by titration for each new system); (d) target denaturation (DNA) that can be performed at this stage or after the probe is applied to the tissue sections; (e) postfixation to retain target nucleic acid; and (f) acetylation of tissues to reduce nonspecific binding of probe.

5. Conditions for hybridization and washing need to be carefully chosen for high specificity of the resultant signal. The temperature of hybridization is critical to determine the stringency of the formation of hybrids in any experiment. Optimal hybridization temperatures have a broad around 25°C below the theoretical melting temperature (T_m) *(24)*. High-stringency hybridization conditions are performed at temperatures of 15–25°C below the theoretical T_m of the hybrids. Low-stringency conditions are between 25 and 30°C below T_m.

Calculation of melting temperature (T_m) of nucleic acid hybrids:

DNA : DNA hybrids *(25,26)*
$T_m = 16.6 \log [Na^+] + 0.41\ (\%\ G+C) + 81.5 - 0.72\ (\%\ FA)$
RNA : RNA hybrids *(25,27)*
$T_m = 18.5 \log [Na^+] + 0.584\ (\%\ G+C) + 79.8 + 0.0012\ (\%\ G+C)^2 - 0.35\ (\%\ FA)$.

Terms
[Na+] = Sodium ion concentration, $2 \times SSC = 0.33\ M$, $5 \times SSC = 0.825$
% G+C = % Guanosine plus cytosine content of DNA (usually 50%)
% FA = % Formamide concentration in final probe mix.

6. Examples of hybridization and washing conditions

   a. **DNA : DNA hybrids**
   Hybridize in 2× SSC, 50% FA, 50% G+C content; $T_m = 58°C$.
   Hybridization temperature = 37°C = T_m-21°C (high stringency).
   Hybridize in 5× SSC, 50% FA, 50% G+C content; $T_m = 64.6°C$.
   Hybridization temperature = 37°C = $T_m - 27.6°C$.
   High-stringency wash in 0.1× SSC, 0% FA; $T_m = 73°C$.
   Washing temperature = 55°C = $T_m - 18°C$.
   Specificity control to remove hybrids in 50% FA, $2 \times SSC$; $T_m = 58°C$.
   Washing temperature = 65°C = $T_m + 7°C$.

   b. **RNA : RNA hybrids**
   Hybridize in 2× SSC, 50% FA, 50% G+C content; $T_m = 86°C$.
   Hybridization temperature = 50°C = $T_m - 36°C$.
   High-stringency wash in 0.1× SSC, 0% FA; $T_m = 79°C$.
   Washing temperature = 60°C = $T_m - 19°C$.

   c. **DNA : RNA hybrids**
   Hybridize in 2× SSC, 50% FA, 50% G+C; $T_m \sim 75°C$.
   Hybridization temperature = 42°C ≈ $T_m - 33°C$.
   High-stringency wash in 0.1× SSC, 0% FA; $T_m \approx 75°C$.
   Washing temperature = 56°C ≈ $T_m - 19°C$.

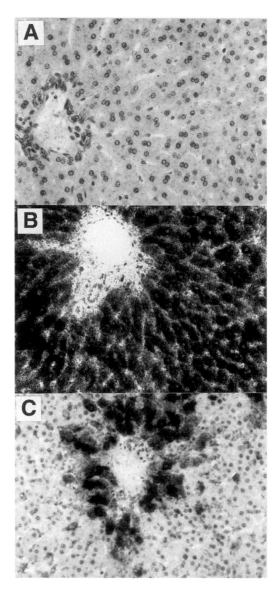

Fig. 2. Detection of woodchuck hepatitis virus (WHV) DNA in EAA-fixed liver tissue by nonradioactive *in situ* hybridization using a digoxigenin-dUTP labeled DNA probe and color development with NBT and X-Phosphate. Liver biopsy tissue was collected from woodchuck no. 38 preinfection (**A**), and at 11 (**B**) and 17 wk (**C**) postinfection with the woodchuck hepatitis B virus as described by Kajino et al. *(8)*. A low level of background staining can be seen in the preinfection sample (**A**). High levels of viral DNA were detected in the cytoplasm of > 95% of hepatocytes (**B**), and

(continued)

7. Analysis of *in situ* hybridization signals: Determining the specificity of resultant signals and learning to recognize nonspecific effects is often the hardest task and requires experience and careful choice of control tissues and probes. It is essential to include negative control tissues that have been fixed and pretreated in the same way as the test samples in every experiment (*see* **Note 3**). These will allow identification of any nonspecific binding of probe to the tissues. Positive control tissues also help in recognition of specific signals but are not always available. Negative control probes allow study of consecutive sections from the same tissue block or sample and identification of nonspecific binding.

8. Localization of viral DNA and antigen. *In situ* hybridization has the advantage that it allows localization of viral DNA to individual cells and consecutive sections can be used to detect antigen and nucleic acid in the same cell. An example of this technique is shown in **Fig. 1**. DHBV surface antigen was dectected by immunoperoxidase staining (**Fig. 1A**) and DHBV DNA has been detected in the same cell by *in situ* hybridization with an ^{125}I-labeled DNA probe and autoradiography (**Fig. 1B**).

Ackowledgments

The author wishes to thank D. S. Miller for technical assistance and development of the plastic coverslips method. Much of the work described in this chapter was performed through collaborative research projects with C. J. Burrell, E. J. Gowans, and W. S. Mason and was supported by funding from the National Health and Medical Research Council of Australia.

References

1. Gowans, E. J., Burrell, C. J., Jilbert, A. R., and Marmion, B. P. (1981) Detection of hepatitis B virus DNA sequences in infected hepatocytes by *in situ* cytohybridization. *J. Med. Virol.* **8**, 67–78.
2. Gowans, E. J., Burrell, C. J., Jilbert, A. R., and Marmion, B. P. (1983) Patterns of single-and double-stranded hepatitis B virus DNA and viral antigen accumulation in infected liver cells. *J. Gen. Virol.* **64**, 1229–1239.
3. Gowans, E. J., Jilbert, A. R., and Burrell, C. J. (1989) Detection of specific DNA and RNA sequences in tissues and cells by *in situ* hybridization, in *Nucleic Acid Probes* (Symons, R. H., ed.), CRC Press, Boca Raton, FL, pp. 140–158.
4. Jilbert, A. R., Freiman, J. S., Gowans, E. J., Holmes, M., Cossart, Y. E., and Burrell, C. J. (1987) Duck hepatitis B virus DNA in liver, spleen and pancreas: analysis by *in situ* and Southern blot hybridization. *Virology* **158**, 330–338.
5. Jilbert, A. R., Freiman, J. S., Burrell, C. J., Holmes, M., Gowans, E. J., and Cossart, Y. E. (1988) Virus-liver cell interactions in duck hepatitis B virus infection: a study of virus dissemination within the liver. *Gastroenterology* **95**, 1375–1382.

---

in 19% of hepatocytes (**C**) during clearance of WHV infection from the liver. Sections were counterstained with hematoxylin and mounted in glycerol saline. Original magnification ×66.

6. Jilbert, A. R., Wu, T.-T., England, J. M., Hall, P., de la M., Carp, N. Z., O'Connell, A. P., and Mason, W. S. (1992) Rapid resolution of duck hepatitis B virus infections occurs after massive hepatocellular involvement. *J. Virol.* **66**, 1377–1388.

7. Mason, W. S., Cullen, J., Saputelli, J., Wu, T.-T., Liu, C., London, T. W., Lustbader, E., Schaffer, P., O'Connell, A. P., Fourel, I., Aldrich, C. E., and Jilbert, A. R. (1994) Characterisation of the antiviral effects of 2' carbodeoxyguanosine in ducks chronically infected with duck hepatitis B virus. *Hepatology:* **19**, 398–411.

8. Kajino, K., Jilbert, A. R., Saputelli, J., Cullen, J., and Mason, W. S. (1994) Woodchuck hepatitis virus infections: very rapid recovery after a prolonged viremia and infection of virtually every hepatocyte. *J. Virol.* **68**, 5792–5803.

9. Mason, W. S., Cullen, J., Moraleda, G., Saputelli, J., Aldrich, C., Miller, D. S., Tennant, B., Frick, L., Averett, D., Condreay, L., and Jilbert, A. R. (1998) Lamivudine therapy of WHV infected woodchucks. *Virology* **245**, 18–32.

10. Herrington, C. S., Burns, J., Graham, A. K., Evans, M., and McGee, J. O. (1989) Interphase cytogenetics using biotin and digoxigenin labeled probes I: relative sensitivity of both reporter molecules for detection of HPV16 in CaSki cells. *J. Clin. Pathol.* **42**, 592–600.

11. Musiani, M., Zerbini, M., Venturoli, S., Gentilomi, G., Borghi, V., Pietrosemoli, P., Pecorari, M., and LaPlaca, M. (1994) Rapid diagnosis of cytomegalovirus encephalitis in patients with AIDS using *in situ* hybridization. *J. Clin. Pathol.* **47**, 886–891.

12. Burns, J., Graham, A. K., Frank, C., Fleming, K. A., Evans, M. F., and McGee, J. O. (1987) Detection of low copy human papilloma virus DNA and mRNA in routine paraffin sections of cervix by non-isotopic *in situ* hybridization. *J. Clin. Pathol.* **40**, 858–864.

13. Coates, P. J., Hall, P. A., Butler, M. G., and D'Ardenne, A. J. (1987) Rapid technique of DNA–DNA *in situ* hybridization on formalin fixed tissue sections using microwave irradiation. *J. Clin. Pathol.* **40**, 865–869.

14. Salimans, M. M., van de Rijke, F. M., Raap, A. K., and van Elsacker Niele, A. M. (1989) Detection of parvovirus B19 DNA in fetal tissues by *in situ* hybridization and polymerase chain reaction. *J. Clin. Pathol.* **42**, 525–530.

15. Desport, M., Collins, M. E., and Brownlie, J. (1994) Detection of bovine virus diarrhoea virus RNA by *in situ* hybridization with digoxigenin-labeled riboprobes. *Intervirology* **37**, 269–276.

16. Van-Rensburg, E. J., Van Heerden, W. F., Venter, E. H., and Raubenheimer, E. J. (1995) Detection of human papillomavirus DNA with *in situ* hybridization in oral squamous carcinoma in a rural black population. *S. Afr. Med. J.* **85**, 894–896.

17. (1996) *Non-Radioactive* In Situ *Hybridization Application Manual*, 2nd ed. Boehringer Mannheim, Germany.

18. Giaid, A., Hamid, Q., Adams, C., Springall, D. R., Terenghi, G., and Polak, J. M. (1989) Non-isotopic RNA probes. Comparison between different labels and detection systems. *Histochemistry.* **93**, 191.

19. Cubie, H. A., Grzybowski, J., da-Silva, C., Duncan, L., Brown, T., and Smith, N. M. (1995) Synthetic oligonucleotide cocktails as probes for detection of human parvovirus B19. *J. Virol. Methods* **53**, 91–102.

20. Kerstens, H. M., Poddighe, P. J., and Hanselaar, A. G. (1995) A novel *in situ* hybridization signal amplification method based on the deposition of biotinylated tyramine. *J. Histochem. Cytochem.* **43,** 347–352.
21. Deichmann, M., Bentz, M., and Haas, R. (1997) Ultra-sensitive FISH is a useful tool for studying chronic HIV-1 infection. *J. Virol. Methods* **65,** 19–25.
22. Plenat, F., Pichard, E., Antunes, L., Vignaud, J. M., Marie, B., Chalabreysse, P., and Muhale, F. (1997) Amplification of immunologic reactions using catalytic deposition at the reactions sites of tyramine derivatives. A decisive gain in sensitivity in immunohistochemistry and *in situ* hybridisaton. *Ann. Pathol.* **17,** 17–23.
23. Strappe, P. M., Wang, T. H., McKenzie, C. A., Lowrie, S., Simmonds, P., and Bell, J. E. (1997) Enhancement of immunohistochemical detection of HIV-1 p24 antigen in brain by tyramide signal amplification. *J. Virol. Methods* **67,** 103–112.
24. Britten, R. J. and Davidson, E. H. (1985) Hybridization strategy, in *Nucleic Acid Hybridization: A Practical Approach* (Hames, B. D. and Higgins, S. J., eds.), IRL Press, Oxford, pp. 1–16.
25. McConaughy, B. L., Laird, C. D., and McCarty, B. J. (1969) Nucleic acid reassociation in formamide. *Biochemistry* **8,** 3289–3295.
26. Thomas, M., White, R. L., and Davis, R. W. (1976) Hybridization of RNA to double-stranded DNA: formation of R-loops. *Proc. Natl. Acad. Sci. USA* **73,** 2294–2298.
27. Bodkin, D. K. and Knudson, D. L. (1985) Sequence relatedness of Palyam virus genes to cognates of the Palyam serogroup viruses by RNA–RNA blot hybridization. *Virology* **143,** 55–62.
28. Rogers, A. R. (1979) *Techniques in Autoradiography,* 3rd ed. Elsevier, North Holland.
29. Henderson, C. (1989) Aminoalkylsilane: an inexpensive, simple preparation for slide adhesion. *J. Histochem.* **12,** 123–124.
30. Wilkinson, D. G. (ed.) (1992) In Situ *Hybridization: A Practical Approach.* IRL Press, Oxford.

14

Signal Amplification for DNA and mRNA

Ernst J. M. Speel, Anton H.N. Hopman, and Paul Komminoth

1. Introduction

In situ hybridization (ISH) permits the localization of specific unique or repeated DNA and RNA sequences at the level of individual cells *(1–4)*. It has significantly advanced the study of gene structure and expression, and, in addition to morphological identification of cell types involved, ISH also allows some quantification of observations, e.g., with respect to tumor burden or viral load. Despite its high degree of detection specificity, the technique still does not allow the routine detection of DNA sequences less than 5 kb in size, and in the case of tissue sections the detection sensitivity is even more limited. The threshold levels for mRNA detection are more difficult to determine, with the reported sensitivity limits of 1–20 copies of mRNA per cell being achieved only in the most sensitive protocols *(5,6,6a)*.

In recent years, strategies to improve the threshold levels of nucleic acid detection in ISH studies have included (*see* **Fig. 1**):

1. The use of increased amounts of hybridized probes, e.g., cocktails of oligonucle-otide probes or multiple cRNA probes *(6–8)*.
2. Protocols to amplify the target nucleic acid sequences prior to ISH, e.g., by *in situ* polymerase chain reaction (PCR) *(9–11, see also* chapter by Nuovo, this volume), *in situ* self-sustained sequence replication (3SR) *(12,13)*, and cycling or repeated primed *in situ* (PRINS) nucleic acid labeling *(14–16)*, which is not mean-ingfully different from *in situ* PCR.
3. Protocols to amplify the signals produced by the hybridization procedures, e.g., by catalyzed reporter deposition (CARD) *(17–20)*.

Target amplification techniques thus combine PCR and hybridization *in situ*, and have been utilized to visualize specific amplified DNA sequences (repeated, single-copy, viral) as well as RNA sequences within isolated cells

From: *Methods in Molecular Biology*, Vol. 123: In Situ *Hybridization Protocols*
Edited by: I. A. Darby © Humana Press Inc., Totowa, NJ

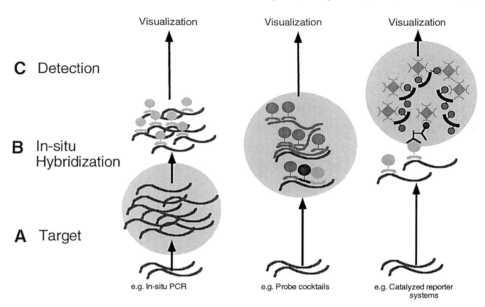

Fig. 1. Schematic drawing of the key steps (*darker circles*) of signal enhancement strategies for the detection of nucleic acid sequences *in situ* (*see* **Introduction**).

and tissue sections and of DNA sequences in chromosomal preparations *(9,21–25)*. Theoretically *in situ* PCR techniques are straightforward: fixation of cells or tissue samples, generating semipermeable cell membranes allowing the primers, nucleotides, and enzymes to enter the cell but avoiding the loss of the generated amplicants, in cell PCR amplification, and direct or indirect (by ISH) detection of the amplicants *(26)*. The practical procedure, however, is associated with several obstacles, such as low amplification efficiency (restricted sensitivity), poor reproducibility (restricted specificity), and difficulties in quantification of the results *(11,27)*. These findings are caused by a number of artifacts as a consequence of PCR amplification *in situ*, such as diffusion of PCR products during and after denaturation from the site of synthesis inside and/or outside the cells, extracellular generation of amplicants, and, in the case of direct incorporation of labeled nucleotides during direct *in situ* PCR, the generation of nonspecific PCR products resulting from mispriming, from fragmented DNA undergoing "repair" by DNA polymerase ("repair" artifacts), or from priming of nonspecific DNA or cDNA fragments ("endogenous priming" artifacts). Repair artifacts may also occur in apoptotic cells or samples that have been pretreated with DNase prior to *in situ* reverse transcriptase (RT)–PCR for mRNA detection *(26,28,29)*. Therefore, it has been recommended to use a multitude of different controls to allow adequate inter-

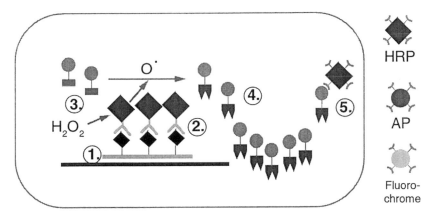

Fig. 2. Principles of ISH combined with CARD signal amplification using biotinylated tyramides. (1) Hybridization with digoxigenin-labeled probe, (2) anti-digoxigenin HRP detection of the probe, (3) peroxidase-tyramide reaction (CARD) with H_2O_2 and biotinylated tyramides (tyramide and oxide radical formation), (4) deposition of activated tyramide molecules onto protein moieties, (5) detection of biotin groups with (strept)avidin conjugates using fluorescence of enzyme cytochemical visualization. HRP, horseradish peroxidase; AP, alkaline phosphatase.

pretation of results *(30)*. In summary, it appears that the theoretical potential of *in situ* PCR is still much greater than its current practical impact, both in research and clinical studies. However, interesting alternatives are being explored, such as the rolling-circle amplification approach *(30a)*.

On the other hand, more and more literature is becoming available that describes approaches to amplify the signals after ISH, of which the most promising is the CARD technique. This method, which was introduced by Bobrow et al. *(31)* for use in immunoblotting and enzyme-linked immunosorbent assay (ELISA) assays, is based on the deposition of haptenized tyramide molecules by peroxidase activity (**Fig. 2**). The highly reactive intermediates of this reaction are bound to electron-rich moieties of proteins, such as tyrosine, phenylalanine or tryptophan, at or near the site of the peroxidase binding site. Visualization of deposited tyramides *in situ* can be either directly after the CARD reaction with fluorescence microscopy, if fluorochrome-labeled tyramides are used, or indirectly with either fluorescence or brightfield microscopy, if biotin, digoxigenin, di- or trinitrophenyl are used as haptens, which can act as further binding sites for anti-hapten antibodies or (strept)avidin conjugates (in the case of biotinylated tyramides) (*see* **Table 1**). Also fluorescein and rhodamine can be used as haptens, as specific antibodies against these fluorochromes are commercially available (from Dako, Glostrup, Denmark,

Table 1
Cytochemical Detection Systems that Are Frequently Used for ISH[a]

Label	First layer	Detection	
		Second layer	Third layer
Biotin	Avidin*,b,c		
Biotin	Avidin*	Biotin-labeled anti-avidin Ab	Avidin*
Hapten[d]	Anti-hapten Ab*		
Hapten	Mouse[e] anti-hapten Ab	Anti-mouse Ab*	
Hapten	Mouse anti-hapten Ab	Rabbit anti-mouse Ab*	Anti-rabbit Ab*
Hapten	Mouse anti-hapten Ab	Biotin-labeled anti-mouse Ab	ABC
Hapten	Mouse anti-hapten Ab	Digoxigenin-labeled anti-mouse Ab	Anti-digoxigenin Ab*

[a]Further amplification of ISH signals can be achieved by combining these detection systems with signal amplification using labeled tyramides (*see* **Subheading 3.5.**).

[b]*Abbreviations used:* Ab, antibody; ABC, avidin biotinylated enzyme (horseradish peroxidase or alkaline phosphatase) complex; *, fluorochrome (e.g., coumarin, fluorescein, Cy2, rhodamine, Texas red, Cy3, Cy5, Cy5.5, Cy7, Alexa dyes) or enzyme (horseradish peroxidase or alkaline phosphatase).

[c]Detection conjugate dilutions should be optimized for one's own experiments. Usually the optimal dilutions of the commercially available conjugates are indicated by the manufacturer (*see* also **refs. 6, 51–53**).

[d]Hapten = biotin, digoxigenin, dinitrophenyl (DNP), fluorescein, rhodamine, or Alexa dyes.

[e]Anti-hapten Ab raised in another species (e.g., rabbit, goat, swine) can also be used as primary Ab in ISH detection schemes.

and Molecular Probes, Eugene, OR, USA). CARD signal amplification has been easily implemented in immunohistochemistry, allowing an enormous (up to 1000-fold) increase in sensitivity when compared with conventional avidin biotinylated enzyme complex (ABC) procedures *(32–36)*. Since 1995, CARD has further been introduced in detection procedures for both DNA and RNA ISH on cell preparations and tissue sections *(17–20,37–41)*. With signal amplification the ISH sensitivity could be improved in the range of two- to 100-fold, enabling the detection of (1) single-copy DNA sequences up to the level of 5 kb in cell preparations, (2) up to three different DNA sequences (repetitive as well as single-copy) simultaneously in cell preparations, (3) low and single-copy human papillomaviruses in cell and tissue preparations, and (4) mRNA ranging from high to low abundancy in cell and tissue preparations (*see*, e.g., **Figs. 3** and **4** for a comparison of insulin mRNA signal intensity after conventional as well as CARD signal amplification). Closer examination of the literature, however, shows that a number of different combinations of cytochemical probe detection (one to three detection conjugate layers) and CARD signal amplification (utilizing different tyramides, CARD amplification buffer, reaction time, and temperature) systems were applied in these studies, and that optimization of each detection system was necessary to obtain a high signal-to-noise ratio. Because with CARD signal amplification both specific and non-specific (background) ISH signals will be amplified, it is essential that background staining after conventional probe detection should be very low to apply this procedure successfully. Therefore, we recommend to always optimize cytochemical probe detection and CARD signal amplification for one's own experiment. For this purpose, an appropriate detection system can be selected from **Table 1** and combined with either a commercially available signal amplification kit (available from NEN Life Science Products [Boston, MA, USA] as Tyramide Signal Amplification [TSA] kits or from Dako [Glostrup, Denmark] as Genpoint kit), used according to the manufacturer's instructions, or CARD signal amplification using freshly synthesized hapten- or fluoro-chrome-labeled tyramides (*see* **ref. *42*** and **Subheading 3.1.**) in a phosphate-buffered saline (PBS) buffer of pH 7.6 containing 0.1 *M* imidazole and 0.001% H_2O_2. In our hands, discretely localized ISH signals of high intensity can be obtained by adjusting the number of cytochemical detection layers, the dilution of detection conjugates (usually the first detection layer can be diluted 2- to 10-fold further than in conventional detection systems), the tyramide concentration in the CARD amplification buffer, and the reaction time (at 37°C) *(20)*. Although probe concentrations have also been diluted in combination with CARD signal amplification, this seems to be advantageous mainly in cases in which complex DNA probes are used (e.g., chromosome painting, yeast artificial chromosome [YAC], P1 or cosmid probes containing repetitive elements

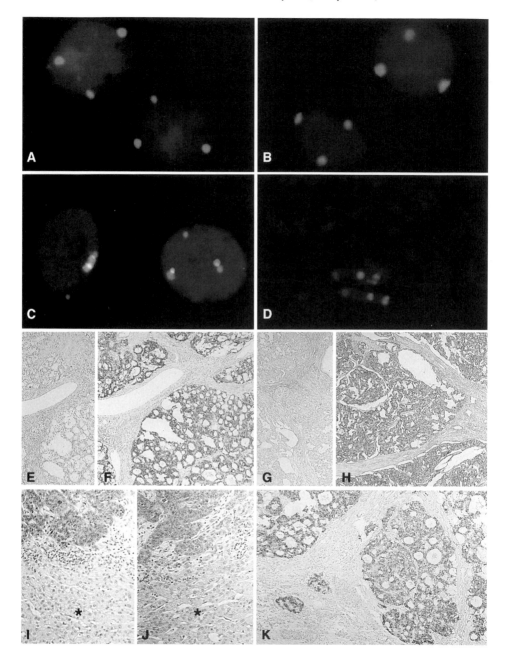

Fig. 3. (*See* Color Plate 5 following p. 176.) Results after ISH and CARD signal amplification on human cell preparations (**A–D**) and routinely fixed, paraffin-embedded tissue sections (**E–K**) (*see* **Tables 1** and **2**). (**A,B**) Fluorescence detection of three centromere 1 copies in   *(continued)*

that need to be blocked by competitor DNA, such as Cot-I DNA) *(20,40)*. An even quicker way to label nucleic acids *in situ* is the use of peroxidase-labeled probes, which can be detected directly with signal amplification *(39,43)*. However, these procedures proved to be clearly less sensitive than the indirect procedures described earlier.

The application of CARD signal amplification to ISH on tissue sections from routinely fixed, paraffin-embedded surgical samples is of particular interest for clinical diagnosis. In this respect we have recently utilized the CARD approach to increase the sensitivity of our diagnostic mRNA ISH procedure and to shorten the overall turnaround time of the assay (*[19]* and *[19a]*; *see also* **Fig. 3E–H**, and **Fig. 4**). This approach allows, e.g., the detection of peptide hormone mRNA within one working day and makes the assay suitable for routine diagnostic purposes. Furthermore, it allows the use of diaminobenzidine (DAB) as a chromogen and as a consequence the application of conventional counterstains and the mounting of slides in xylene-based mounting solutions, making the procedure more acceptable for the technician to perform and, e.g., the pathologist to interpret. In addition, the possibility to synthesize differently la-

---

bladder tumor cell line T24 using a chromosome 1-specific centromere probe (biotin-labeled), AvPO/fluorescein-tyramide (**A**) or AvPO/rhodamine-tyramide (**B**) visualization, and PI (**A**) or DAPI (**B**) counterstaining. (**C**) Fluorescence detection of three centromere 11 and two 11p15 copies in bladder tumor TCC 9 using a chromosome 11-specific centromere and 11p15 probe (digoxigenin- and biotin-labeled), ShADigPO/fluorescein-tyramide and MADig-GAMPO/rhodamine-tyramide visualization, and DAPI counterstaining. (**D**) Fluorescence detection of chromosome 1q42–43, 1p36 and centromere loci on a lymphocyte metaphase spread using specific biotin-, digoxigenin-, and fluorescein-labeled probes, and AvPO-BioGAA-AvPO/coumarin-tyramide, MADig-GAMPO/rhodamine-tyramide and AFluPO/fluorescein-tyramide visualization without counterstaining. (**E,F**) Brightfield detection of insulin mRNA in an insulinoma with (**F**) or without (**E**) antisense oligonucleotide probe (digoxigenin-labeled) using ShADigPO/digoxigenin-tyramide/ShADigPO detection, PO–DAB visualization, and hematoxylin counterstaining. (**G,H**) Brightfield detection of vaso-active intestinal polypeptide (VIP) mRNA in a VIPoma with (**H**) or without (**G**) signal amplification using the same visualization procedure as in (**E,F**). (**I,J**) Brightfield detection of insulin mRNA in a liver metastasis of an insulinoma, showing digoxigenin-tyramides (**I**) providing less background than biotin-tyramides (**J**) after signal amplification due to the presence of endogenous biotin in liver tissue. Visualization in (**I**) as described in (**E,F**), in (**J**) with ShADigPO/biotin-tyramide/streptavidin PO (NEN), followed by the PO–DAB reaction and hematoxylin counterstaining. (**K**) Brightfield detection of insulin mRNA in an insulinoma as in (**E,F**), now using trinitrophenyl-tyramides and peroxidase-conjugated rabbit anti-trinitrophenyl (Dako) for detection.

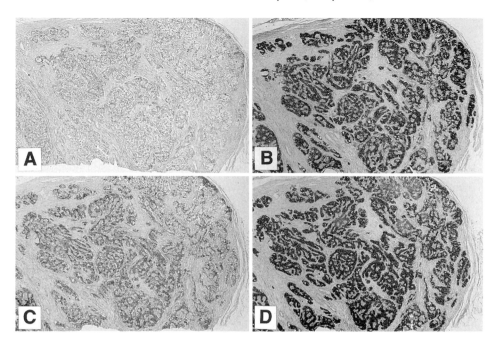

Fig. 4. Comparison of CARD signal amplification (**B,D**) with conventional detection (**A,C**) of insulin mRNA in an insulinoma, showing higher signal-to-noise ratios with than without the utilization of CARD. The hybridized antisense digoxigenin-labeled oligonucleotide probe was visualized by the PO–DAB reaction and tissues were counterstained by hematoxylin (*see* **Tables 1** and **2**). Detection with (**A**) ShADigPO, (**B**) ShADigPO/digoxigenin-tyramide/ShAdigPO, (**C**) MADig-Biotinylated horse anti-mouse-avidin biotinylated peroxidase complex (ABC), and (**D**) ABC (*see* [**C**])/digoxigenin-tyramide/ShADigPO (*see* **Tables 1** and **2**).

beled tyramides enables the substitution of biotinylated tyramides with, e.g., digoxigenin or di- or trinitriphenyl labeled tyramides, in cases where the detection of endogenous biotin can cause background problems, such as in liver and kidney tissue (*19a*; *see* **Fig. 1I–K**).

In conclusion, CARD signal amplification using labeled tyramides holds enormous potential to significantly influence future directions in the development of ISH. It is anticipated that especially efforts to signal amplification are the key strategy for the future to make ISH an easy to perform, fast, highly sensitive, and efficient method and therefore suitable for diagnostic laboratories.

2. Materials

2.1. Synthesis of Biotinylated Tyramides for Use in CARD Signal Amplification

1. N-hydroxysuccinimide ester of biotin (sulfosuccinimidyl-6-[biotinimide]hexanoate) (mol wt 557; Pierce, Rockford, IL, USA).
2. Dimethylformamide (DMF; Pierce).
3. Tyramine-HCl (mol wt 173; Sigma, St. Louis, MO, USA).
4. Triethylamine (TEA, 7.2 *M*; Pierce).
5. 100% Ethanol (Merck, Darmstadt, Germany).

2.2. Multiple-Target DNA In Situ Hybridization on Cell Preparations

1. Slides with fixed cells and/or chromosomes.
2. RNase A (Boehringer [now Roche], Mannheim, Germany).
3. 20× SSC: 3 *M* NaCl, 300 m*M* trisodium citrate, pH 7.0.
4. 2× SSC (diluted from stock 20× SSC).
5. Pepsin from porcine stomach mucosa (2500–3500 U/mg) (Sigma).
6. PBS: 140 m*M* NaCl, 2.7 m*M* KCl, 10 m*M* Na_2HPO_4, 1.8 m*M* KH_2PO_4, pH 7.6.
7. 1% Formaldehyde (diluted from 37% formaldehyde [Merck]) in PBS with 50 m*M* $MgCl_2$ (for chromosome preparations), without magnesium in the case of cell preparations.
8. Dehydration series of 70%, 96%, and 100% ethanol.
9. DNA probes (*see* **ref. 20**).
10. Nick-translation mix for biotin, digoxigenin, and/or fluorochrome labeling (Boehringer), Bionick labeling system (Life Technologies, Basel, Switzerland), or separate labeling components *(44)*.
11. Hybridization mixture, pH 7.0: 50–60% formamide, 10% dextran sulfate (Sigma), 2× SSC, 0.05–0.5 mg/mL of sonicated herring sperm DNA (Sigma) as well as yeast tRNA (50× excess; Boehringer).
12. Human Cot-I DNA (Life Technologies).
13. Genomic human placenta DNA (Sigma).
14. 70% Formamide/2× SSC at 70°C.
15. Metal box or heating plate.
16. Wash buffer I: 50–60% formamide, 2× SSC (diluted from 20× SSC), 0.05% Tween-20.
17. Wash buffer II: 2× SSC, 0.05% Tween-20 (repetitive probes) or 0.1× SSC (single-copy probes).
18. Rubber cement.
19. Water bath at 37°C.
20. Water bath at 70–75°C.
21. Water bath at 42°C.
22. Humid chamber.
23. Incubator at 37°C.

2.3. mRNA In Situ Hybridization on Formalin-Fixed, Paraffin-Embedded Tissue Sections

1. Tissue blocks of formalin-fixed, paraffin-embedded samples.
2. Super Frost Plus object slides (Merck).
3. Xylene.
4. Dehydration series of 70%, 96%, and 100% ethanol.
5. 0.1% Diethylpyrocarbonate (DEPC; Sigma)-treated Milli-Q.
6. PBS (*see* **Subheading 2.2., item 6**).
7. PBS/0.5% SDS: PBS containing 0.5% sodium dodecyl sulfate (SDS).
8. Pepsin (*see* **Subheading 2.2., item 5**).
9. 4% Paraformaldehyde (Merck) in PBS at 4°C.
10. 0.1 M Glycine in PBS.
11. 0.3% H_2O_2 in methanol (diluted from 30% H_2O_2 [Merck]).
12. 4× SSC (diluted from stock 20× SSC, *see* **Subheading 2.2., item 3**).
13. Oligonucleotides synthesized without or with internally amino-dTTP-labeled nucleotides. The amino groups can be coupled to different haptens, such as biotin, digoxigenin, or fluorochromes.
14. DIG oligonucleotide tailing kit (Boehringer).
15. EasyHyb (Boehringer): 4× SSC, 10% dextran sulfate, 0.5% SDS, 0.5 mg/mL of salmon testis ssDNA (Sigma), 0.25 mg/mL of yeast tRNA (Boehringer), 100 µg/mL of polyadenylic acid (Sigma), 0.125 U/mL of polydeoxyadenylic acid (Sigma), and 50 pmol/mL Randomer (NEN Life Science Products, Boston, MA, USA).
16. 2× SSC, 1× SSC, and 0.5× SSC (diluted from stock 20× SSC, *see* **Subheading 2.2., item 3**).
17. Plastic hydrophobic coverslips (Gelbond Film; FMC Bioproducts, Vallensbaek Strand, Denmark).
18. Humid chamber.
19. Water bath at 37°C.

2.4. Cytochemical Probe Detection

1. Nonfat dry milk powder or blocking reagent (Boehringer).
2. Normal goat serum (NGS).
3. 20× SSC (*see* **Subheading 2.2., item 3**).
4. PBS (*see* **Subheading 2.2., item 6**).
5. Tween-20 (Merck).
6. 0.1 M Phosphate buffer, pH 7.3.
7. Detection conjugates (*see* **Tables 1** and **2**).
8. Dehydration series of 70%, 96%, and 100% ethanol.
9. 30% H_2O_2 (Merck).
10. Diaminobenzidine (DAB) (Walter, Kiel, Germany).
11. 1% Cobalt (II) chloride ($CoCl_2 \cdot 6H_2O$) (Merck).
12. 1% Nickel (II) sulfamatetetrahydrate ($H_4N_2NiO_6S_2 \cdot 4H_2O$) (Merck).
13. PO–DAB buffer: 50 mM Tris-HCl, pH 7.6.

Table 2
Protocol for the Fluorescence Detection of Three Nucleic Acid Sequences in Cell Preparations, Hybridized with a Biotin-, Digoxigenin- and Fluorescein-Labeled DNA Probe, Using Multiple CARD Reactions[a]

Detection step	Time	Temperature
1. Detect biotin with AvPO[b,c] (diluted 1:50–200).	30 min	37°C
2. Visualize PO activity in blue (coumarin-tyramide) as described in **Subheading 3.5.**, **step 2b.**	5–15 min	37°C
3. Inactivate residual PO activity with 0.01 *N* HCl.	10 min	RT
4. Detect digoxigenin with ShADigPO[c] (diluted 1:100–200).	30 min	37°C
5. Detect PO activity in red (rhodamine-tyramide) as described in **Subheading 3.5.**, **step 2b.**	5–15 min	37°C
6. Inactivate residual PO activity with 0.01 *N* HCl.	10 min	37°C
7. Detect fluorescein with AFluPO[c] (diluted 1:2000).	30 min	37°C
8. Detect PO activity in green (fluorescein-tyramide) as described in **Subheading 3.5.**, **step 2b.**	5–15 min	37°C
9. Embed in Vectashield without DNA counterstain.	10 min	RT

[a]For details of detection systems, *see* **Table 1** and **ref. 20.**
[b]*Abbreviations used*: AFluPO, PO-conjugated anti-fluorescein (NEN Life Science Products); AvPO, PO-conjugated avidin (Dako); PO, horseradish peroxidase; RT, room temperature; ShADigPO, PO-conjugated sheep anti-digoxigenin Fab fragments (Boehringer).
[c]For smaller DNA targets, an amplification step with biotinylated goat anti-avidin (1:200; Vector) and AvPO (1:50–200) can be applied in **step 1**, with monoclonal mouse anti-digoxigenin (1:20,000; Sigma) and PO-conjugated goat anti-mouse IgG (1:200; Dako) in **step 4**, and with rabbit anti-fluorescein (1:2000; Dako) and PO-conjugated swine anti-rabbit IgG (1:200; Dako) in **step 7**.

14. Vectashield fluorescence embedding medium (Vector, Burlingame, CA, USA).
15. DAPI: 4',6-Diamidino-2-phenyl indole (Sigma).
16. PI: Propidium iodide (Sigma).
17. YOYO (Molecular Probes, Eugene, OR, USA).
18. Hematoxylin: Hematoxylin (Solution Gill no. 3) (Sigma): distilled water (1:2–4).
19. Nuclear fast red.
20. Embedding media for brightfield microscopy: we usually use 0.2 M Tris, pH 7.6:glycerol (1:9, v/v) as aqueous-based and TissueTek, Entellan (Merck) or immersion oil (Zeiss) as organic-based embedding medium.
21. DNA blocking buffer: 4× SSC (diluted from stock 20× SSC), 5% nonfat dry milk.
22. RNA blocking buffer: 0.5% blocking reagent in RNA detection buffer.
23. RNA detection buffer: 50 mM Tris-HCl, pH 7.4, 300 mM NaCl.
24. DNA washing buffers: 4× SSC (diluted from stock 20× SSC), 0.05% Tween-20, PBS, 0.05% Tween-20.
25. RNA washing buffer: RNA detection buffer containing 0.1% Tween-20.
26. Incubator at 37°C.
27. Fluorescence and brightfield microscope (Zeiss Axiophot).
28. Photographic film (64–100 ASA film for brightfield images and 400–640 ASA for fluorescent images).
29. CCD camera and image processing software (e.g., Vysis Quips Genetic Workstation (Vysis, Stuttgart, Germany) or Metasystems Image Pro System (Metasystems, Sandhausen, Germany).

2.5. CARD Signal Amplification

1. Detection conjugates labeled with peroxidase (*see* **Subheading 2.4.** and **Tables 1** and **2**).
2. PBS (*see* **Subheading 2.2.**, **item 6**).
3. RNA detection buffer (*see* **Subheading 2.4.**, **item 23**).
4. Imidazole (Merck).
5. 30% H_2O_2 (Merck).
6. CARD reaction buffer: 1× Amplification Diluent (NEN Life Science Products, Brussels, Belgium) or PBS containing 0.1 M imidazole, pH 7.6, and 0.001% H_2O_2.
7. Tyramide stock solutions, freshly synthesized in DMF (*see* **Subheading 3.1.** and **ref. *42***) and diluted to stock solutions of 1 mg/mL with ethanol, or DMF: biotin-, digoxigenin-, trinitrophenyl-, coumarin-, fluorescein-, rhodamine and Cy3-labeled tyramides (biotin-labeled tyramides are also commercially available from NEN and Dako, and fluorochrome-labeled tyramides from NEN).
8. Washing buffer: PBS containing 0.05% Tween-20.
9. RNA washing buffer: RNA detection buffer containing 0.1% Tween-20.
10. Fluorescence and brightfield microscope (Zeiss Axiophot).
11. Photographic film (64–100 ASA film for brightfield images and 400–640 ASA for fluorescence images).
12. Charged-couple device (CCD) camera and image processing software, e.g., Vysis Quips Genetic Workstation (Vysis) or Metasystems Image Pro System (Metasystems).

3. Methods

3.1. Synthesis of Biotinylated Tyramides for Use in CARD Signal Amplification

1. Dissolve 1 mg of the *N*-hydroxysuccinimide ester of biotin in 100 µL of DMF (= 1.8 µmol) (*see* **Notes 1** and **2**).
2. Dissolve 10 mg of tyramine-HCl in 1 mL of DMF (= 58 µmol) and add a 1.25× equivalent amount of TEA (= 10 µL) (*see* **Note 3**).
3. 28.4 µL Tyramine/TEA stock solution (= 1.76 µmol, *see* **step 2**) is added to 100 µL of biotin ester stock solution (= 1.8 µmol, *see* **step 1**), mixed, and left at room temperature in the dark for 2 h (**Notes 3** and **4**).
4. The synthesized biotinylated tyramides are then further diluted with 872 µL of ethanol to obtain a stock solution of 1 mg/mL, which can be stored for several years at 4°C and used without further purification (*see* **Note 5**).

3.2. Multiple-Target DNA In Situ Hybridization on Cell Preparations

1a. Metaphase chromosomes are freshly prepared from peripheral blood lymphocytes by standard methods (*see* e.g., **refs. 45–48**), fixed in methanol:acetic acid (3:1), and spread on acid/alcohol cleaned slides (*see* **Note 6**).
1b. Prepare preparations from routinely cultured normal diploid cells and tumor cell lines by: (i) Trypsinization (if necessary), harvesting, washing in PBS, and fixation in 70% ethanol (–20°C) (= ethanol suspension) (*see* **Note 7**). (ii) Growing cells on glass cover slips and fixation in cold methanol (–20°C) for 5 s and acetone (4°C) for 3 × 5 s. Air dry and store at –20°C. (Optionally other fixatives may be utilized, such as 2–4% paraformaldehyde.) (iii) Cytospinning floating cells onto glass slides (1000 rpm for 5 min), air drying for 1 h at room temperature, and fixation and storage as described for cells on coverslips.
2. Incubate cell preparations optionally with 100 µL of 100 µg/mL RNase A in 2× SSC under a coverslip for 1 h at 37°C to digest RNA in the cells; wash 3 × 5 min with 2× SSC.
3. Incubate with 50–100 µg/mL of pepsin (Sigma) in 0.01 *M* HCl for 10–20 min at 37°C to improve probe accessibility. A stock of pepsin can be made in Milli-Q (10 mg/mL) and stored at –20°C. This stock can then be diluted with 0.01 *M* HCl to the desired final concentration. Wash slides 1 × 2 min with 0.01 *M* HCl, and 2 × 5 min with PBS (*see* **Note 8**).
4. Post-fix slides for 20 min at 4°C (or 10 min at room temperature) in 1% formaldehyde in PBS with (chromosomes) or without (cells) 50 m*M* MgCl$_2$, wash 2 × 5 min with PBS, and dehydrate through a series of ethanol (70%, 96%, 100%).
5. Label the DNA probes of interest (repetitive, unique, or repetitive element-containing probes) with hapten (biotin, digoxigenin, dinitrophenyl, or fluorochrome)-labeled dUTP according to the nick-translation procedure *(44,49,50)* (*see* **Notes 9** and **10**).
6. Place 10 µL of hybridization mixture containing 1–2 ng/µL of labeled repetitive probe DNA under a 20 × 20 mm coverslip on the slide, and seal optionally with

rubber cement. In case of unique and repetitive element-containing DNA probes usually 2–10 ng/µL of probe DNA is used, in the latter case together with an excess of Cot-I or total human DNA (100- to 500-fold).

7. Denature probe and cellular DNA simultaneously at 70–75°C for 3–5 min on the bottom of a metal box or on a metal heating plate. Optionally, in the case of repetitive element-containing probes, denature probe DNA and repetitive components in hybridization mixture for 5 min at 75°C, preanneal for 0.5–2 h at 37°C, and apply to cell preparations that have been denatured separately in 70% formamide/2× SSC at 70–75°C, followed by dehydration in chilled ethanol and air drying.

8. Perform *in situ* hybridization for 2–16 h at 37°C in a humid chamber.

9. Perform stringent posthybridization washes at 42°C in wash buffer I for 2 × 5 min, and wash buffer II (2× SSC, 0.05% Tween-20) for 2 × 5 min. In the case of single-copy DNA probes washing with wash buffer II (0.1× SSC) should be at 60°C (*see* **Subheading 2.2.**, **item 17**).

3.3. mRNA In Situ *Hybridization on Formalin-Fixed, Paraffin-Embedded Tissue Sections*

1. Cut serial sections (4–5 µm) from tissue blocks that were formalin fixed and paraffin-embedded according to standard procedures and mount them on Superfrost Plus object slides.

2. Deparaffinize tissue sections in three changes of fresh xylene, and rehydrate in graded alcohols (100%, 96%, and 70% ethanol, 5 min each) and DEPC-treated distilled water (Milli-Q, 2 × 5 min) (*see* **Note 11**).

3. Permeabilize for 15 min in PBS/0.5% SDS at room temperature and for 5–30 min in 0.1% pepsin in 0.2 N HCl at 37°C with short Milli-Q washes in between (*see* **Note 12**).

4. Wash 2× shortly with PBS at 4°C, postfix the sections in 4% paraformaldehyde in PBS for 5 min at 4°C, and incubate in 0.1 M glycine in PBS for 2 × 3 min at room temperature to stop the reaction.

5. Incubate sections for 15 min in PBS/0.5% SDS at room temperature and rinse 2 × 3 min with Milli-Q.

6. Treat sections for 20 min at room temperature with 0.3% H_2O_2 in methanol to inactivate endogenous peroxidase, and rinse again 2 × 3 min in Milli-Q, and 2 × 3 min in 4× SSC at room temperature.

7. Label the oligonucleotide probes of interest with digoxigenin either at their 3'-ends using a DIG-oligonucleotide-labeling kit, or chemically by incorporating amino-dT nucleotides in the oligonucleotide during synthesis which can be coupled to digoxigenin afterwards (*see* **Note 12**).

8. Prehybridize the sections with 30 µL of EasyHyb under a plastic hydrophobic coverslip for 1 h at room temperature, and dip the slides shortly in 4× SSC to remove the coverslips.

9. Hybridize with 30 µL of EasyHyb containing 2–10 ng of DIG-labeled oligonucleotide probe for 2–16 h at room temperature.

10. Perform a quick wash of the slides in 4× SSC to remove ISH coverslips and excess EasyHyb, and follow with stringent washes in 2× SSC, 1× SSC, and 0.5× SSC for 2×15 min each at 37°C.

3.4. Cytochemical Probe Detection

1. Place 50–100 μL of DNA or RNA blocking buffer on the slide and leave for 5–15 min at room temperature to reduce background staining in the detection procedures. For DNA detection in cell preparations 5% nonfat milk powder in 4× SSC is a sufficient blocking buffer, whereas for RNA detection in tissue sections 4% nonfat dry milk powder in RNA detection buffer is needed for sufficient blocking (*see* **Note 12**). Instead of nonfat dry milk powder, 0.5% blocking reagent from Boehringer in RNA detection buffer has proven to be an appropriate blocking buffer, which we prefer for standardization (*see* **Note 13**).
2. Dilute detection conjugates for DNA detection as follows: Dilute avidin conjugates in DNA blocking buffer, and antibody conjugates in PBS, 0.05% Tween-20, 2% NGS. Dilute all conjugates for RNA detection in RNA blocking buffer.
3. For single-target probe detection, incubate the slides for 30–60 min at 37°C with the first detection layer (**Table 1**) and wash 2× 5–10 min in the appropriate DNA washing buffers (4× SSC, 0.05% Tween-20 for avidin conjugates; PBS, 0.05% Tween-20 for antibody conjugates) or RNA washing buffer. Repeat this step with the next detection layer(s) until all incubations are complete.
4. After the last detection layer, wash samples with the appropriate washing buffer for 2× 5–10 min, followed by PBS or RNA detection buffer for 5–10 min at room temperature, in the case of DNA or RNA detection, respectively. For fluorescence probe detection, dehydrate samples in an ethanol series (70%, 96%, and 100% ethanol) and embed as described in **step 6**. For enzyme cytochemical probe detection, visualize the nucleic acid targets by an appropriate enzyme reaction (see, e.g., the peroxidase-diaminobenzidine [PO–DAB] reaction with or without nickel–cobalt intensification (**step 5**), or other peroxidase or alkaline phosphatase reactions as described elsewhere *(51–53)* and embed as described in **step 6**.
5. Visualize the nucleic acid targets with the horseradish peroxidase–diaminobenzidine (PO–DAB) reaction: Mix 120 mg of DAB in 200 mL of PO–DAB buffer and add 80 μL of 30% H_2O_2 just before use. Incubate the slides for 5–10 min at room temperature and rinse with tap water. For nickel–cobalt intensification of ISH signals: Mix 100 mg DAB in 200 mL of 0.1 M phosphate buffer, pH 7.3, and slowly add 5 mL of 1% cobalt chloride and 4 mL of 1% nickel sulfamate during light vortex mixing. Incubate sections for 5 min at room temperature in this mixture, and then add 66 μL of 30% H_2O_2. Incubate sections for another 5–10 min at room temperature and rinse with tap water (*see* **Note 14**).
6. After all detection layers have been applied to the slides, dehydrate the slides in the case of fluorescence detection in an ethanol series (70%, 96%, and 100% ethanol) and embed them in Vectashield with or without the addition of 0.5 μg/mL of DAPI (blue fluorescent), PI (red fluorescent), or YOYO (green fluorescent) as DNA counterstain. Chromosomes, nuclei, or tissue sections with nucleic

acid targets visualized by enzyme precipitates are counterstained by hematoxylin (e.g., in the case of PO–DAB visualization) or other appropriate DNA counterstains (e.g., 0.1% nuclear fast red in the case of PO–DAB visualization with nickel–cobalt intensification, or (m)ethyl green). Dehydrate slides with PO–DAB precipitates in an ethanol series (70%, 96%, and 100% ethanol) and embed in Tissue Tek, or any other aqueous- or organic-based mounting medium. Sections stained with other enzyme precipitates should be counterstained with the appropriate DNA counterstain and embedded in an aqueous- or organic-based mounting medium dependent on the solubility of the enzyme precipitate used (see, e.g., **refs. *52,53***).

7. To detect multiple nucleic acid targets labeled with different haptens, choose a combination of cytochemical detection systems from **Table 1**. Multiple DNA targets can be visualized simultaneously by suitable combinations of detection systems using conjugated fluorochromes, or enzymes in combination with different enzyme precipitation reactions (**refs. *52,53*** and **Table 2**).

8. Examine slides under a fluorescence (fluorochromes) or brightfield (enzyme precipitates) microscope with appropriate filter sets. Microscope images can be recorded directly on photographic film, or with a CCD camera.

3.5. CARD Signal Amplification

1. To use CARD signal amplification, *in situ* hybridized probes should be detected with peroxidase conjugates (*see* **Tables 1** and **2**, and **Subheading 3.4.**), applied subsequently to the slides. Instead of performing an enzyme precipitation reaction, the peroxidase activity is now utilized to deposit labeled tyramide molecules in the vicinity of the enzyme molecules. These deposited tyramide molecules then can be detected directly, if labeled with fluorochromes, or indirectly with subsequent detection layers (*see* **Fig. 2** and **Table 1**), if labeled with haptens.

2a. For single-target ISH, wash slides with PBS or RNA detection buffer for 5–10 min at room temperature, after application of the last peroxidase detection layer (*see* **Subheading 3.4.**, **step 4**, and **Note 15**). Detect peroxidase activity on the slides by application of 50–100 μL of reaction mixture containing the desired labeled tyramides for 5–15 min at 37°C. For DNA detection on cell preparations, optimal reaction mixtures contain 1:250–1:1000 dilutions of 1 mg/mL of stock solutions of, respectively, coumarin-, fluorescein-, rhodamine-, Cy3-, biotin-, digoxigenin-, and trinitrophenyl-tyramides in CARD reaction buffer (*see* **Note 16**). For RNA detection on tissue sections, optimal reaction mixtures contain 1:50–100 dilutions of the stock solutions in CARD reaction buffer. Thereafter, wash slides 2 × 5 min with washing buffer or RNA washing buffer. For direct fluorescence detection, wash slides 1 × 5 min with PBS or RNA detection buffer and embed in Vectashield (*see* **step 3**). For indirect detection (fluorescence or brightfield), apply the next anti-hapten detection layer and follow **protocol 3.4**, from **step 3** on.

2b. To detect multiple DNA targets *in situ* with different labeled tyramides, hybridize with differently labeled probes and visualize them with a combination of cytochemical detection systems from **Table 1**, all utilizing peroxidase-conjugates. Visualize the probes consecutively with the appropriate detection system and CARD amplification reaction. As an example, **Table 2** shows the detection of three DNA targets with CARD signal amplification using three different fluorochrome-labeled tyramides.A similar protocol may be utilized for multiple-target mRNA ISH.

3. Embed slides with fluorochrome-labeled tyramides in Vectashield with or without the addition of 0.5 μg/mL of DAPI, PI, or YOYO as DNA counterstain. Chromosomes or nuclei with DNA targets visualized by enzyme precipitates are counterstained by and embedded as described in **Subheading 3.4., step 6**.
4. Examine slides under a fluorescence (fluorochromes) or brightfield (enzyme precipitates) microscope with appropriate filter sets. Microscope images can be recorded directly on photographic film, or with a CCD camera.

4. Notes

1. Synthesis of other tyramide conjugates (e.g., with digoxigenin, trinitrophenyl, coumarin, fluorescein, rhodamine, and Cy3) that have been used for *in situ* nucleic acid detection are described elsewhere *(42)*. All synthesized tyramides could be applied at the same concentrations as commercially available tyramides (2–10 μ*M*).
2. Because *N*-hydroxysuccinimide esters are prone to hydrolysis and are light sensitive, they need to be freshly dissolved only shortly before tyramide synthesis.
3. Hydrolysis of the biotin ester can be circumvented by performing the tyramide synthesis reaction in water-free medium (e.g., DMF) and by adding TEA in an 1.25× equivalent amount to deprotonate the amino group of tyramine-HCl (final pH should be between 7 and 8).
4. For an efficient biotin coupling, the biotin ester was added in a 1.1× molar excess compared to the tyramine/TEA.
5. Synthesized tyramides can also be further diluted in DMF or dimethylsulfoxide.
6. After dehydration and air drying, slides should be used directly, stored for up to 4 wk at room temperature, or stored in a dry box (sealed with a plastic bag) at –70°C for up to 2 years. Cell suspensions can be stored for up to 3 months at –20°C.
7. Ethanol suspensions can be stored in this way for several years.
8. After the 0.01 *M* HCl wash, preparations can be dehydrated optionally in 70% ethanol/0.01 *M* HCl, pH 2.0, 96% and 100% ethanol followed by air drying to preserve chromosomal and cellular morphology.
9. Apart from nick-translation other specialized labeling reactions are available to label DNA probes with modified nucleotides, including random primed labeling and PCR (*see* also **refs.** *53,54*).
10. For good *in situ* hybridization, labeled probe molecules should be in the range of 100–500 nucleotides long.
11. All solutions containing water were prepared with 0.1% DEPC-treated water.
12. A number of steps in our mRNA ISH protocol proved to be essential to obtain high signal-to-noise ratios, including treatments with SDS and pepsin for optimal tissue pretreatment, the use of blocking reagents during detection, and the use of chemically labeled oligonucleotide probes instead of 3'-end tailed oligonucleotides *(19a)*.

13. Blocking with 5% nonfat dry milk in 4× SSC proved to be less efficient than with 4% in RNA detection buffer to obtain high signal-to-noise ratios for mRNA ISH.

14. It is recommended to follow every enzyme reaction under the microscope to ensure discrete localization of the *in situ* signals.

15. In comparison with conventional ISH procedures, the same standard probe concentrations were generally used for signal amplification, whereas the detection conjugates, in particular applied in the first layer, were 2- to 10-fold further diluted to obtain high signal-to-noise ratios *(19a,20)*. Dependent on the size and amount of nucleic acid target to be detected, one to three detection layers were applied to the preparations before CARD. Generally, one or two layers were sufficient for repetitive DNA sequences (e.g., chromosome centromeres) and high abundant mRNA, whereas two or three layers were needed for single-copy DNA sequences and intermediate/low abundant mRNA. With respect to repetitive-element containing probes, such as cosmid probes containing Alu sequences, a high excess of competitor DNA (total genomic or Cot-I DNA) need to be hybridized simultaneously with the probe to prevent reduction of the signal-to-noise ratio. In the light of reducing experimental costs (e.g., if expensive probes are used), optimal amplification conditions can also be found by reducing the probe concentrations in combination with more detection layers and/or higher tyramide concentrations (*see* also **Note 16**).

16. Besides establishing probe concentration and detection system for efficient signal amplification (*see* **Note 15**), the tyramide and H_2O_2 concentration as well as the CARD reaction time are further factors influencing the final result. We recommend to optimize the tyramide concentration in a PBS/0.1 M imidazole, pH 7.6, buffer containing 0.001% H_2O_2 during a CARD reaction time of 5–15 min at room temperature or 37°C as starting point.

Acknowledgments

The authors thank P. Saremaslani for excellent technical assistance, H. Neff and N. Wey for photographic and computer-assisted reproductions, Prof. M. Werner for helpful discussions, and Profs. J. Roth, Ph. U. Heitz, and F. C. S. Ramaekers for continuous support.

References

1. Joos, S., Fink, T. M., Rätsch, A., and Lichter, P. (1994) Mapping and chromosome analysis: the potential of fluorescence *in situ* hybridization. *J. Biotechnol.* **35**, 135–153.

2. Dirks, R. W. (1996) RNA molecules lighting up under the microscope. *Histochem. Cell Biol.* **106**, 151–166.

3. Hopman, A. H. N., Voorter, C. E. M., Speel, E. J. M., and Ramaekers, F. C. S. (1997) *In situ* hybridization and comparative genomic hybridization, in *Cytogenetic Cancer Markers* (Wolman, S. R. and Sell, S., eds.), Humana Press, Totowa, NJ, pp. 45–69.

4. McNicol, A. M. and Farquharson, M. A. (1997) *In situ* hybridization and its diagnostic applications in pathology. *J. Pathol.* **182**, 250–261.

5. Höfler, H., Childers, H., Montminy, M. R., Lechan, R. M., Goodmann, R. H., and Wolfe, H. J. (1986) *In situ* hybridization methods for the detection of somatostatin mRNA in tissue sections using antisense RNA probes. *Histochem. J.* **18**, 597–604.

6. Komminoth, P. (1996) Detection of mRNA in tissue sections using digoxigenin-labeled RNA and oligonucleotide probes, in *Nonradioactive* In Situ *Hybridization Application Manual*. Boehringer, Mannheim, Germany, pp. 126–135.

6a. Femino, A. M., Fay, F. S., Fogarty, K., and Singer, R. H. (1998) Visualization of single RNA transcripts in situ. *Science* **280**, 585–590.

7. Lloyd, R. V. and Jin, L. (1995) *In situ* hybridization analysis of chromogranin A and B mRNAs in neuroendocrine tumors with digoxigenin-labeled oligonucleotide probe cocktails. *Diagn. Mol. Pathol.* **4**, 143–151.

8. Trembleau, A. and Bloom, F. E. (1995) Enhanced sensitivity for light and electron microscopic *in situ* hybridization with multiple simultaneous non-radioactive oligodeoxynucleotide probes. *J. Histochem. Cytochem.* **43**, 829–841.

9. Haase, A. T., Retzel, E. F., and Staskus, K. A. (1990) Amplification and detection of lentiviral DNA inside cells. *Proc. Natl. Acad. Sci. USA* **87**, 4971–4975.

10. Nuovo, G. (1992) PCR *In Situ* Hybridization. Raven Press, New York.

11. Komminoth, P. and Long, A. A. (1993) In-situ polymerase chain reaction. An overview of methods, applications and limitations of a new molecular technique. *Virchows Arch. B Cell Pathol.* **64**, 67–73.

12. Zehbe, I., Hacker, G. W., Sällström, J. F., Rylander, E., and Wilander, E. (1994) Self-sustained sequence replication-based amplification (3SR) for the in-situ detection of mRNA in cultured cells. *Cell Vision* **1**, 20–24.

13. Höfler, H., Pütz, B., Mueller, J., Neubert, W., Sutter, G., and Gais, P. (1995) *In situ* amplification of measles virus RNA by the self-sustained sequence replication reaction. *Lab. Invest.* **73**, 577–585.

14. Gosden, J. and Hanratty, D. (1993) PCR *in situ*: a rapid alternative to *in situ* hybridization for mapping short, low copy number sequences without isotopes. *BioTechniques* **15**, 78–80.

15. Terkelsen, C., Koch, J., Kolvraa, S., Hindkjaer, J., Pedersen, S., and Bolund, L. (1993) Repeated primed *in situ* labeling: formation and labeling of specific DNA sequences in chromosomes and nuclei. *Cytogenet. Cell Genet.* **63**, 235–237.

16. Troyer, D. L., Goad, D. W., Xie, H., Rohrer, G. A., Alexander, L. J., and Beattle, C. W. (1994) Use of direct *in situ* single-copy (DISC) PCR to physically map five porcine microsatellites. *Cytogenet. Cell Genet.* **67**, 199–204.

17. Kerstens, H. M., Poddighe, P. J., and Hanselaar, A. G. (1995) A novel *in situ* hybridization signal amplification method based on the deposition of biotinylated tyramine. *J. Histochem. Cytochem.* **43**, 347–352.

18. Raap, A. K., Van De Corput, M. P. C., Vervenne, R. A. W., Van Gijlswijk, R. P. M., Tanke, H. J., and Wiegant, J. (1995) Ultra-sensitive FISH using peroxidase-mediated deposition of biotin- or fluorochrome tyramides. *Hum. Mol. Genet.* **4**, 529–534.

19. Komminoth, P. and Werner, M. (1997) Target and signal amplification: approaches to increase the sensitivity of *in situ* hybridization. *Histochem. Cell Biol.* **108**, 325–333.

19a. Speel, E. J. M., Saremaslani, P., Roth, J., Hopman, A. H. N., and Komminoth, P. (1998) Improved mRNA in situ hybridization on formaldehyde-fixed and paraffin-embedded tissue using signal amplification with different haptenized tyramides. Histochem. Cell Biol. 110, 571–577.

19b. Speel, E. J. M., Hopman, A. H. N., and Komminoth, P. (1999) Amplification methods to increase the sensitivity of in situ hybridization: Play CARD(S). *J. Histochem. Cytochem.* **47,** 281–288.

20. Speel, E. J. M., Ramaekers, F. C. S., and Hopman, A. H. N. (1997) Sensitive multicolor fluorescence *in situ* hybridization using catalyzed reporter deposition (CARD) amplification. *J. Histochem. Cytochem.* **45,** 1439–1446.

21. Gosden, J. and Lawson, D. (1995) In-situ cyclic amplification of oligonucleotide primed synthesis (cycling PRINS), in *PCR Application Manual.* Boehringer, Mannheim, Germany, pp. 115–118.

22. Komminoth, P., Long, A. A., Ray, R., and Wolfe, H. J. (1992) *In situ* polymerase chain reaction detection of viral DNA, single copy genes and gene rearrangements in cell suspensions and cytospins. *Diagn. Mol. Pathol.* **1,** 85–97.

23. Komminoth, P., Adams, V., Long, A. A., Roth, J., Saremaslani, P., Flury, R., Schmid, M., and Heitz, P. U. (1994) Evaluation of methods for hepatitis C virus (HCV) detection in liver biopsies: comparison of histology, immunohistochemistry, in-situ hybridization, reverse transcriptase (RT) PCR and in-situ RT PCR. *Pathol. Res. Pract.* **190,** 1017–1025.

24. Staskus, K., Couch, L., Bitterman, P., Retzel, E., Zupancic, M., List, J., and Haase, A. (1991) *In situ* amplification of visna virus DNA in tissue sections reveals a reservoir of latently infected cells. *Microb. Pathogen.* **11,** 67–76.

25. Zehbe, I., Sällström, J. F., Hacker, G. W., Hauser-Kronberger, C., Rylander, E., and Wilander, E. (1994) Indirect and direct *in situ* PCR for the detection of human papillomavirus. An evaluation of two methods and a double staining technique. *Cell Vision* **1,** 163–167.

26. Komminoth, P. and Long, A. A. (1995) In-situ polymerase chain reaction — methodology, applications and non-specific pathways, in *PCR Application Manual.* Boehringer, Mannheim, Germany, pp. 97–106.

27. Höfler, H. (1993) *In situ* polymerase chain reaction: toy or tool? *Histochemistry* **99,** 103–104.

28. Long, A. A., Komminoth, P., Lee, E., and Wolfe, H. J. (1993) Comparison of indirect and direct in-situ polymerase chain reaction in cell preparations and tissue sections. Detection of viral DNA, gene rearrangements and chromosomal translocations. *Histochemistry* **99,** 151–162.

29. Sällström, J. F., Zehbe, I., Alemi, M., and Wilander, E. (1993) Pitfalls of *in situ* polymerase chain reaction (PCR) using direct incorporation of labeled nucleotides. *Anticancer Res.* **13,** 1153.

30. Long, A. A. and Komminoth, P. (1996) *In situ* polymerase chain reaction: an overview, in *Methods in Molecular Biology,* vol. XX, *Protocols for PRINS and In Situ PCR* (Gosden, J. R., ed.), Humana Press, Totowa, NJ, pp. 141–161.

30a. Lizardi, P. M., Huang, X., Zhu, Z., Bray-Ward, P., Thomas, D. C., and Ward, D. C. (1998) Mutation detection and single-molecule counting using insothermal rolling-circle amplification. *Nat. Genet.* **199,** 225–232.

31. Bobrow, M. N., Harris, T. D., Shaughnessy, K. J., and Litt, G. J. (1989) Catalyzed reporter deposition, a novel method of signal amplification. Application to immunoassays. *J. Immunol. Methods* **125,** 279–285.

32. Adams, J. C. (1992) Biotin amplification of biotin and horseradish peroxidase signals in histochemical stains. *J. Histochem. Cytochem.* **40**, 1457–1463.
33. Berghorn, K. A., Bonnett, J. H., and Hoffman, G. E. (1994) cFos immunoreactivity is enhanced with biotin amplification. *J. Histochem. Cytochem.* **42**, 1635–1642.
34. Merz, H., Malisius, R., Mannweiler, S., Zhou, R., Hartmann, W., Orscheschek, K., Moubayed, P., and Feller, A. C. (1995) ImmunoMax. A maximized immunohistochemical method for the retrieval and enhancement of hidden antigens. *Lab. Invest.* **73**, 149–156.
35. Werner, M., Von Wasielewski, R., and Komminoth, P. (1996) Antigen retrieval, signal amplification and intensification in immunohistochemistry. *Histochem. Cell Biol.* **105**, 253–260.
36. Von Wasielewski, R., Mengel, M., Gignac, S., Wilkens, L., Werner, M., and Georgii, A. (1997) Tyramine amplification technique in routine immunohistochemistry. *J. Histochem. Cytochem.* **45**, 1455–1459.
37. Poddighe, P. J., Bulten, J., Kerstens, H. M. J., Robben, J. C. M., Melchers, W. J. G., and Hanselaar, A. G. J. M. (1996) Human papilloma virus detection by *in situ* hybridisation signal amplification based on biotinylated tyramine deposition. *Clin. Mol. Pathol.* **49**, M340–344.
38. Macechko, P. T., Krueger, L., Hirsch, B., and Erlandsen, S. L. (1997) Comparison of immunologic amplification vs enzymatic deposition of fluorochrome-conjugated tyramide as detection systems for FISH. *J. Histochem. Cytochem.* **45**, 359–363.
39. Schmidt, B. F., Chao, J., Zhu, Z., Debiasio, R. L., and Fisher, G. (1997) Signal amplification in the detection of single-copy DNA and RNA by enzyme-catalyzed deposition (CARD) of the novel fluorescent reporter substrate Cy3. 29-tyramide. *J. Histochem. Cytochem.* **45**, 365–373.
40. Van Gijlswijk, R. P. M., Zijlmans, H. J. M. A. A., Wiegant, J., Bobrow, M. N., Erickson, T. J., Adler, K. E., Tanke, H. J., and Raap, A. K. (1997) Fluorochrome-labeled tyramides: use in immunocytochemistry and fluorescence *in situ* hybridization. *J. Histochem. Cytochem.* **45**, 375–382.
41. Zehbe, I., Hacker, G. W., Su, H., Hauser-Kronberger, C., Hainfeld, J. F., and Tubbs, R. (1997) Sensitive *in situ* hybridization with catalyzed reporter deposition, streptavidin-nanogold, and silver acetate autometallography. *Am. J. Pathol.* **150**, 1553–1561.
42. Hopman, A. H. N., Ramaekers, F. C. S., and Speel, E. J. M. (1998) Rapid synthesis of biotin-, digoxigenin-, trinitrophenyl-, and fluorochrome-labeled tyramides and their application for *in situ* hybridization using CARD-amplification. *J. Histochem. Cytochem.* **46**, 771–777.
43. Van Gijlswijk, R. P. M., Wiegant, J., Vervenne, R., Lasan, R., Tanke, H. J., and Raap, A. K. (1996) Horseradish peroxidase-labeled oligonucleotides and fluorescent tyramides for rapid detection of chromosome-specific repeat sequences. *Cytogenet. Cell Genet.* **75**, 258–262.
44. Lichter, P. and Ried, T. (1994) Molecular analysis of chromosome aberrations: *in situ* hybridization, in *Methods in Molecular Biology*, vol. 29. *Chromosome Analysis Protocols* (Gosden, J. R., ed.), Humana Press, Totowa, NJ, pp. 449–478.
45. Dutrillaux, B. and Viegas-Pequignot, E. (1985) High resolution R- and G-banding in the same preparation. *Hum. Genet.* **57**, 93–95.

46. Speel, E. J. M., Schutte, B., Wiegant, J., Ramaekers, F. C. S., and Hopman, A. H. N. (1992) A novel fluorescence detection method for *in situ* hybridization, based on the alkaline phosphatase-fast red reaction. *J. Histochem. Cytochem.* **40**, 1299–1308.

47. Spowart, G. (1994) Mitotic metaphase chromosome preparation from peripheral blood for high resolution, in *Methods in Molecular Biology*, vol. 29. *Chromosome Analysis Protocols* (Gosden, J. R., ed.), Humana Press, Totowa, NJ, pp. 1–10.

48. Ross, F. (1994) Chromosome preparation from hematological malignancies, in *Methods in Molecular Biology*, vol. 29. *Chromosome Analysis Protocols* (Gosden, J. R., ed.), Humana Press, Totowa, NJ, pp. 11–25.

49. Langer, P. R., Waldrop, A. A., and Ward, D. C. (1981) Enzymatic synthesis of biotin-labeled polynucleotides: novel nucleic acid affinity probes. *Proc. Natl. Acad. Sci. USA* **78**, 6633–6637.

50. Boehringer (1996) *Nonradioactive* In Situ *Hybridization Application Manual.* 2nd ed. Boehringer, Mannheim, Germany.

51. Speel, E. J. M., Kamps, M., Bonnet, J., Ramaekers, F. C. S., and Hopman, A. H. N. (1993) Multicolour preparations for *in situ* hybridization using precipitating enzyme cytochemistry in combination with reflection contrast microscopy. *Histochemistry* **100**, 357–366.

52. Speel, E. J. M., Jansen, M. P. H. M., Ramaekers, F. C. S., and Hopman, A. H. N. (1994) A novel triple-color detection procedure for brightfield microscopy, combining *in situ* hybridization with immunocytochemistry. *J. Histochem. Cytochem.* **42**, 1299–1307.

53. Speel, E. J. M., Ramaekers, F. C. S., and Hopman, A. H. N. (1995) Detection systems for *in situ* hybridization, and the combination with immunocytochemistry. Who is still afraid of red, green and blue? *Histochem. J.* **27**, 833–858.

54. Hopman, A. H. N., Speel, E. J. M., Voorter, C. E. M., and Ramaekers, F. C. S. (1995) Probe labelling methods, in *Non-Isotopic Methods in Molecular Biology: A Practical Approach* (Levy, E. R. and Herrington, C. S., eds.), IRL Press, Oxford University Press, Oxford, UK, pp. 1–24.

15

In Situ Localization of PCR-Amplified DNA and cDNA

Gerard J. Nuovo

1. Introduction

In situ hybridization is the only DNA- or RNA-based molecular biology based test that allows for the direct correlation of the results with the histologic and cytologic features of the sample. The DNA/RNA extraction that precedes filter hybridization (slot blot, hybrid capture, or Southern blot) and polymerase chain reaction (PCR) precludes this analysis. The relative sensitivities of the three different assays are presented in **Table 1**. It is evident that *in situ* hybridization is a relatively insensitive test. A reflection of this relative insensitivity is seen in occult or latent infection by a virus where the copy number is low. In such situations, the virus is rarely detected by *in situ* hybridization even though it was detected by either PCR or filter hybridization *(1–6)*. This is not to say that the technique of *in situ* hybridization has remained static. The detection threshold of this assay has improved substantially in the last 10 years. Another point worth emphasizing about *in situ* hybridization is that one does not need to use radiolabeled probes (usually ^{35}S or ^{3}H) to maximize its sensitivity. Although true 10 years ago, recent and dramatic advances in nonisotopic labeling and, more importantly, detection systems has greatly enhanced the sensitivity when using such common labels as biotin and digoxigenin *(7–12)*. Still, only the most aggressive salesman would claim (and incorrectly at that) that any given *in situ* system can routinely detect one DNA or RNA copy per cell. In my experience, this statement applies to the newer generation of posthybridization "signal amplification systems" (such as the cascade amplification system) which are not able to routinely detect one copy per cell. **Figure 1** shows that the one copy of HPV 16 DNA present in SiHa cells cannot be detected by *in situ* hybridization without prior PCR-amplification. The detec-

From: *Methods in Molecular Biology*, Vol. 123: In Situ *Hybridization Protocols*
Edited by: I. A. Darby © Humana Press Inc., Totowa, NJ

Table 1
Sensitivities of the Molecular Based Assays

Assay	Detection threshold
In situ hybridization	10–20 copies per cell
Southern blot hybridization	1 copy per 50 cells
Slot blot hybridization	1 copy per 100 cells
PCR (standard)	1 copy per 1000 cells
PCR (with hot start modification)	1 copy per 1,000,000 cells

tion threshold of *in situ* hybridization is 10–20 copies per cell; note the signal after *in situ* hybridization analysis of the 20 copies of HPV 18 present in HeLa cells (**Fig. 1**).

Despite the widespread use of both PCR and *in situ* hybridization in the last several years it has proved difficult to combine the two. If this combination could be accomplished, DNA would be amplified in intact cells and then visualized with *in situ* hybridization. The ability to accomplish PCR *in situ* hybridization in paraffin-embedded tissue has been difficult for several reasons. One must expose the target DNA without destroying tissue morphology. Optimal concentrations of the essential reagents such as the primers, Mg^{2+}, and the DNA polymerase must be determined. Further, and perhaps most importantly, if the reaction is to be carried out directly on glass slides, loss of tissue adherence and tissue drying would have to be circumvented. These problems have been overcome to the point that one may reliably amplify both DNA and, for RNA, cDNA in paraffin-embedded, fixed tissues. The purpose of this chapter is to provide readers with the protocols this laboratory has developed for the *in situ* localization of PCR-amplified DNA and cDNA. This will include a discussion of the key components of successful *in situ* hybridization as, of course, an in-depth knowledge of the mechanics of *in situ* hybridization is essential for performing PCR *in situ* hybridization.

2. Materials

2.1. Slide Preparation

1. Use silane coated slides which may be purchased from ONCOR Corporation (Gaithersburg, MD) (*see* **Note 1**).

2.2. Fixative

1. Use 10% buffered formalin and fix the cells or tissue for 15–24 h (*see* **Note 2**).

2.3. Protease Digestion

1. Use 2 mg/mL of pepsin or trypsin at room temperature for 5–30 min (*see* **Note 3**).

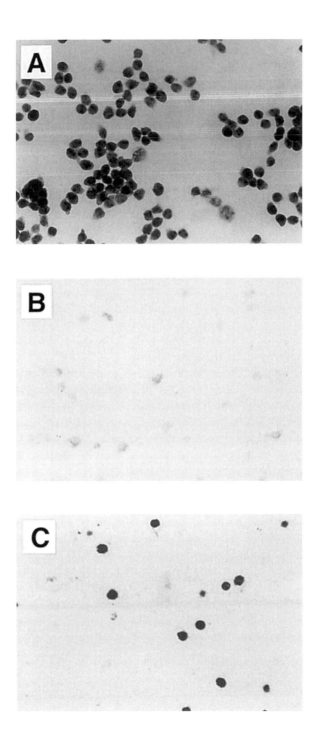

Fig. 1. Detection of HPV DNA by *in situ* hybridization. The 20 copies of HPV 18 DNA are detectable in HeLa cells using standard *in situ* hybridization (**A**) whereas the one copy of HPV 16 DNA in SiHa cells is not (**B**) unless amplified first by PCR (**C**).

2.4. Probe Cocktail

Add together the following ingredients:

1. 50 µL Deionized formamide (use 10 µL for oligoprobes and add 40 µL water).
2. 30 µL 25% Dextran sulfate.
3. 10 µL 20× Saline sodium citrate (SSC).
4. 10 µL Labeled probe (stock solution of 5 to 10 µg/mL) (*see* also **Note 4**).

2.5. Posthybridization Wash

1. The wash solution contains 2.5% bovine serum albumin and 0.2× SSC (or 30 mM sodium chloride) and is heated to 45°C (oligoprobes) or 60°C (for full-length probes). The slides should be kept in the wash solution for 10 min (*see* **Note 5**).

2.6. Detection Systems

1. (For biotin system) Use a streptavidin–alkaline phosphatase (AP) conjugate (ONCOR or Enzo Diagnostics, Farmingdale, NY, USA; use solution as is); for digoxigenin use the antidigoxigenin–AP conjugate (1:200 dilution, Enzo Diagnostics, prepare immediately prior to use).
2. After the AP conjugate is used, the slides should be placed in a solution of 0.1 M Tris-HCl, pH 9.0–9.5, and 0.1 M NaCl (detection reagent solution).
3. The chromagen 4-nitroblue tetrazolium chloride/5-bromo-4-chloro-3-indolyl-phosphate (NBT/BCIP) is added to the detection reagent.
4. Counterstain is nuclear fast red (ONCOR).

2.7. The PCR Step of PCR In Situ Hybridization

1. Use silane-coated glass slides.
2. Fix the cells or tissue in 10% buffered formalin for 24–48 h.
3. Deparaffinize tissue in xylene (5 min) and 100% ethanol (5 min).
4. Digest in 2 mg/mL of pepsin or trypsin at room temperature for 5–30 min.
5. The amplifying solution needs to contain 4.5 mM MgCl$_2$, 1 µM each of primers 1 and 2, 2.5 U/25 µL of *Taq* polymerase (assuming bovine serum albumin is added), 200 µM of the dNTPs, and the GeneAmp kit buffer (Perkin-Elmer) (*see* **Note 7**).

3. Methods

3.1. Running the Reaction

1. Place several 4-µm paraffin-embedded sections or two cytospins on a silane-coated glass slide.
2. Wash paraffin-embedded sections in xylene for 5 min; then in 100% ethanol for 5 min, then air dry (for tissue sections only).
4. Digest in pepsin; inactivate protease by washing in 0.1 M Tris, pH 7.5, and 0.1 M NaCl for 3 min.
5. Wash slides in 100% ethanol for 3 min, then air dry. For PCR step (if no PCR step, proceed to **step 13**).

6. Add 2.5 µL of GeneAmp buffer, 4.5 µL of $MgCl_2$ (25 mM stock), 4 µL of dNTPs (2.5 mM stock), 1 µL of primer 1 and primer 2 (each 20 µM stock), and 11.2 µL of sterile water. Remove 4 µL (keep on ice for hot start).
7. Place solution on two separate sections; add plastic coverslip, anchor with nail polish. (Alternatives that do not require mineral oil include the ampliclip/cover system of Perkin Elmer or the SelfSeal reagent from MJ Research.)
8. Time delay file — 82°C for 7 min.
9. At the onset of this file add 0.8 µL of *Taq* polymerase to the tube on ice.
10. At 75°C, lift one edge of the coverslip gently and add 2.4 µL to each section; overlay with preheated mineral oil. (If using the ampliclip/amplicover system or SelfSeal reagent, the *Taq* polymerase should be added to the slide at 55°C.)
11. Switch to time delay file — 94°C for 3 min.
12. Link this time delay file to a cycling file of 55°C — 2 min and 94°C — 1 min for 35 cycles; at conclusion link to soak file of 4°C; if using aluminum boat method, remove mineral oil with xylene and ethanol washes and air dry.
13. Add 5–10 µL of the probe cocktail to a given tissue section.
14. Overlay with plastic coverslip cut slightly larger than tissue section.
15. Place slide on hot plate — 95–100°C — for 5 min.
16. Remove bubbles over tissue gently with a toothpick.
17. Place slides in humidity chamber at 37°C for 2 h.
18. Remove coverslips — hold down one end with fingernail and lift off coverslip with toothpick.
19. Place in wash solution for 10 min at 60°C (assuming full-length probe).
20. Wipe off excess wash solution and put slides in a humidity chamber; do not let slides dry out.
21. Add appropriate alkaline phosphatase conjugate and place slides in humidity chamber.
22. Incubate for 30 min at 37°C.
23. Wash slides at room temperature for 3 min in a solution of 0.1 M Tris-HCl (pH 9.0–9.5) and 0.1 M NaCl (detection reagent solution).
24. Place slides in detection reagent solution to which NBT/BCIP has been added (ONCOR or Enzo kit).
25. Incubate slides for 30 min to 2 h, checking results periodically under microscope.
26. Counterstain with nuclear fast red and coverslip; view under microscope.

3.2. Comments and Troubleshooting

3.2.1. Background or No Signal

The most common problems encountered with *in situ* hybridization are background and the absence of a signal. Background may be defined as the presence of a hybridization signal with a specific probe in areas of the tissue where the signal should not be present (i.e., normal endocervical cells or in basal cells with HPV). Of course, in some instances one may not be sure where the *in situ* signal should localize. A more strict definition of

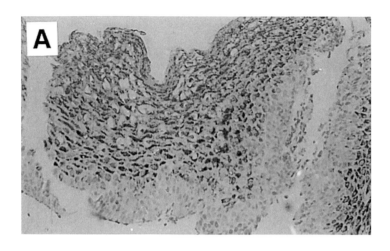

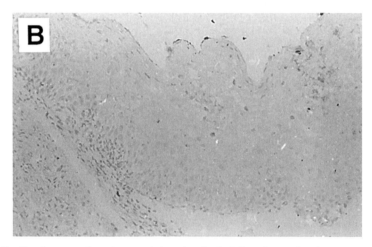

Fig. 2. Background noise and *in situ* hybridization: the importance of the posthybridization wash. A large amount of nonspecific staining is seen with the HPV 6/11 probe (**A**) if the posthybridization wash was done for 15 min at 20°C in a solution with 1.5 *M* NaCl. Note how the background is primarily cytoplasmic; HPV DNA typically localizes to the nucleus. The background is eradicated if the posthybridization wash is 15 min at 50°C in a solution that contains 150 m*M* NaCl (**B**); this lesion actually contained HPV 33.

background would be a hybridization signal in tissues or cells known not to contain the target of interest (this may be determined by solution phase PCR or, for some viral infections, the lack of the diagnostic histologic changes). Background is the result of nonspecific binding of the probe to nontarget molecules. Two simple and logical ways to deal with background are to decrease the concentration of the probe and/or to increase the stringency of the posthybridization wash. The result of doing the latter is depicted in **Fig. 2**. If background is a problem, first try decreasing the concentration of the probe 10-fold. If background persists, decrease the concentration of the sodium in the posthybridization wash 10-fold or increase the temperature of the posthybridization wash by 15°C.

An obvious problem with *in situ* hybridization is the absence of a hybridization signal. I recommend following a flow-chart type of problem-solving tree, which is presented in **Fig. 3**.

3.2.2. Probe Size

The probe size for standard *in situ* hybridization is from 90 to 250 base pairs in size; optimal probe size is from 90 to 140 bp. However, one may use much smaller (20–40 bp) probes called oligoprobes. There are two main reasons to use oligoprobes: they are more readily available than the larger probes which require a cloned sequence of DNA; one only needs to know the sequence of the target of interest, readily available in the literature, to generate an oligoprobe. Second, one is obliged to use an oligoprobe internal to the sequence being amplified in solution phase PCR to assure oneself that the signal is indeed the PCR product (as opposed to a signal due to primer oligomerization). Because oligoprobes are often at least five times shorter than a "standard" probe, there is a substantial reduction in the number of basepair matches and thus the strength of the hybridized complex compared to the larger probes. The practical consequence is that the wash conditions must be carefully chosen so as to minimize background but not to lose the signal. In practical terms, I have seen the signal lost for a 20-mer oligoprobe with a posthybridization wash in 30 m*M* salt at 45°C; under these conditions the signal for a larger homologous probe would remain intact. Hence, I use different probe cocktail and posthybridization wash conditions for oligoprobes.

It should be stressed that primer oligomerization does NOT appear to occur inside nuclei during *in situ* PCR *(33)*. This may reflect the relatively high protein concentration inside the nucleus; single-stranded binding proteins can inhibit primer oligomerization during solution phase PCR *(33)*. This observation has important practical implications for PCR *in situ* hybridization. Specifically, one may use the full-length probe with PCR *in situ* hybridization, even though it includes the region of the primers, and still detect target specific signal assuming that other potential causes of background have been eliminated. Full length probes permit much more stringent washes owing to their wide range of signal to background. Thus, it is much easier to eliminate background and still preserve signal with a full length probe relative to an oligoprobe.

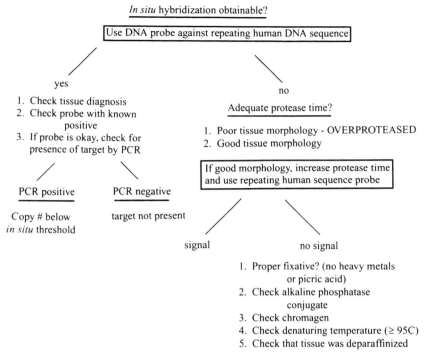

Fig. 3. Flow chart for a negative result with the *in situ* hybridization. The figure details a step-by-step approach to follow if a hybridization signal is not evident with *in situ* analysis.

3.2.3. Labels, Detection and General Considerations

Direct incorporation of reporter molecules is possible for DNA targets with PCR *in situ* but only under strictly defined conditions. The inclusion of a labeled nucleotide in the amplifying solution is the major modification in this technique compared to PCR *in situ* hybridization, in which the PCR product is not directly labeled but rather detected with a labeled probe. Most of our work with *in situ* PCR has focused on the labeled nucleotide digoxigenin dUTP (DIG-dUTP; Boehringer Mannheim, Indianapolis, IN, USA) *(26–30)*. There are only two modifications to the protocol listed for PCR *in situ* hybridization. First, 10 µM DIG-dUTP is added to the amplifying mixture. Second, after completing the amplifying reaction and removing of the coverslip and mineral oil, we wash the slides for 10 min at 65°C in a solution of 2% bovine serum albumin and 0.1× SSC to remove unincorporated DIG-dUTP and, more importantly, the labeled primer oligomers that may have formed in the overlying amplifying solution (as compared to inside the cell). After this wash, the digoxigenin that has been incorporated into the amplified DNA may be detected according to the manufacturer's protocol (Boehringer Mannheim; I use a 1:200 dilution of the antibody). It is important to note that one must use the hot start maneuver to perform target-specific incorporation of the labeled nucleotide during *in situ* PCR for DNA targets. Further, one must use cells or tissue that has *not* been exposed to dry heat prior to *in situ* PCR, which induces a primer independent signal due to DNA nicking *(33)*. This negates the use of *in situ* PCR for DNA targets with paraffin-embedded tissue, which is heated at 65°C for 4 h prior to embedding. For paraffin-embedded tissues, one must use PCR *in situ* hybridization with a probe step after the PCR. The DNase digestion step that is done after protease digestion for reverse transcriptase (RT) *in situ* PCR eliminates all the DNA synthesis pathways (mispriming, target specific and primer independent). This allows for the target-specific incorporation of the labeled nucleotide in the amplified cDNA *(33)*. The sensitivity of *in situ* PCR is dependent on the exact fixation, protease digestion, and reagent conditions listed previously (**Fig. 4**). A summary of the PCR *in situ* procedure is listed in **Fig. 5**.

With regard to specificity, in solution phase PCR, two pathways compete with target-specific DNA synthesis. These are mispriming and primer oligomerization *(30)*. If the hot start modification is not employed, the mispriming and primer oligomerization pathways can easily overwhelm target specific DNA synthesis such that a large amount of DNA is synthesized but it is *mostly nonspecific (30,31)*. This is not surprising when one considers that there is far more nontarget and primer DNA in a reaction mixture relative to target DNA. It has been shown that under nonhot start conditions the detection threshold for the target of interest with solution phase PCR (assuming 1 μg of nontarget DNA) may be greater than *several thousand copies*, not the 1–100 copies most articles quote *(30,31)*. However, nonspecific incorporation is greatly curtailed by the hot start modification. The end result is that one can reliably detect 10 copies per tissue with the hot start modification of PCR *(30,31)*. Similarly, the hot start modification of PCR *in situ* hybridization (or with direct labeling using *in situ* PCR), by inhibiting mispriming, allows for the reproducible detection of one target copy per cell using a single primer pair *(30,33)*.

Two different approaches may be used to demonstrate the specificity of hot start *in situ* PCR. First, different cell populations can be mixed and a primer pair is used that is able to amplify a target in only one of the two populations. Second, one may use "irrelevant primers," primer pairs that could not possibly find targets in the cells being studied. The latter is simply accomplished by making primers based on the cDNA of an RNA virus such as measles, which cannot make its cDNA in vivo.

SiHa cells are from a human cervical cancer cell line that, as expected, contain HPV DNA although only one viral copy per cell. These were mixed with

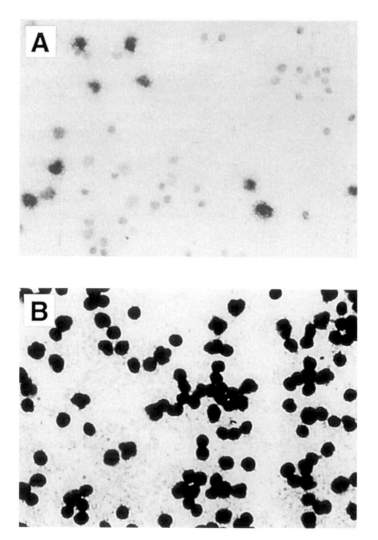

Fig. 4. Effect of fixation and protease digestion on detection of amplified bcl-2 DNA in peripheral blood leukocytes (PBMs). A signal was evident in about 25% of the cells after PCR *in situ* with direct incorporation of digoxigenin dUTP for cells fixed in acetone for 5 min with no digestion (**A**). If the PBMs were fixed for 15 h in buffered formalin and pretreated with trypsin then an intense signal is evident in each cell (**B**).

peripheral blood leukocytes and mononuclear cells (PBMs) from a noninfected individual, although HPV cannot infect PBMs if an HPV infection is present at some other site. Standard conditions refer to experiments in which all reagents were added *before* raising the temperature of the heating block. Whereas SiHa cells might give specific or nonspecific product, amplified DNA from PBMs

1. Paraffin-embedded tissue or cytospin on silane-coated glass slide; digest with protease solution

2. Add amplifying solution, plastic coverslip and anchor with nail polish

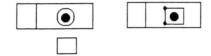

3. Place in aluminum boat, then on Thermal Cycler

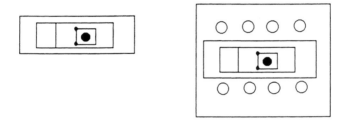

4. Add Taq DNA at 60C by lifting coverslip, overlay with mineral oil, denature, cycle

5. Standard in situ hybridization

Fig. 5. A graphic representation for doing PCR in situ hybridization.

must be nonspecific when the HPV specific primers are employed. Under hot start conditions with a single HPV 16 primer pair only some of the cells incorporated digoxigenin. The negative cells proved to be the leukocytes as they reacted with an antibody against leukocyte common antigen in a double labeling technique. Under standard conditions all of the cells, including the leukocytes, incorporated digoxigenin (*see* **Fig. 2** in **ref. *30***). These results provided reassurance that the hot start modification greatly inhibited nonspecific pathways (*30*).

In an analogous experiment, measles-infected HeLa cells were mixed with the PBMs of an noninfected individual. The measles-infected HeLa cells are

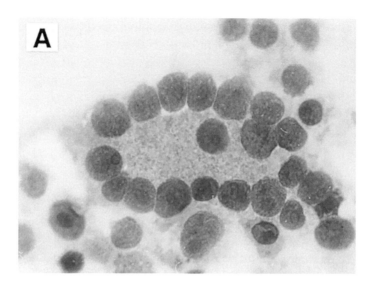

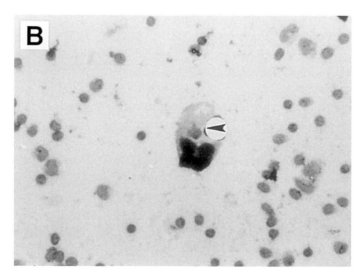

Fig. 6. Cytological and molecular analysis of measles-infected HeLa cells. The multinucleation and nuclear inclusions characteristic of measles infected cells are evident on routine HandE stain **(A)**. Only the measles-infected HeLa cells were positive using direct incorporation when these cells were mixed with peripheral blood leukocytes; note the cytoplasmic signal (*arrow*, **B**) demonstrating the specificity of direct incorporation using the hot start modification in cytospin preparations.

multinucleated and thus easily differentiated cytologically from the single nucleated leukocytes. These mixing experiments proved that only the measles-infected HeLa cells had a detectable signal with direct incorporation and measles-specific primers if the hot start modification was employed with the RT *in situ* PCR technique (**Fig. 6**) *(32,33)*. If the hot start modification was omitted then both the HeLa cells and PBMs had detectable signal.

In experiments with irrelevant primers, we analyzed two different cellular samples — PBMs and a cervical swab. HIV-1 DNA was routinely found in CD4 positive PBMs from patients with AIDS *(26)*. Similarly, HPV DNA was regularly detected in dysplastic cells from patients with squamous intraepithelial lesions *(29,33)*.

Thus far we have analyzed approx 30,000 PBMs from people who are HIV-1 negative by PCR and who lack any of the risk factors for AIDS. Further, we have analyzed about 5000 PBMs from AIDS patients by *in situ* PCR using primers for measles cDNA. In either instance, any positive cell would presumably represent nonspecific uptake of dUTP. Each of the 35,000 PBMs was negative. Of the 3000 cervical epithelial cells analyzed by *in situ* PCR with primers for measles cDNA, 4 (0.1%) have been positive. We have obtained similar results using PBMs fixed for 15 h after protease digestion and *in situ* PCR using the measles-specific primers. Recall that the rate of nonspecific incorporation under these conditions for standard *in situ* PCR approaches 100%. Clearly, nonspecific uptake is a rare event with the hot start modification although very common under standard conditions *(23)*.

4. Brief mention will be made of our RT PCR *in situ* hybridization protocol and results. More detailed information is available *(32,33)*. The model system I have described is with the RNA virus hepatitis C. All of these reactions were performed directly on glass slides. The key point to emphasize about our protocol is the overnight pretreatment in RNase-free DNase. The end point of determining the proper DNase digestion time is: *the inability to synthesize genomic based DNA with either target-specific or irrelevant primers or via the primer-independent pathway.*

The inability to synthesize DNA in the liver tissue after DNase digestion permits target-specific incorporation of the labeled nucleotide into the cDNA made during the RT step of the RT *in situ* PCR reaction (**Fig. 7**). Also note that one may exploit the fact that there is invariably DNA synthesis in the non-DNase-treated paraffin-embedded tissue section by using it as a positive PCR control and the DNase non-RT section as the negative control (**Fig. 7**).

In the RT step, the solution is covered with a plastic coverslip, anchored with nail polish, and overlaid with mineral oil as described and illustrated previously for PCR *in situ* hybridization. We use the same reagent concentrations for RT as listed in the RT PCR kit (Perkin Elmer, Norwalk, CT, USA). After RT, we usually do *in situ* PCR *(32,33)*. Alternatively, one may do the RT and PCR steps under the same amplifying solution using

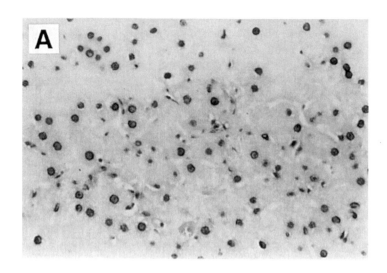

Fig. 7. Direct incorporation *in situ* PCR; effect of DNase pretreatment on tissue sections. This liver biopsy did not contain hepatitis C RNA, yet a signal is evident with hot start *in situ* PCR using hepatitis C-specific primers if the DNase step is omitted **(A)**. The signal is eradicated if DNase pretreatment is done **(B)**.

the enzyme rTth *(33)*; this one-step protocol simplifies the procedure and is the one I recommend.

Amplified hepatitis C cDNA was detectable in several of the liver biopsies, mostly from patients with serological evidence of the infection. The specificity was demonstrated by omitting the direct incorporation of DIG-d-UTP in the

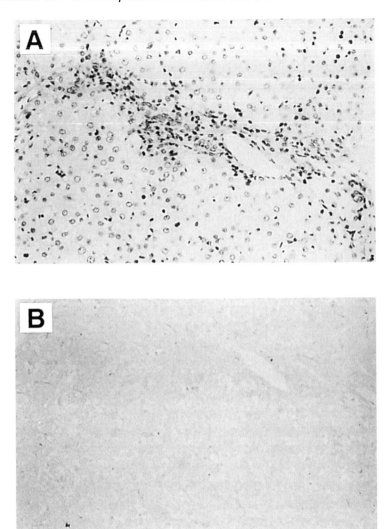

Fig. 8. Detection of hepatitis C in a liver biopsy. Hepatitis C was detected in this liver biopsy after DNase pretreatment using RT *in situ* PCR (**A**); the signal was lost if the RT step was omitted (**B**); omission of the RT is an important negative control in RT *in situ* PCR.

PCR step and using a labeled internal oligoprobe and demonstrating in serial sections that the same cells were positive (**Fig. 8**). Note the inclusion of the essential negative and positive controls. A common cause of a false-positive result with RT *in situ* PCR is the persistence of genomic based DNA synthesis due to inadequate protease digestion which does not permit sufficient access of the native DNA to the DNase. This will be recognized by a nuclear based signal in the negative control and can be rectified by increasing the time of protease digestion. Another common problem with liver biopsies is protease overdigestion. This can be recognized as a weak to no signal with the test and the positive control and the concomitant loss of the tissue structure. The latter may be recognized as poor staining of the nucleus and cytoplasm and a prominence of the basement membrane of the cells.

4. Notes

1. Perhaps the major technical advancement in the field dealt simply with the preparation of the glass slide. About 10 years ago, it was routine to pretreat the slides with materials such as poly-L-lysine, glue, or other adhesives to improve adherence *(13–16)*. Although such pretreatments certainly worked better than untreated slides where most sections would fall off, in my experience the sections would, at best, remain on about 75% of the time. It should be noted adherence is much less a problem with cytologic preparations, such as Pap smears. One can do *in situ* hybridization or *in situ* PCR using destained Pap smears on routine slides with excellent adherence. Loss of adherence using tissue sections was circumvented by the use of organosilane, a chemical used in industry to treat glass. I have stored silane-coated slides at room temperature in closed boxes and used these slides successfully as much as *8 years* after pretreatment. If tissue sections fall off at a rate greater than 5% and one is using commercially prepared silane slides, the problem is probably air bubbles under the tissue which usually reflects inexperience of the technician placing the sections on the slides.

2. I do not recommend using frozen tissues for *in situ* hybridization. The morphology is at best poor, which defeats the purpose of the test. Further, some claim that protease digestion is still required with *in situ* hybridization with the use of frozen tissues so even this step cannot necessarily be omitted *(10,17)*. There are many different fixatives that various laboratories use to process tissues. In my own experience, I have seen buffered formalin, unbuffered formalin, Bouin's solution, and B5 (each of which contains picric acid), Zenker's solution (which contains mercury), and 95% ethanol used for a variety of purposes. When using tissue sections, buffered formalin, pH 7.0, is the best fixative for *in situ* hybridization and PCR *in situ* hybridization, although other crosslinking fixatives such as glutaraldehyde and paraformaldehyde are acceptable. Fixatives that contain either picric acid or heavy metals may allow for successful *in situ* hybridization but this is dependent on the length of time of fixation (15, 18-22). Two hours' fixation in Bouin's solution has minimal effect on the intensity of the hybridization signal, which may be completely eradicated after overnight (15 hour) fixation; intermediate results are seen after 8 h of Bouin's fixation *(15,20)*. However,

fixation in a solution that contains a heavy metal or picric acid will not permit either PCR or PCR *in situ* hybridization *(19,23)*.

If only frozen tissue is available, it is recommended that the tissue be slowly thawed at room temperature and then fixed in 10% buffered formalin overnight. Although some freezing artifact will be evident, both *in situ* hybridization and PCR *in situ* hybridization may be done with good results (Nuovo, G. J., *unpublished observations*).

Cellular preparations, such as Pap smears and fine needle aspirates, are usually fixed in an alcohol-based solution immediately after being obtained. One can use such specimens for successful *in situ* hybridization or PCR *in situ* hybridization. The key is to use a relatively weak protease digestion; I recommend proteinase K at 10 μg/mL of water for 10 min.

3. Fixatives, especially those whose primary mode of action is crosslinking such as formalin, hinder penetration of the probe to the target nucleic acid molecule *(15,22–24)*. Different methods have been used to facilitate probe entry including treatment with various chemicals such as HCl, photofluor, detergents, heating (microwave), and sodium sulfite *(12)*, to name but a few. However, most of the interest has focused on pretreatment with proteases.

Many different proteases have been used for *in situ* hybridization and immunohistochemistry. These include trypsin, pepsin, proteinase K, and pronase. We have found that these are equivalent for *in situ* hybridization but each requires determining the optimal time and concentration, which may vary considerably for different tissues. For example, tissues such as biopsies of kidney, liver, and lymph nodes are relatively sensitive to protease digestion whereas cervical tissue and, especially, autopsy material often requires relatively more prolonged digestion *(22)*.

Insufficient protease treatment is recognizable as a diminished or completely absent hybridization signal (**Fig. 9**). Too much protease treatment is easy to recognize as the tissue morphology will be destroyed (**Fig. 10**). It is important to emphasize that in my experience protease overtreatment is the *most likely* cause of poor morphology for *in situ* hybridization and PCR *in situ* hybridization. If poor tissue morphology is a problem, then decrease either the time of protease digestion or the protease concentration 10-fold. To determine if the protease time was adequate, it is strongly recommend that, if using human tissues, one use a labeled "consensus" human DNA probe. When the protease time is correct, all the cells should give an intense hybridization signal with the use of these probes; many of these probes use repetitive *alu* sequences which are present in numbers much greater than 10,000 copies per cell.

The protease digestion time recommended above of 5–30 min, in my experience, will be adequate for about 95% of surgical biopsies. Brief mention should be made of what may be called "hyperfixed" samples. In my experience, these are usually cellular preparations which have been fixed for a long period of time (e.g., weeks), sometimes at elevated temperatures. In such cases, I have needed to use up to 100 min of digestion in 2 mg/mL of pepsin to obtain the strong, optimal signal with *in situ* hybridization or *in situ* PCR (Nuovo, G. J., *unpublished observations*).

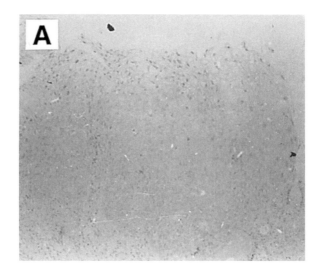

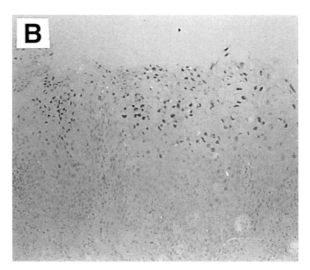

Fig. 9. Effect of protease digestion on the intensity of the hybridization signal. With no protease pretreatment, a hybridization signal was not evident in this low-grade cervical squamous intraepithelial lesion (SIL) which contained HPV 51 (**A**). A signal was evident in the serial section when the hybridization was preceded by treatment of the tissue in trypsin for 13 min (**B**).

4. The function of the formamide and relatively low salt concentration is to facilitate denaturing of the probe and target DNA at 100°C, approx 40°C above the melting temperature (T_m) of homologous hybridized DNA for hybrids of 100 bp, as is typical for so-called full-length probes (probes made from templates at least

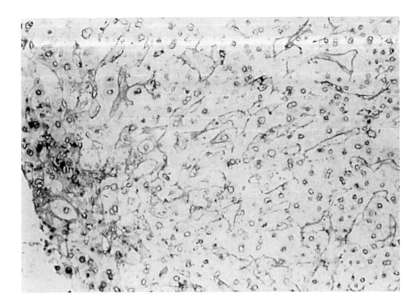

Fig. 10. Effect of protease digestion: poor tissue morphology. This liver biopsy was overdigested in pepsin as evidenced by the lack of nuclear detail toward the center and concomitant loss of the hybridization signal. Note the signal in a few of the cells toward the periphery which apparently are more resistant to protease digestion.

200 bp in size). Bromley et al. have done extensive work correlating the concentration of the probe and the intensity of the hybridization signal under a wide variety of conditions *(25)*. The probe concentration we start with is 1 µg/mL. This amount is rarely associated with background problems for the nonisotopic-labeled probes. However if background is a problem the concentration of the probe may be decreased 10-fold, with excellent preservation of the hybridization signal (*see* below) *(25)*.

5. Background is easily corrected with the most common nonisotopic systems — biotin and digoxigenin (**Fig. 2**). The conditions listed previously usually readily disallow the relatively few hydrogen bonds between the large (100–200 bp) probe and nontarget molecules but still maintain the probe/target complex. However, it is important to emphasize that for oligoprobes (20–40 bp), one needs to decrease the stringency by increasing the salt concentration to 150–300 mM and, as noted previously, decreasing the temperature of the posthybridization wash, or risk losing the entire signal.

6. After the posthybridization wash one is left with a target/probe complex. Labeled nucleotides are incorporated into the probe. A key component of the biotin or digoxigenin systems is the enzyme alkaline phosphatase, which will be attached to the probe/target complex. For biotin this is readily accomplished with a streptavidin–alkaline phosphatase conjugate. An advantage of this system is that

any immunohistochemistry laboratory will have extensive experience with such conjugates and thus be aware of its nuances. For the digoxigenin system one employs an antibody against digoxigenin that is conjugated to the alkaline phosphatase. Although a wide variety of chromagens are available, I recommend 5-bromo-4-chloro-3-indolyl phosphate which in the presence of nitroblue tetrazolium (NBT/BCIP) yields a blue precipitate. The counterstain, nuclear fast red, which stains the cytoplasm and negative nuclei pale pink, allows for easy interpretation of the histologic features and improves the contrast of photomicroscopy.

7. Note that the optimal concentrations of the Mg^{2+} and *Taq* polymerase are greater than those for standard PCR. This may reflect difficulty in entry of these reagents to the site of DNA amplification and/or sequestration of the Mg^{2+} by cellular components. Consistent with this hypothesis is our observation that one may use 10-fold less *Taq* polymerase with the addition of 1 mg/mL of bovine serum albumin (BSA) to the amplifying solution in PCR *in situ* hybridization; BSA can block adsorption of the enzyme to the glass slide or plastic coverslip *(23)*.

Acknowledgments

The authors greatly appreciate the technical and material assistance of Drs. John Atwood, Will Bloch, Deborah French, Dennis Groff, Larry Haff, Brian Holaway, John Sninsky, Dr. Selaki, and Eric Spitzer. Ms. Angella Forde, Frances Gallery, and Kim Rhatigan provided expert technical assistance. Ms. Phyllis MacConnell has been especially helpful with the technical aspects of the *in situ* PCR procedures. We are especially indebted to Mr. S. B. Lewis for his financial support. This work was supported by a grant from the Lewis Foundation to G. J. N.

References

1. de Villiers, E. M., Schneider, A., and Miklaw, H. (1987) Human papillomavirus infections in women with and without abnormal cervical cytology. *Lancet* **ii,** 703–706.
2. Nuovo, G. J. and Cottral, S. (1989) Occult infection of the uterine cervix by human papillomavirus in postmenopausal women. *Am. J. Obstet. Gynecol.* **160,** 340–344.
3. Nuovo, G. J., Darfler, M. M., Impraim, C. C., and Bromley, S. E. (1991) Occurrence of multiple types of human papillomavirus in genital tract lesions: analysis by *in situ* hybridization and the polymerase chain reaction. *Am. J. Pathol.* **58,** 518–523.
4. Nuovo, G. J., Hochman, H., Eliezri, Y. D., Comite, S., Lastarria, D., and Silvers, D. N. (1990) Detection of human papillomavirus DNA in penile lesions histologically negative for condylomata: analysis by *in situ* hybridization and the polymerase chain reaction. *Am. J. Surg. Pathol.* **14,** 829–836.
5. Nuovo, G. J. (1990) Human papillomavirus DNA in genital tract lesions histologically negative for condylomata, analysis by *in situ*, Southern blot hybridization and the polymerase chain reaction. *Am. J. Surg. Pathol.* **14,** 643–651.
6. Nuovo, G. J. (1989) A comparison of different methodologies (biotin based and 35S based.), for the detection of human papillomavirus DNA. *Lab. Invest.* **61,** 471–476.

7. Crum, C. P., Nuovo, G. J., Friedman, D., and Silverstein, S. J. (1988) A comparison of biotin and isotope labeled ribonucleic acid probes for *in situ* detection of HPV 16 ribonucleic acid in genital precancers. *Lab. Invest.* **58,** 354–359.

8. Nuovo, G. J. (1989) A comparison of slot blot, Southern blot and *in situ* hybridization analyses for human papillomavirus DNA in genital tract lesions. *Obstet. Gynecol.* **74,** 673–677.

9. Ostrow, R. S., Manias, D. A., Clark, B. A., Okagaki, T., Twiggs, L. B., and Faras, A. J. (1987) Detection of human papillomavirus DNA in invasive carcinomas of the cervix by *in situ* hybridization. *Cancer Res.* **47,** 649–653.

10. Walboomers, J. M. M., Melchers, W. J. G., and Mullink, H. (1988) Sensitivity of *in situ* detection with biotinylated probes of human papillomavirus type 16 DNA in frozen tissue sections of squamous cell carcinoma of the cervix. *Am. J. Pathol.* **131,** 587–594.

11. Crum, C. P., Symbula, M., and Ward, B. E. (1989) Topography of early HPV 16 transcription in high-grade genital precancers. *Am. J. Pathol.* **134,** 1183–1188.

12. Nagai, N., Nuovo, G. J., Friedman, D., and Crum, C. P. (1987) Detection of papillomavirus nucleic acids in genital precancers with the *in situ* hybridization technique, A review. *Int. J. Gynecol. Pathol.* **6,** 366–379.

13. Crum, C. P., Nuovo, G., Friedman, D., and Silverstein, S. J. (1988) Accumulation of RNA homologous to human papillomavirus type 16 open reading frames in genital precancers. *J. Virol.* **62,** 84–90.

14. Crum, C. P., Nagai, N., Levine, R. U., and Silverstein, S. J. (1986) *In situ* hybridization analysis of human papillomavirus 16 DNA sequences in early cervical neoplasia. *Am. J. Pathol.* **123,** 174–182.

15. Nuovo, G. J. and Silverstein, S. J. (1988) Comparison of formalin, buffered formalin, and Bouin's fixation on the detection of human papillomavirus DNA from genital lesions. *Lab. Invest.* **59,** 720–724.

16. Nuovo, G. J., Friedman, D., Silverstein, S. J., and Crum, C. P. (1987) Transcription of human papillomavirus type 16 in genital precancers. *Cancer Cells.* **5,** 337–343.

17. Lebargy, F., Bulle, F., Siegrist, S., Guellaen, G., and Bernaudin, J. (1990) Localization by *in situ* hybridization of γ glutamyl transpeptidase mRNA in the rat kidney using 35S-labeled RNA probes. *Lab. Invest.* **62,** 731–735.

18. McAllister, H. A. and Rock, D. L. (1985) Comparative usefulness of tissue fixatives for *in situ* viral nucleic acid hybridization. *J. Histochem. Cytochem.* **33,** 1026–1032.

19. Greer, C. E., Peterson, S. L., Kiviat, N. B. and Manos, M. M. (1991) PCR amplification from paraffin-embedded tissues: effects of fixative and fixative times. *Am. J. Clin. Pathol.* 95,117–124.

20. Nuovo, G. J. (1989) Buffered formalin is the superior fixative for the detection of human papillomavirus DNA by *in situ* hybridization analysis. *Am. J. Pathol.* **134,** 837–842.

21. Tournier, I., Bernuau, D., Poliard, A., Schoevaert, D., and Feldmann, G. (1987) Detection of albumin mRNAs in rat liver by *in situ* hybridization, usefulness of paraffin embedding and comparison of various fixation procedures. *J. Histochem. Cytochem.* **35,** 453–459.

22. Nuovo, G. J. (1991) Comparison of Bouin's solution and buffered formalin fixation on the detection rate by *in situ* hybridization of human papillomavirus DNA in genital tract lesions. *J. Histotech.* **14,** 13–18.

23. Nuovo, G. J., Gallery, F., Hom, R., MacConnell, P., and Bloch, W. (1993) Importance of different variables for optimizing *in situ* detection of PCR-amplified DNA. *PCR. Method Appl.* **2,** 305–312.

24. Goelz, S. E., Hamilton, S. R., and Vogelstein, B. (1985) Purification of DNA from formaldehyde fixed and paraffin-embedded tissue. *Biochem. Biophys. Res. Commun.* **130,** 118–124.

25. Bromley, S. E., Darfler, M. M., Hammer, M. L., Jones-Trower, A., Primus, M. A., and Kreider, J. W. (1990) *In situ* hybridization to human papillomavirus DNA in fixed tissue samples, Comparison of detection methods, in *Papillomaviruses* (Crum, C. P., ed.), Wiley-Liss, New York, 46–53.

26. Nuovo, G. J., MacConnell, P. and Becker, J. (1992) Rapid *in situ* detection of PCR-amplified HIV-1 DNA. *Diagn. Mol. Pathol.* **1,** 98–102.

27. Nuovo, G. J., Gallery, F., and MacConnell, P. (1992) Analysis of the distribution pattern of PCR-amplified HPV 6 DNA in vulvar warts by *in situ* hybridization. *Mod. Pathol.* **5,** 444–448.

28. Nuovo, G. J., Becker, J., MacConnell, P., Comite, S., and Hochman, H. (1992) Histological distribution of PCR-amplified HPV 6 and 11 DNA in penile lesions. *Am. J. Surg. Pathol.* **16,** 269–275.

29. Nuovo, G. J., MacConnell, P., Forde, A., and Delvenne, P. (1991) Detection of human papillomavirus DNA in formalin fixed tissues by *in situ* hybridization after amplification by PCR. *Am. J. Pathol.* **139,** 847–854.

30. Nuovo, G. J., Gallery, F., MacConnell, P., Becker, J., and Bloch, W. (1991) An improved technique for the detection of DNA by *in situ* hybridization after PCR-amplification. *Am. J. Pathol.* **139,** 1239–1244.

31. Chou, Q., Russell, M., Birch, D. E., Raymond, J., and Bloch, W. (1992) Prevention of pre-PCR mis-priming and primer dimerization improves low-copy-number amplifications. *Nucleic Acids Res.* **20,** 1717–1723.

32. Nuovo, G. J., Gorgone, G., MacConnell, P., and Goravic, P. (1992) *In situ* localization of human and viral cDNAs after PCR-amplification. *PCR Method Appl.* **2,** 117–123.

33. Nuovo, G. J. (1997) *PCR In Situ Hybridization, Protocols and Applications*, 3rd ed., Lippinicott-Raven, New York.

16

Ultrastructural Detection of Nucleic Acids on Thin Sections of Tissue Embedded in Hydrophilic Resin

Dominique Le Guellec

1. Introduction

Detection of nucleic acid sequences at the ultrastructural level allows the localization of RNA and DNA molecules to specific subcellular compartments. At the light microscope level, it is already evident that DNA and RNA sequences do not have a random distribution, suggesting that their localization inside the cell is related to their role, or the role of their transcribed proteins.

The detection of mRNA at the ultrastructural level by *in situ* hybridization was first described by Jacob et al. *(1)* only 2 years after the first detection of mRNA at the light microscope level by Gall and Pardue *(2)*. The current phase of development of this technique began in 1986, with the utilization of nonradioactive probes by Binder et al. *(3)*. Since then, the areas of application of ultrastructural *in situ* hybridization have greatly expanded (*see* **Note 1**).

Three main methods have now been developed to detect nucleic acid sequences at the ultrastructural level by *in situ* hybridization: preembedding *(4–6)*, ultrathin frozen sections *(7–10)*, and postembedding *(3,11–14)* methods. These all differ in sensitivity, ultrastructural preservation, and detection system. In a previous study *(15)*, we compared these three methods in terms of specificity, sensitivity, resolution, and cytological preservation. The preembedding method gave the best ultrastructural preservation and an intermediate sensitivity, but its resolution is low, as hybrids are detected by an enzymatic detection system (gold particles could not be used without pretreatment which decreases the cytological preservation). The method using frozen sections gave the best sensitivity but the poorest ultrastructural preservation. Using the postembedding method, resolution was as high as with the frozen

From: *Methods in Molecular Biology*, Vol. 123: In Situ *Hybridization Protocols*
Edited by: I. A. Darby © Humana Press Inc., Totowa, NJ

section method, ultrastructural conservation was intermediate, and sensitivity was low because only nucleic acid sequences located at the surface of sections are accessible to the probe and can thus be detected. This method seems to be a good compromise between sensitivity and ultrastructural preservation. Its other advantages are the possibility of storing specimens, and its simplicity. For these reasons, we have chosen to develop the postembedding method.

The principle of the postembedding method is to embed tissue or cells in a hydrophilic resin such as Lowicryl resin *(3,12,14,16,17)*, LR white *(11,18–21)*, LR gold *(22,23)*, or HM20 *(24)* and to incubate thin sections with a radioactive or nonradioactively labeled probe. Hybrids are detected by autoradiography or by immunocytology using gold particles, which give the best resolution. Ultrastructural preservation can be improved by rapid freezing and freeze substitution followed by low-temperature embedding *(25)*.

2. Materials

2.1. Chemicals

1. Paraformaldehyde.
2. Phosphate buffer.
3. Ethanol.
4. Methanol.
5. Lowicryl resin (Chemi Werke Lowe, Germany).
6. Probe labeling kit.
7. Hapten-dUTP or hapten-UTP.
8. Formamide.
9. Mixed bed resin (200–400 mesh).
10. Sodium chloride.
11. Tri-sodium citrate.
12. Dextran sulfate.
13. Bovine serum albumin, fraction V (Sigma, St. Louis, MO, USA).
14. Polyvinylpyrrolidone.
15. Ficoll 400.
16. Salmon or herring sperm DNA.
17. Dithiothreitol.
18. RNase A.
19. Tween-20.
20. Ovalbumin.
21. Anti-biotin antiserum.
22. Tris-HCl.
23. Colloidal gold particles.
24. Glutaraldehyde.
25. Uranyl acetate.

2.2. Stock Solutions

2.2.1. Deionized Formamide

1. In a flask mix 20% w/v of mixed bed resin (200–400 mesh) (Bio-Rad) and 80% v/v of formamide (Merck).
2. Stir gently for 30 min in darkness.
3. After decanting or filtration, keep the supernatant.
4. Deionized formamide can be stored at –20°C.

2.2.2. 20× Standard Saline Citrate Buffer (20× SSC)

1. Dissolve 3 M NaCl and 0.3 M sodium citrate in water.
2. Adjust pH to 7.0.
3. Sterilize by autoclaving.
4. Conserve the solution at room temperature.

2.2.3. 50% Dextran Sulfate

1. To prepare 100 µL of a 50% solution, weigh 50 mg of dextran sulfate (MW = 500,000) into an Eppendorf vial.
2. Add 65 µL of autoclaved distilled water.
3. Dissolve the powder completely with a Pasteur pipet tip.
4. Centrifuge a few seconds at 10,000g to eliminate bubbles.
5. It is preferable to prepare this solution just before use.

2.2.4. 100× Denhardt's Solution

1. Dissolve 2 g of bovine serum albumin type V (BSA) (Sigma), 2 g of polyvinylpyrrolidone (Sigma), and 2 g of Ficoll 400 (Sigma) in 100 mL of autoclaved distilled water.
2. Store at –20°C.

2.2.5. Denatured Salmon or Herring Sperm DNA

1. Dissolve the DNA in autoclaved distilled water at a concentration of 25 mg/mL.
2. Store at –20°C.
3. Before use, denature the DNA for 10 min at 95°C and chill it quickly on ice.

2.2.6. 1 M Dithiothreitol (1 M DTT)

1. Dissolve 1.5 g in 10 mL of 0.01 M sterilized sodium acetate, pH 7.2.
2. Store aliquots at –20°C. Avoid thawing and refreezing.

2.3. Hybridization Buffer

The viscosity of the dextran solution is very high, making it difficult to pipet accurately. Therefore, it is preferable to prepare this solution first, just before use, adding the other solutions directly to the same tube.

Initial concentrations	Final concentrations
a. 50% Dextran sulfate	10%
b. Pure deionized formamide	50%
c. 20× SSC	4×
d. 100× Denhardt's solution	1×
e. 1 *M* DTT	10 m*M*
f. 25 mg/mL denatured sperm DNA	100 µg/mL

2.4. Equipment

1. –20°C Freezer.
2. Freezing apparatus such as KF80 (Reichert-Jung).
3. Cryosubstitution apparatus such as CS Auto Substitution Apparatus (Reichert-Jung).
4. Beem plastic capsules.
5. Ultraviolet light source (360 mm long wavelength).
6. Gold or nickel grids.
7. Centrifuge.
8. Ceramic plate with depressions for incubation/washing steps.
9. Incubator.

3. Methods

3.1. Fixation

In all *in situ* hybridization protocols, tissues or cells must be fixed to preserve the cellular structures. However, ideally the fixation should not inhibit hybridization of probes to target RNA or DNA. The standard fixative solution is 1–4% paraformaldehyde in phosphate buffer. Depending on the abundance of the target nucleic sequences, low concentrations of glutaraldehyde can be added to the fixative solution, but when the number of target molecules to detect is low, glutaraldehyde decreases the hybridization signal dramatically *(25)*. Some experiments have described fixation using only glutaraldehyde at 0.2% *(26)* or 1.6% *(27)* but in these cases the target RNA or DNA was abundant.

When chemical fixation cannot be used, or to improve the ultrastructural preservation and maintain good hybridization, the best option is cryofixation of specimens followed by cryosubstitution. The freezing process must achieve very fast cooling rates so as to minimize ice crystal formation (*see* **Note 2**). Ice is then removed by substitution with an organic solvent that is miscible with the embedding resin (e.g., acetone, methanol, or ethanol). To improve the ultrastructural preservation, 1% osmium tetroxide can be added to the organic solvent, but this decreases the level of hybridization signal *(28)*.

3.2. Embedding in Lowicryl

The majority of studies of ultrastructural *in situ* hybridization have used Lowicryl resins. Other hydrophilic resins have also been tested, such as LR white *(19–21)*, LR gold *(22,23)* or HM 20 *(24)*.

Lowicryl resins (*see* **Note 3**) consist of highly crosslinked acrylate- and methacrylate-based embedding media that can be polymerized by long-wavelength (360 nm) ultraviolet light. Because the initiation of the photopolymerization process is largely independent of temperature, blocks may be polymerized at the same temperatures that are used for infiltration. Lowicryl K4M and K11M are polar resins formulated to provide low viscosity at low temperature. The resins are assumed to maintain intra- and intermolecular hydrogen bonds, replacing water by the resin solvent solution. Both resins have similar properties. K4M is usable down to –35°C. Lowicryl K11M has a much lower viscosity and can be applied at temperatures 20–30°C lower than those used for K4M.

3.2.1. Embedding in Lowicryl K4M

Monomer	Side chain (R)	Weight percent
a. Hydroxypropyl methacrylate	CH_2–CH_2–CH_2OH	48.4
b. Hydroxyethyl acrylate	CH_2–CH_2–OH	23.7
c. *n*-Hexyl methacrylate	CH_2–$(CH_2)4$–CH_3	9.0
d. Crosslinker		13.9

3.2.2. Lowicryl K11M

Monomer	Weight percent
a. *n*-Butyl methacrylate	2.2
b. 2-Methoxy ethyl methacrylate	10.2
c. 2-Ethoxy ethyl methacrylate	11.2
d. 2-Hydroxy ethyl acrylate	20.6
e. 2-Hydroxy propyl methacrylate	41.0
f. 1,3-Butanediol dimethacrylate	4.8
g. Benzoin methyl ether	0.5

3.2.3. Preparation of Lowicryl Resins (see **Note 4**)

The commercial kit contains the resin as three components to be mixed in the following proportions:

3.2.3.1. LOWICRYL K4M

a. Monomer B	13.7 g	
b. Crosslinker A	2.7 g	
c. Photo-initiator C	0.1 g	

3.2.3.2. LOWICRYL K11M

a. Monomer I	19 g
b. Crosslinker H	1 g
c. Photoinitiator C	0.1 g

As the resins are of low viscosity, they require little mixing. Weigh the first two components of the resin directly into a flask. Mix gently to avoid bubbles, because incorporation of too much oxygen into the mixture may interfere with polymerization. Add the photoinitiator and mix gently until it is completely dissolved.

3.3. Dehydration and Impregnation

Cells and tissue samples are prepared for Lowicryl resin embedding following the Progressive Lowering of Temperature Process (PLT) using Lowicryl K4M, or the freeze-substitution process, using Lowicryl K11M (*see* **Note 5**).

3.3.1. Dehydration

3.3.1.1. PROTOCOL FOR DEHYDRATION AT LOW TEMPERATURE WITH THE PLT METHOD

1. 30% Ethanol 30 min at 0°C.
2. 50% Ethanol 60 min at –20°C.
3. 70% Ethanol 60 min at –20°C.
4. 90% Ethanol 60 min at –20°C.
5. 100% Ethanol 60 min at –20°C.
6. 100% Ethanol 60 min at –20°C.

3.3.1.2. PROTOCOL FOR FREEZE-SUBSTITUTION

The ice is removed by substitution with an organic solvent, ethanol or methanol. Methanol is better because it is less dehydrating than ethanol and substitutes ice even in the presence of 10% water.

1. 100% Methanol at –90°C for 48 h.
2. The temperature is raised to –80°C for 10 h.
3. 100% Methanol at –80°C for 48 h.
4. The temperature is raised to –60°C for 20 h.
5. 100% Methanol at –60°C for 48 h.
6. The temperature is raised to 50°C for 10 h.

The duration of each dehydration step may differ depending on the nature of the sample to be embedded. Larger samples will require longer incubations at each step.

3.3.2. Infiltration and Embedding

The duration and the dilution of the resin for each step of infiltration and embedding are similar for the PLT and freeze substitution methods, but for the PLT method, the temperature is –20°C, and the resin used is Lowicryl K4M.

For the freeze-substitution method, the temperature is −50°C, and the resin used is Lowicryl K11M. In both cases, the organic solvent used is ethanol.

The samples are put into Beem plastic capsules which should be completely filled to the top (with fresh precooled resin) to minimize the air space over the resin which may inhibit polymerization.

3.4. Ultraviolet Polymerization

Before polymerization, the samples are put into a fresh change of 100% Lowicryl resin. Polymerization is carried out by indirect UV light (360 nm long wavelength) for 2 or 3 d at −20°C for K4M and −50°C for K11M. If possible, the temperature should be raised slowly (10°C/h) to room temperature and polymerization continued for 24–48 h at room temperature.

3.5. Sectioning

When the Lowicryl resins have been correctly polymerized, they are easily sectioned with either glass or diamond knives. Usually, a sectioning speed of 2 to 5 mm/s is recommended but with some experiments, a similar speed to that used for epoxy resins can be used (1–1.5 mm/s). The thickness of the sections produced is approx 60–80 nm. If the block of resin is difficult to section (*see* **Note 6**), because polymerization is insufficient and the resin is too soft, it is possible to polymerize further at room temperature for 24–48 h.

3.6. Grids

Thin sections are mounted on gold or nickel grids. Copper grids are not appropriate because the formamide present in the hybridization buffer is highly reactive with copper.

3.7. Prehybridization Treatments

Before the hybridization step, in general it is not necessary to treat the sections with chemical agents or proteolytic enzymes as a means of improving the accessibility of the target DNA or RNA to the probe. In special cases, such as the detection of viral sequences, denaturation, proteolytic steps, or DNase or RNase treatment can be performed *(29,30)*. Before hybridization, incubation of the thin sections in hybridization buffer without the probe may be used to diminish the nonspecific adsorption of the probe onto the tissue section. Usually, this incubation is not necessary because nonspecific interactions between the Lowicryl resins and the probes are few.

3.8. Choice and Labeling of the Probe

Three types of probes, oligodeoxynucleotides *(6,15)*, cDNAs *(1,3,4,12,14)*, and cRNAs *(20,21,28,31,32)*, have been tested and all can be used to detect nucleic acid sequences at the ultrastructural level with similar degrees of success and sensitivity (*see* **Note 7**).

3.8.1. cDNA Labeling by Random Priming

1. Denature 25 ng of linearized cDNA at 95°C for 5–10 min (depending on the size of the cDNA).
2. Cool on ice.
3. Add the following to an Eppendorf vial on ice and make up to a final volume of 20 μL
   a. 2 μL of 10× labeling buffer: 500 mM Tris-HCl; 50 mM MgCl$_2$; 20 mM dithiothreitol; 2 mM HEPES, pH 7.5; 52 A_{260} U/mL of random hexadeoxyribonucleotides.
   b. 1 μL of 0.5 M dCTP.
   c. 1 μL of 0.5 M dATP.
   d. 1 μL of 0.5 M dGTP.
   e. 1 μL of mixture containing 0.65 mM dTTP and 0.35 mM hapten (labeled) dUTP.
   f. 2 U of Klenow fragment of DNA polymerase I.
4. Mix gently and centrifuge briefly in a microfuge.
5. Incubate 2 h at 37°C or 20 h at room temperature (*see* **Note 8**).
6. Purify the labeled probe by ethanol precipitation.
7. Resuspend the probe in water. The labeled probe can be stored at –20°C for several months.

3.8.2. Oligodeoxynucleotide Labeling by Tailing at the 3' End

1. Mix the following in an Eppendorf vial:
   a. 100 pmol of oligonucleotide.
   b. 1 nmol of hapten (labeled)-11-dUTP.
   c. 9 nmol of dATP.
   d. 55 U of terminal transferase (Promega).
   e. 4 μL of 5× terminal transferase labeling buffer (500 mM cacodylate buffer, pH 6.8; 5 mM CoCl$_2$, 0.5 mM dithiothreitol).
   f. 2 μL of 25 mM CoCl$_2$.
2. Incubate at 37°C for 45 min in a final volume of 20 μL.
3. Purify the labeled probe by ethanol precipitation.
4. Resuspend the probe in water. The labeled probe can be stored at –20°C for several months.

3.8.3. cRNA Labeling by Transcription

1. Combine the following in an Eppendorf tube on ice, and make up to a final volume of 20 μL:
   a. 1 μL of linearized template.
   b. 2 μL 10× transcription buffer: 400 mM Tris-HCl, 60 mM MgCl$_2$, 100 mM dithiothreitol, 20 mM spermidine, pH 8.0.

   c. 1 µL 10 mM ATP.
   d. 1 µL 10 mM GTP.
   e. 1 µL 10 mM CTP.
   f. 1 µL of UTP mixture (6 mM UTP with 4 mM hapten (labeled)-11 UTP).
   g. 20 U of RNase inhibitor.
   h. 25 U of T3 or T7 or SP6 RNA polymerase.
2. Mix gently and centrifuge briefly in a microfuge.
3. Incubate for 2 h at 37°C.
4. Add 10 U of DNase I to eliminate the DNA template.
5. Incubate for 15 min at 37°C.
6. Purify the labeled probe by ethanol precipitation.
7. Resuspend the probe in RNase free water and store at –20°C.

3.9. Hybridization

As in classical *in situ* hybridization, the hybridization process at the ultrastructural level consists of depositing the labeled probe on a tissue section and the formation of hydrogen bonds with the target RNA or DNA. The stability of hybrids depends on the temperature, the monovalent cation concentration, the percentage of G-C bases, and the hybrid size. The melting temperature (T_m) is the temperature at which 50% of the two DNA strands are denatured. The T_m increases by 16.6°C for each 10-fold increase in sodium concentration and by 0.41°C for every 1% increase in the percentage of G–C bases.

For a hybrid size of more than 200 bp, the T_m follows the following equation:

$$T_m = 81.5°C + 16.6 \log(Na^+) + 0.41\ (\%\ G–C) – 0.65\ (\%F)$$

and for a hybrid size of less than 200 base pairs :

$$T_m = 81.5°C + 16.6 \log(Na^+) + 0.41\ (\%\ G–C) – 820/L – 0.65\ (\%F)$$

where Na^+ is the ionic concentration (in moles per liter) of sodium salt, L the size in base pairs of the probe, and F the percentage of formamide in the hybridization buffer. The nearer the hybridization temperature is to the T_m, the more specific the hybridization reaction will be. Usually, hybridization is carried out 20–25°C below the T_m. In fact, this temperature is too high and inconsistent with good structural preservation of cells. For this reason formamide is added to the hybridization buffer to diminish the hybridization temperature. In the presence of 45–50% formamide, the hybridization temperature is often approx 37°C. Because the stability of RNA/RNA hybrids is better than that of DNA/RNA and DNA/DNA hybrids, the temperature of hybridization can be increased to 50°C when riboprobes are used.

The composition of the hybridization solution must provide optimal conditions for the formation of hybrids at the incubation temperature. The two most important components are salt and deionized formamide. Dextran sulfate is usually added to accelerate the speed of the hybridization process, but the effect of this molecule in ultrastructural hybridization has not been demonstrated. Denhardt's solution contains high mol wt molecules that favor hybridization and bovine serum albumin (BSA) to decrease the nonspecific hybridization. Macromolecules such as tRNA and DNA are added to reduce nonspecific adsorption between the probe and the tissue section, but the nonspecific interactions between probe and Lowicryl resin have been found to be low.

The concentration of the probe varies from 1 to 10 µg/mL. In the same way, the duration of incubation of sections with the probe varies from 1 h to overnight. It is preferable to incubate for a short period because sections can be damaged by long incubation periods. Escaig-Haye et al. *(33)* have observed that 50% of hybrids are formed at the end of the first hour of incubation. From 4 to 5 h of incubation is a good compromise between obtaining good hybridization and good ultrastructure preservation.

3.9.1. Hybridization Procedure

Thin sections are incubated in a drop of 1–5 µL of the hybridization buffer deposited on a ceramic plate (Pelanne Instruments), containing 5 µg of nonradioactive denatured probe per milliliter (for cDNA probes), 5 µg of probe per milliliter (for cRNA), or 100 pmol of oligodeoxynucleotide probe per milliliter. The plate should be humidified with the hybridization buffer. The temperature and the duration of the incubation are as described previously.

3.9.2. Washing

The aim of washing is to eliminate the nonspecific hybridization and nonspecific adsorption of the probe to the tissue section. The four parameters affecting washing are temperature, ionic force, time, and agitation. The first two parameters allow the elimination of mismatches between the probe and nonspecific nucleic acids. During washing, temperature is increased and salt concentration reduced. The second and third parameters allow elimination of nonspecific adsorption of the probe to the tissue section. All of these parameters must be determined experimentally. Washing can be carried out from 4× SSC to 0.1× SSC, from 10 min to 1 h in each, from room temperature up to 50°C, but as indicated previously, the nonspecific binding between the probe and the tissue section is low, so washing may be reduced to short rinses in 1× SSC or PBS *(3)*.

When riboprobes are used, it is possible to reduce the background by tissue section incubation in 10 µg/mL RNase A for 30 min at 37°C to eliminate single-stranded probe molecules which remain unhybridized but bound to the section.

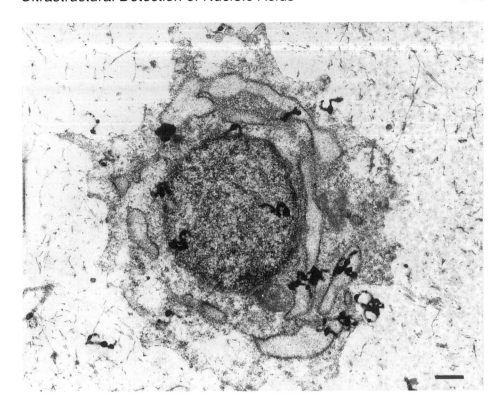

Fig. 1. *In situ* hybridization using ^{35}S-labeled chick fibronectin cDNA on a thin section of sclera (stage 32 chick embryo) dehydrated by progressively lowering temperature method and embedded in Lowicryl K4M. Labeling is observed principally over the cytoplasm, associated with the rough endoplasmic reticulum. The resolution is very low when radioisotopes are used as the label and autoradiography used as the detection system. Bar = 500 nm.

3.10. Detection

The choice of the detection system depends on the probe label used: radiolabeled probes are detected by autoradiography (**Fig. 1**) whereas probes labeled with antigens (e.g., biotin) are detected by immunohistochemistry (**Figs. 2** and **3**). Because nonradioactively labeled probes are now more frequently used, only the detection method by immunohistochemistry is described here. The most successful of the nonradioactive nucleotides is biotin conjugated to nucleotides (**Fig. 2**). Digoxigenin- and fluorescein-labeled (**Fig. 3**) probes have been tested but the data to date do not show if these molecules give the same sensitivity as biotin. As in standard immunhistochemistry, antigen molecule detec-

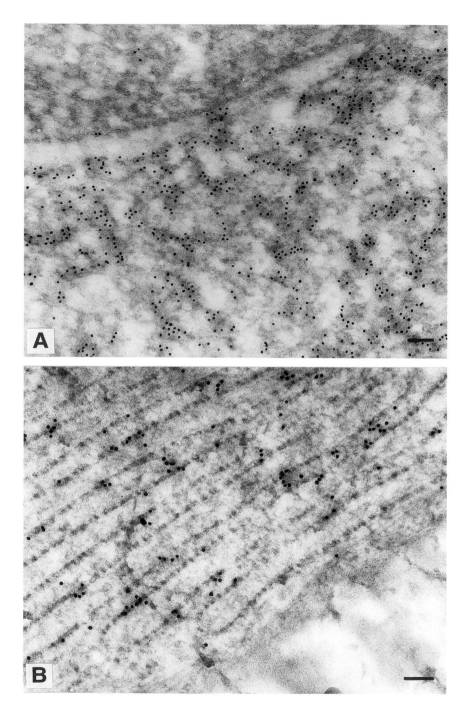

Fig. 2. *In situ* hybridization using biotin-11-UTP labeled type I collagen cRNA on a thin section of fish scale dehydrated by a progressively lowering temperature method and embedded in Lowicryl K4M (**A**). The same tissue after quick freezing   *(continued)*

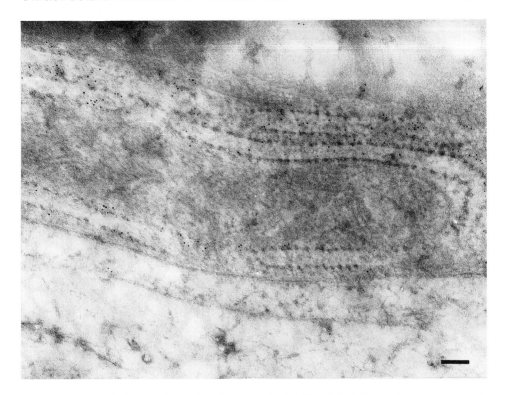

Fig. 3. *In situ* hybridization using fluorescein-11-UTP labeled type I collagen cRNA on a thin section of fish scale after quick freezing and freeze substitution in the presence of 1% osmium tetroxide and embedding in LR white. Fluorescein is detected by the "indirect" method and the marker used is colloidal gold. Bar = 100 nm.

tion can be carried out either by the "direct" or the "indirect" method. The primary antibody can be replaced by a molecule with a high affinity for the antigen (e.g., streptavidin for biotin). The marker used must be electron dense. Colloidal gold (5, 10, or 15 nm) is the most widely used marker because it gives the highest resolution. It is conjugated to the primary or the secondary antibody, to protein A (used as a secondary antibody), or to streptavidin. Ferritin has been tested by Brangeon and Sossountzov *(32)*, but this marker is difficult to observe under the electron microscope because of its small size.

---

and freeze substitution in the presence of 1% osmium tetroxide and embedding in LR white **(B)**. Biotin is detected by the "indirect" method and the marker is colloidal gold. **(A)** A strong signal is observed over the rough endoplasmic reticulum membranes with high resolution. **(B)** The cellular ultrastructural preservation is better when the quick freezing method is used, but fewer gold particles are observed. Bar = 100 nm.

In a previous study *(15)*, we tested three different systems to detect biotin-labeled probes; anti-biotin antibody conjugated to gold particles ("direct" method), unconjugated anti-biotin antibody detected by a secondary antibody conjugated to gold particles ("indirect" method), and streptavidin coupled to colloidal gold. The best sensitivity was obtained with the "indirect" method as follows (**Figs. 2** and **3**):

1. Incubate thin sections for 15 min in buffer A (0.1 *M* phosphate buffer, pH 7.4; 0.65 *M* NaCl; 0.05% Tween-20; and 0.5% ovalbumin).
2. Incubate for 2 h in rabbit anti-biotin serum (Enzo).
3. Rinse for 15 min in buffer A.
4. Incubate for 15 min in buffer B (0.02 *M* Tris-HCl, pH 7.6; 0.65 *M* NaCl; 0.05% Tween-20; 0.5% ovalbumin).
5. Incubate for 30 min in anti-rabbit serum conjugated with gold particles of 5 or 15 nm (Biocell) diluted 1:50 in buffer B.
6. Rinse for 15 min in buffer B.
7. Rinse for 2 min in 2× SSC.
8. Fix by contact with a drop of 1–2% glutaraldehyde for 5 min.
9. Wash in 2× SSC, 2 × 5 min each.
10. Wash in distilled water for 2 min.

3.11. Section Staining

Sections should be stained with either saturated aqueous or alcoholic solutions of uranyl acetate following the standard methods.

3.12. Controls

During the hybridization process, nonspecific signals can be observed owing to adsorption of the probe onto cell structures or because of mismatches hybridizing due to insufficiently stringent conditions. This can be controlled for by using different tissues or different probes or by inhibition of the hybridization signal (*see* **Note 9**). In a heterogeneous tissue, only some cells will express the gene of interest and these should be positive. The absence of signal in other cells is a good check for specificity. To show that the hybridization is not due to a nonspecific adsorption of the probe, the tissue can be hybridized with a heterologous probe (sense riboprobe, sense oligodeoxynucleotide, or any unrelated cDNA that is not expressed in the studied cells). The control probe should have characteristics similar to those of the test probe and the hybridization procedure should be carried out under identical conditions to those in which the test probe is used.

Specific hybridization can be abolished by competition between the labeled probe and an excess of unlabeled probe (100 times greater concentration). This control is only available when oligodeoxynucleotide or cRNA (riboprobes) are used, and not with cDNA probes.

As a negative control, the target nucleic acid can be destroyed by enzymatic pretreatment with DNase or RNase, prior to the hybridization process. However, these enzymes should not degrade the probe itself. For example, it is not suitable to pretreat sections with RNase when riboprobes are to be used. This negative control demonstrates that the test probe hybridizes nucleic acids and is not adsorbed on cell constituents.

In addition to the controls attesting to the specificity of the hybridization, all steps of the immunohistochemical detection method should be controlled by the standard methods. The number of gold particles detected can also be an indication of the specificity. When cDNA or cRNA probes are used, hybrids are visible as a cluster of three or more particles, whereas the background is characterized by isolated gold particles. As indicated previously, this evaluation of the specificity cannot be done with oligodeoxynucleotides because the few antigens attached at the 3'-end may result in detection by only one gold particle.

3.13. Quantitation

It is important to validate the results by quantitation to demonstrate the specificity of the hybridization signal and to detect the cells and cellular compartments that contain the nucleic acid studied. The use of gold particles as a marker allows easy quantitation compared to the other markers such as enzyme substrate methods or radioisotopic detection by autoradiography. Quantitation should be made randomly in a given grid square. The primary purpose of quantitation is to determine the level of background. Gold particles are also counted on cells that are known to contain the nucleic acid studied, on the extracellular matrix, and on the grid. Grains should also be counted on negative control sections. Gold particles are then counted on positive cells. The density of gold particles is determined per unit area of cells. Standard statistical techniques can then be used to examine differences in hybridization between sections or treatments.

4. Notes

1. The areas of application of ultrastructural *in situ* hybridization have greatly expanded since the first utilization of nonradioactive probes by Binder et al. *(3)*. The main fields of application of this method are the localization of viral nucleic acids *(4,12,20,27,29,30,34,35)*, nuclear RNA *(17,36,37)*, and mitochondrial RNA *(33,38)*, and cytoplasmic mRNA *(5,9,11,13,15,21,25,28,39,40)*. The interactions between cells and viruses have also been extensively studied. Viral nucleic acids in infected cells seem to be the best material for *in situ* hybridization because the target nucleic acid is present in abundance. It has been shown that it is possible to specifically detect both double- and single-stranded viral DNA sequences *(12,42,43)*, single-stranded DNA segments only *(41)*, both DNA and RNA single-

stranded sequences *(30)*, or viral RNA molecules only *(42)* using sections embedded in Lowicryl.

The localization of nucleic acid sequences in the nucleolus has allowed a better understanding of the structure and function of this complex organelle *(17,36,37)*. Several studies have described the spatial relationship between mRNA and its cellular environment. An association between the mRNA encoding cytoskeletal molecules and the specific translated protein has also been observed *(9)*. The presence of mRNA encoding neuropeptide has been described in axons *(44–46)*, which altered the previously accepted idea that in neurons, mRNA was located only in the cell body.

The data mentioned here are only a few examples of ultrastructural *in situ* hybridization among many that have contributed to the understanding of cell biology.

2. The cell ultrastructure is well preserved only to a depth of a few micrometers. It is impossible to prevent ice crystal formation deeper in the section. After the freezing process, specimens can be stored in liquid nitrogen before they are embedded in resin.

3. The general composition of Lowicryl resin is:

Methacrylate	$CH_2 = C(CH_3)-CO-O-R$
Acrylate	$CH_2 = CH-CO-O-R$
Dimethacrylate crosslinker	$CH_2 = C(CH_3)-CO-O-R-O-CO-(CH_3)C = CH_2$

4. Liquid Lowicryl and Lowicryl vapors are toxic and may cause skin irritations in sensitive individuals. For these reasons, it is imperative to take precautions when handling Lowicryl. It is necessary to work in a good fume hood, to avoid any direct contact with liquid or vapors and to wear proper gloves, mask, and safety glasses.

5. The specimen preparation following the PLT method can be carried out inside a simple freezer, but the freeze-substitution process requires specific apparatus.

6. Precautions should be taken to ensure that the block face does not become wet during sectioning. This can be accomplished by using a level of fluid in the knife trough that is slightly less than would be used for araldite or Epon resins.

7. For light microscopic hybridization, the choice of probe is important in terms of sensitivity: oligodeoxynucleotide probes are not generally very sensitive, while cRNA probes give the highest sensitivity. At the ultrastructural level, the difference in sensitivity between the three probes is less marked, because hybrids are visualized with gold particles. This visualization is not dependent on the intensity of staining or on the quantity of β-rays emitted by a radioisotope. When hybrids are detected by autoradiography or by antibodies coupled to enzymes, the number of silver grains or the intensity of staining is directly correlated to the specific activity of the probe and to the number of hybrids. The specific activity of oligodeoxynucleotides, and their sensitivity, is low compared to the other probes, because of the small number of labeled nucleotides incorporated at the 3'-

end. At the ultrastructural level, in theory, a single hybrid molecule can be detected by only one gold particle when oligodeoxynucleotide probes are used, whereas long probes, such as cDNA or cRNA probes, give clusters of gold particles. In each case, every gold particle is visible at the electron microscopic level.

Other factors need to be considered for the choice of probe. Oligodeoxynucleotide probes allow a choice of a specific sequence of a gene, but the full length of the mRNA is not accessible to the probe, and only the part that is in contact with the surface of the section can be hybridized. Thus it is more likely that inaccessible sequences will represent a problem. Several oligodeoxynucleotide probes can be used but the hybridization parameters must be identical for each probe. For this reason, it is preferable to use long probes such as cDNA or cRNA probes, but these also have some drawbacks. Part of the long probe may not be hybridized, which reduces the stability of the hybrid formed. cDNA probes have a low hybridization efficiency because they are double-stranded probes, and cRNA probes can be easily hydrolyzed by RNases.

8. The incorporation of the hapten-dUTP is often more effective after a 20-h incubation at room temperature.

9. A single negative control is insufficient, several are necessary to demonstrate the specificity of the hybridization. Troubleshooting may be necessary with some controls such as in the case of positivity of a negative control. For example, a positive signal can be observed after RNase pretreatment because of insufficient washing of sections after enzyme incubation and a failure to eliminate short RNA sequences.

References

1. Jacob, J., Todd, K., Binstiel, M. L., and Bird, A. (1971) Molecular hybridization of ³H-labelled ribosomal RNA in ultrathin sections prepared for electron microscopy. *Biochem. Biophys. Acta* **228**, 761–766.
2. Gall, J. G. and Pardue, M. L. (1969) Formation and detection of RNA-DNA hybrids molecules in cytological preparations. *Proc. Natl. Acad. Sci. USA* **63**, 378–381.
3. Binder, M., Tourmente, S., Roth, J., Renaud, M., and Gehring, W. J. (1986) *In situ* hybridization at the electron microscope level: localization of transcripts on ultrathin sections of Lowicryl K4M-embedded tissue using biotinylated probes and protein-A gold complexes. *J. Cell Biol.* **102**, 1646–1653.
4. Croissant, O., Dauguet, C., Jeanteur, P., and Orth, G. (1972) Application de la technique d'hybridation moléculaire *in situ* à la mise en évidence au microscope électronique, de la réplication végétative de l'ADN viral dans les papillomes provoqués par le virus de Shope chez le lapin cottontail. *C. R. Acad. Sci. Paris* **274**, 614–617.
5. Guitteny, A. F. and Bloch, B. (1989) Ultrastructural detection of the vasopressin messenger RNA in the normal and Brattleboro rat. *Histochemistry* **92**, 277–281.
6. Trembleau, A. (1993) Pre-embedding *in situ* hybridization, in *Hybridization Techniques for Electron Microscopy* (Morel, G., ed.), CRC Press, Boca Raton, FL, pp. 139–162.

7. Hutchison, N. J., Langer-Safer, P., Ward, D., and Hamkalo, B. (1982) *In situ* hybridization at the electron microscope level: hybrid detection by autoradiography and colloidal gold. *J. Cell Biol.* **95**, 609–618.

8. Hutchison, N. J. (1984) Hybridization histochemistry: *in situ* hybridization at the electron microscope level, in *Immunolabeling for Electron Microscopy* (Polak, J. M. and Varndell, I. M., eds.), Elsevier, Amsterdam, pp. 341–351.

9. Singer, R. H., Langevin, G. L., and Lawrence, J. B. (1989) Ultrastructural visualization of cytoskeletal messenger RNAs and their associated proteins using double label *in situ* hybridization. *J. Cell Biol.* **108**, 2343–2353.

10. Morel, G. (ed.) (1993) *Hybridization Techniques for Electron Microscopy.* CRC Press, Boca Raton, FL.

11. Webster, H. F., Lamperth, L., Favilla, J. T., Lemke, G., Tesin, D., and Manuelidis, L. (1987) Use of a biotinylated probe and *in situ* hybridization for light and electron microscopic localization of P° mRNA in myelin-forming Schwann cells. *Histochemistry* **86**, 441–444.

12. Puvion-Dutilleul, F. and Puvion, E. (1989) Ultrastructural localization of viral DNA in thin sections of herpes simplex virus type I infected cells by *in situ* hybridization. *Eur. J. Cell Biol.* **49**, 99–109.

13. Le Guellec, D., Frappart, L., and Willems, R. (1990) Ultrastructural localization of fibronectin mRNA in chick embryo by *in situ* hybridization using 35S or biotin labeled cDNA probes. *Biol. Cell.* **70**, 159–165.

14. Frappart, L. and Le Guellec, D. (1993) Ultrastructural *in situ* hybridization of mRNA on Lowicryl sections using quick-freezing, freeze-substitution, and low temperature embedding, in *Hybridization Techniques for Electron Microscopy* (Morel, G., ed.), CRC Press, Boca Raton, FL, pp. 221–242.

15. Le Guellec, D., Trembleau, A., Pechoux, C., Gossard, F., and Morel, G. (1992) Ultrastructural non-radioactive *in situ* hybridization of GH mRNA in rat pituitary gland: pre-embedding vs ultra-thin frozen sections vs post-embedding. *J. Histochem. Cytochem.* **40**, 979–986.

16. Escaig-Haye, F., Grigoriev, V., and Fournier, J. G. (1989) Détection ultrastructurale d'ARN ribosomal par hybridation *in situ* à l'aide d'une sonde biotinylée sur coupes ultrafines de cellules animales en culture. *C. R. Acad. Sci. Paris* **309**, 429–434.

17. Thiry, M. and Thiry-Blaise, L. (1990) Locating transcribed and non-transcribed rDNA spacer sequences within the nucleolus by *in situ* hybridization and immunoelectron microscopy. *Nucleic Acids Res.* **19**, 11–15.

18. Brangeon, J., Prioul, J. L., and Forchioni, A. (1988) Localization of mRNAs for the small and large subunits of rubesco using electron microscope *in situ* hybridization. *Plant Physiol.* **86**, 990–992.

19. Brangeon, J., Nato, A., and Forchioni, A. (1989) Ultrastructural detection of ribulose-1,5-biphosphate carboxylase protein and its subunit mRNAs in wild-type and holoenzyme-deficient Nicotiana using immunogold and *in situ* hybridization techniques. *Planta* **177**, 151–159.

20. Troxler, M., Pasamontes, L., Egger, D., and Brenz, K. (1990) *In situ* hybridization for light and electron microscopy: a comparison of methods for the localization of viral RNA using biotinylated DNA and RNA probes. *J. Virol. Methods* **30**, 1–14.

21. Wenderoth, M. P. and Eisenberg, B. R. (1991) Ultrastructural distribution of myosin heavy chain mRNA in cardiac tissue: a comparison of frozen LR White embedment. *J. Histochem. Cytochem.* **39**, 1025–1033.

22. Mc Fadden, G. I., Cornish, E. C., Boning, I., and Clarke, A. E. (1988) A simple fixation and embedding method for use *in situ* hybridization histochemistry of plant tissues. *Histochem. J.* **20**, 575–586.

23. Mc Fadden, G. I. (1989) *In situ* hybridization in plants: from macroscopic to ultrastructural resolution. *Cell Biol. Int. Rep.* **13**, 3–21.

24. Fournier, J. G. (ed.) (1994) *Histologie Moléculaire.* Lavoisier, Paris.

25. Le Guellec, D., Frappart, L., and Desprez, P. Y. (1991) Ultrastructural localization of mRNA encoding for the EGF receptor in human breast cell cancer line BT20 by *in situ* hybridization. *J. Histochem. Cytochem.* **39**, 1–6.

26. Thiry, M. and Thiry-Blaise, L. (1989) *In situ* hybridization at the electron microscopic level: an improved method for precise localization of ribosomal DNA and RNA. *Eur. J. Cell Biol.* **50**, 235–243.

27. Puvion-Dutilleul, F. and Puvion, E. (1991) Ultrastructural localization of defined sequences of viral RNA and DNA by *in situ* hybridization of biotinylated DNA probes on sections of herpes simplex virus type 1 infected cells. *J. Electron Microsc. Tech.* **18**, 336–353.

28. Le Guellec, D. and Zylberberg, L. (1998) Expression of type I and type V collagen mRNAs in the elasmoid scales of a teleost fish as revealed by *in situ* hybridization. *Connect. Tissue Res.* **39,** 257–267.

29. Puvion-Dutilleul, F. and Puvion, E. (1990) Replicating single-strand adenovirus type 5 DNA molecules accumulate within well-delimited area of lytically infected Hela cells. *Eur. J. Cell Biol.* **52**, 379–388.

30. Puvion-Dutilleul, F., Roussev, R., and Puvion, E. (1992) Distribution of viral RNA molecules during the adenovirus type 5 infectious cycle in Hela cells. *J. Struct. Biol.* **108**, 209–220.

31. Fisher, D., Weisenberg, D., and Scheer, U. (1992) *In situ* hybridization of DIG-labeled rRNA probes to mouse liver ultrathin sections with a DIG-labeled RNA probe, in *Nonradioactive* In Situ *Hybridization. Application Manual.* Boehringer Mannheim, Germany, pp. 56–61.

32. Brangeon, J. and Sossountzov, L. (1993) Electron microscopic *in situ* hybridization to RNA or DNA in plant cells, in *Hybridization Techniques for Electron Microscopy* (Morel, G., ed.), CRC Press, Boca Raton, FL, pp. 301–348.

33. Escaig-Haye, F., Grigoriev, V., Sharova, I. A., Rudneva, V., Buckrinskaya, A., and Fournier, J. G. (1991) Analysis of human mitochondrial transcripts using electron microscopic *in situ* hybridization. *J. Cell Sci.* **100**, 851–862.

34. Escaig-Haye, F., Grigoriev, I., Sharova, I., Rudneva, V., Buckrinskaya, A., and Fournier, J. G. (1992) Ultrastructural localization of HIV-1 RNA and core proteins. Simultaneous visualization using double immunogold labelling after *in situ* hybridization and immunocytochemistry. *J. Submicrosc. Cytol. Pathol.* **24**, 437–443.

35. Multhaupt, H. A. B., Rafferty, P. A., and Warhol, M. J. (1992) Ultrastructural localization of human papilloma virus by nonradioactive *in situ* hybridization on tissue of human cervical intraepithelial neoplasia. *Lab. Invest.* **67**, 512–518.

36. Wachtler, F., Schöfer, C., Mosgöller, K., Schwarzacher, H. G., Guichaoua, M., Hartung, M., Stahl, A., Bergé-Lefranc, J. L., Gonzalez, I., and Sylvester, J. (1992) Human ribosomal RNA gene repeats are localized in the dense fibrillar component of nucleoli: light and electron microscopic *in situ* hybridization in human Sertoli cells. *Exp. Cell Res.* **198**, 135–143.

37. Olmedilla, A., Testillano, P. S., Vicente, O., Deseny, M., and Risueno, M. C. (1993) Ultrastructural rRNA localization in plant cell nucleoli. *J. Cell Sci.* **106**, 1333–1346.

38. Tourmente S., Savre-Train, I., Berthier, F., and Renaud, M. (1990) Expression of six mitochondrial genes during *Drosophila* oogenesis: analysis by *in situ* hybridization. *Cell Diff. Dev.* **31**, 137–149.

39. Erickson, P. A., Feinstein, S. C., Lewis G. P., and Fisher, S. K. (1992) Glial fibrillar acidic protein and its mRNA: ultrastructural detection and determination of changes after CNS injury. *J. Struct. Biol.* **108**, 148–161.

40. Li, X., Franceschi, V. R., and Okita, T. W. (1993) Segregation of storage protein mRNAs on the rough endoplasmic reticulum membranes of rice endosperm cells. *Cell* **72**, 869–879.

41. Puvion-Dutilleul, F., Picard, E., Laithier, E., and Puvion, E. (1989) Cytochemical study of parental herpes simplex virus type I DNA during infection of the cell. *Eur. J. Cell Biol.* **50**, 187–200.

42. Puvion-Dutilleul, F. and Puvion, E. (1991) Sites of transcription of adenovirus type 5 genomes in relation to early viral DNA replication in infected HeLa cells. A high resolution *in situ* hybridization and autoradiographical study. *Biol. Cell* **71**, 135–147.

43. Puvion-Dutilleul, F. and Puvion, E. (1990) Analysis by *in situ* hybridization and autoradiography of sites of replication and storage of single-and double-stranded adenovirus type 5 DNA in lytically infected HeLa cells. *J. Struct. Biol.* **103**, 280–289.

44. Jirikowsky, G. F., Sanna, P. P., and Bloom, F. E. (1990) mRNA coding for oxytocin is present in axons of the hypothalamo-neurohypophysial tract. *Proc. Natl. Acad. Sci. USA* **87**, 7400–7404.

45. Trembleau, A., Morales, M., and Bloom, F. E. (1994) Aggregation of vasopressin mRNA in a subset of axonal swellings of the median eminence and posterior pituitary: light and electron microscopic evidence. *J. Neurosci.* **14**, 39–53.

46. Trembleau, A., Melia, K. R., and Bloom, F. E. (1995) BC1 RNA and vasopressin mRNA in rat neurohypophysis: axonal compartmentalization and differential regulation during dehydration and rehydration. *Eur. J. Neurosci.* **7**, 2249–2260.

17

In Situ Hybridization for Electron Microscopy

Ross F. Waller and Geoffrey I. McFadden

1. Introduction

In the great majority of cases *in situ* hybridization is used to localize mRNA species at the tissue level, or DNA at the chromosome level. These approaches are generally best done by light microscopy. There are instances, however, when it becomes important to localize nucleic acids at the subcellular level — this brings us into the domain of the electron microscope. Distribution of nucleic acids within the cell can be an important component of their function. The partitioning, manufacturing, trafficking, and processing of different nucleic acids is critical to the functioning of cells, and is only beginning to be understood.

In situ hybridization for electron microscopy (EM) allows us to visualise this subcellular choreography of molecules. The EM methodologies differ in some respects from other *in situ* protocols, but the principles are essentially the same. A labelled probe is allowed to hybridize with target nucleic acids in an ultrathin section (usually plastic or cryosections) and the bound probe is then detected with a marker (usually colloidal gold) (**Fig. 1**). In this chapter we present an outline of the utility and challenges of EM *in situ* hybridization, as well as protocols used and developed by us for localizing several varieties of RNA in a wide range of cells including animals, plants, protists, and bacteria (**Figs. 2–5**).

1.1. Current Applications and Future Speculations

In situ hybridization studies began before the advent of recombinant DNA technology and therefore probe production was limited to highly abundant molecules that could be harvested from cells by conventional biochemical methods (e.g., ribosomal RNAs, satellite DNA, viral DNA). Molecular clon-

From: *Methods in Molecular Biology*, Vol. 123: In Situ *Hybridization Protocols*
Edited by: I. A. Darby © Humana Press Inc., Totowa, NJ

Direct system   Indirect system

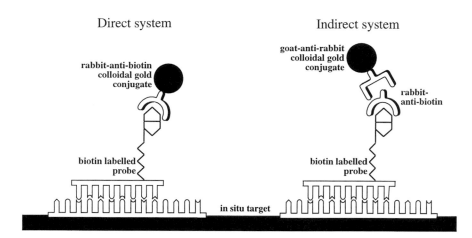

Fig. 1. Diagrammatic representation of *in situ* hybridization. Biotin-tagged probe molecules hybridize to molecules in the cell. Subsequent immunogold detection is directed toward the biotin tag. The detection can be either direct, with colloidal gold attached to the primary antibody, or indirect, with gold attached to a secondary antibody. In the past, radiolabeled probes were detected by microautoradiography. Long exposure times (commonly in the order of months) have made the more recently developed immunobased technology the detection system of choice for most workers. A method for microautoradiography can be found in Fakan and Fakan (*28*). Diagram not to scale.

ing, the polymerase chain reaction (PCR), and oligonucleotide biosynthesis have removed these boundaries and allow virtually no limit to the probes that can be constructed. Furthermore, new probe detection systems continue to be developed and refined, again extending the applications of this technology. This section briefly highlights some of the varied applications of *in situ* hybridization and speculates on the future role of the technology.

1.2. Ancient Alliances

The theory of endosymbiosis describes the formation of organelles by the stable merger of two cells, with one taking up residence inside the other. As a result multiple distinct genetic compartments are created and persist in a cell. *In situ* hybridization has served our laboratory well by enabling us to identify, and define the boundaries of, these discrete cellular spaces. Using rRNA probes, algal cells of the classes Cryptophyceae and Chlorarachniophyceae are revealed as true chimeras of two different eukaryotes (**Fig. 2**) (*1*). *In situ* hybridization has also been recently used to identify a hitherto unrecognized plastid in apicomplexan parasites (that cause several diseases including malaria) by targeting nucleic acid sequences of known plastid similarity (**Fig. 5A**) (*2,3*).

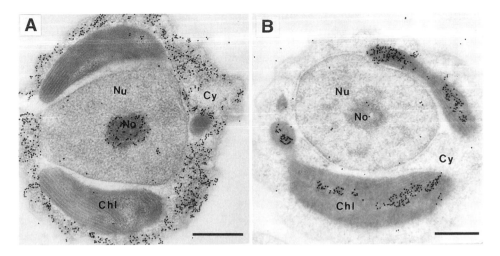

Fig. 2. Identification of different cellular compartments in unicellular algae by *in situ* hybridization. (**A**) Cross section of an alga showing nucleus (Nu) labeled with biotinylated antisense RNA probes specific for nuclear-encoded small subunit ribosomal RNA (ssu rRNA). The probe can hybridize both to mature ribosomes throughout the cytoplasm (Cy) or nascent ssu rRNA molecules at their sites of transcription in the nucleolus (No). (**B**) Similar section labeled with chloroplast rRNA rpobe. The probe only labels the ribosomes in the chloroplast (Chl) and not those in the cytoplasm, which have a different rRNA sequence. Scale bar = 0.5 µm.

1.3. Molecular Invaders

More recent immigrant nucleic acids have also been tracked through their host cells by *in situ* hybridization. Viruses and viroids, through their dependence on their host for replication, must observe specific rendezvous within the host. These rendezvous sites can help illuminate their replicative biology, as well as their pathology. Viroids of plants have been localized to separate cell compartments, either the nucleus or the chloroplast, according to which of two major groupings they belong *(4,5)*. This discovery implicates the nuclear RNA polymerases as the replicative machinery for one group, and the chloroplast RNA polymerases as that for the other group. The study of HIV (human immunodeficiency virus) has sought *in situ* hybridization to further understanding of its pathogenicity and transmission. HIV-1 RNA transcripts have been shown to accumulate in increased amounts in the mitochondria relative to the nucleus and cytoplasm in infected CD4 T lymphocytes *(6)*. Such an accumulation is suggested to compromise mitochondrial function and mediate the cytopathic effect seen in these cells *(6)*. Other work has confirmed that spermatozoa of

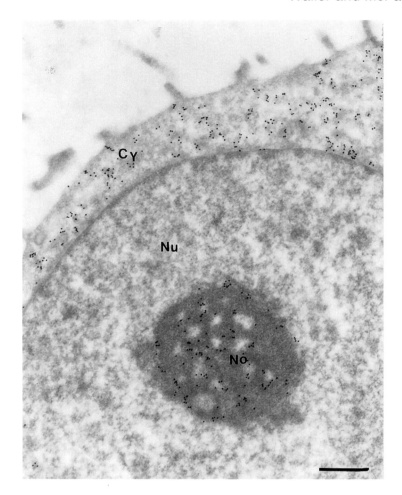

Fig. 3. Human fibroblast showing the nucleus (Nu) labeled with biotinylated antisense RNA probes specific for nuclear-encoded small subunit ribosomal RNA. The probe can hybridize both to mature ribosomes throughout the cytoplasm (Cy) or nascent ssu rRNA molecules at their sites of transcription in the nucleolus (No). Scale bar = 0.5 µm.

infected patients can contain HIV-particles, and these particles can be transmitted to the fertilized oocyte *(7,8)*.

1.4. Molecular Traffic

At an even higher level of nucleic acid dynamics, mRNA behavior, we again find many applications of *in situ* hybridization. Nascent poly(A)$^+$ RNAs have been observed in a speckled pattern in the nucleus, a portion of

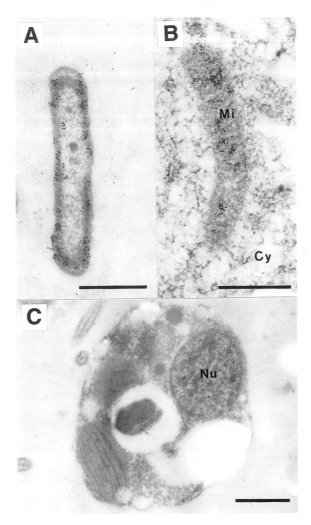

Fig. 4. **(A)** Bacterium labeled with bacterial rRNA probe. The gold label is spread over the cytoplasm at the periphery but absent from the nucleoid. **(B)** Mitochondrion (Mi) of fibroblast labeled with probe for mitochondrial rRNA. **(C)** Cross-section of unicellular algae labeled with probe for U6 small nuclear RNA from the spliceosome. The gold particles are restricted to the nucleus (Nu). Scale bars = 0.5 µm.

which are shown to form a persistent structure apparently important to nuclear function *(9)*. A similar pattern is shown in studies of snRNAs (components of mRNA processing spliceosomes), further building on our knowledge of nuclear function compartmentalization (**Fig. 4B**) *(10,11)*. Following transcript processing, *in situ* hybridization has tracked poly(A)⁺ RNA's pas-

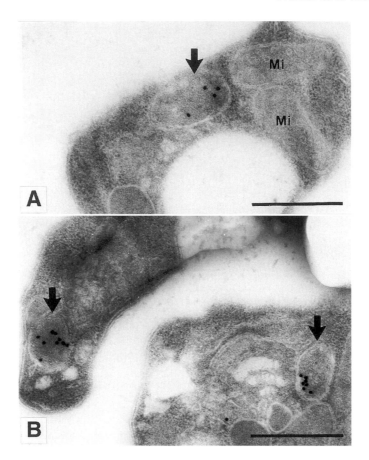

Fig. 5. Identification of a chloroplast in the human pathogen *Toxoplasma*. Probes for the chloroplast-like rRNA gene from the parasite hybridized to a small, ovoid organelle (arrow) distinct from the mitochondrion (Mi). These experiments identified a third genetic compartment in *Toxoplasma* and demonstrated that the parasites had a previous photosynthetic history *(2)*. Scale bar = 0.5 μm.

sage through the nuclear pores *(9)*. Beyond the nucleus, numerous studies are illuminating a complex process of specific mRNA sorting in the cytoplasm. Nonhomogeneous distributions of mRNAs are often seen, some of which show function related targeting *(12,13)*. The cytoskeleton is implicated in this process of mRNA sorting by *in situ* hybridization data from cell wholemounts *(14–17)*.

1.5. Future Challenges

The future challenges of *in situ* hybridization technique are to localize nonabundant nucleic acid targets with ever increasing sensitivity. Two methods of artificially amplifying the *in situ* target levels are PRINS (primed *in situ* labeling) and *in situ* PCR *(16,18)*. These methods are based on the specific hybridization of an oligonucleotide probe followed by enzymatic extension of the anchored probe incorporating labeled nucleotides. Their development has been at the light microscope level but they are now showing promising signs of extension to electron microscopic investigation *(12)*. Should these techniques prove viable they will bolster the role of *in situ* hybridization in the study of cell biology.

2. Materials
2.1. Specimen Preparation
1. Paraformaldehyde and/or glutaraldahyde.
2. 500 mM Stock solution of PIPES (piperazineethanesulfonic acid), pH 7.0.
3. Sucrose.
4. Graded ethanol series (10%, 30%, 50%, 70%).
5. LR Gold (London Resin Company).
6. Benzoyl peroxide paste (London Resin Company).
7. Pioloform (Agar Aids).
8. Nickel or gold mesh EM grids.

2.2. Preparing Probes
2.2.1. Hapten-Labeled Probes
1. Template DNA.
2. 5× Transcription buffer: 200 mM Tris-HCl, pH 7.5, 30 mM MgCl$_2$, 10 mM spermidine, 50 mM NaCl.
3. 100 mM Dithiothreitol (DTT).
4. RNasin (20 U/μL) (Promega, Madison, WI, USA).
5. 10 mM ATP, 10 mM GTP, 10 mM CTP, 10 mM UTP.
6. 2.5 mM Biotin-11-UTP (Sigma, St. Louis, MO, USA) or 2.5 mM biotin-16-UTP (Boehringer Mannheim, Germany) or 2.5 mM DIG-11-UTP (Boehringer Mannheim, Germany).
7. Phage RNA polymerase (T7, T3, or SP6) (20U/μL) (Promega, Madison, WI, USA).
8. Diethylpyrrocarbonate (DEPC)-treated water.
9. 37°C Water bath.
10. 20% Sodium dodecyl sulfate (SDS).
11. BioSpin 30 column (Bio-Rad, Richmond, CA, USA).
12. 4 M LiCl.
13. 3 M Na Acetate (for DNA probes only).
14. *E. coli* transfer RNA (Sigma, St. Louis, MO, USA) (for DNA probes only).
15. Absolute ethanol.
16. Hybridization buffer:
    a. 50 mM PIPES, pH 7.2.
    b. 150 mM NaCl.

   c. 5 m*M* EDTA.

   d. 100 µg/mL of powdered herring sperm DNA (Type D-6898, Sigma).

   e. 0.1% Ficoll (Pharmacia, Uppsala, Sweden).

   f. 0.1% Polyvinylpyrrolidine (PVP-40, Sigma, St. Louis, MO, USA).

   g. 0.1% Bovine serum albumin (BSA) (fraction V, Sigma, St. Louis, MO, USA).

   h. 50% Deionized formamide.

Small aliquots (1 mL) should be stored at −20°C.

2.2.2. Random-Priming Probes

1. Water bath at 37°C.
2. Template DNA.
3. Random primers.
4. 10× Messing buffer: 50 m*M* Tris-HCl, pH 7.6, 100 m*M* MgCl$_2$, 100 m*M* DDT.
5. 1 m*M* dATP, 1 m*M* dGTP, 1 m*M* dCTP.
6. 1 m*M* Biotin-11-dUTP (Sigma, St. Louis, MO, USA) or 1 m*M* DIG-11-dUTP (Boehringer, Mannheim, Germany).
7. DNA polymerase 1 (Klenow subunit) (Promega, Madison, WI, USA).

2.3. Hybridization

1. Hybridization oven (capable of accurately maintaining set temperature).
2. A small sealable container for hybridization. (It is essential that this container can be made vapor tight to ensure that the probe solution does not evaporate during hybridization. A small amount of hybridization buffer is placed in this container, either as a drop or in a small cap, to humidify the chamber.)
3. Hybridization buffer: *see* **Subheading 2.2**.
4. 4× Standard saline citrate (SSC): 1× SSC = 150 m*M* sodium chloride; 15 m*M* sodium citrate; sodium phosphate buffer, pH 7.0.

2.4. Detection of Hybrids

2.4.1. Buffers

1. SC buffer: 50 m*M* PIPES, 0.5 *M* NaCl, 0.5% Tween-20.
2. BSA (type A-7030) (Sigma, St. Louis, MO, USA).

2.4.2. Direct Method

Anti-hapten gold-conjugate. A range of gold sizes conjugated with different anti-biotin or anti-DIG antibodies are commercially available from the following suppliers: British Biocell International, Cardiff, UK; Boehringer Mannheim, Germany; Enzo Biochemicals, New York, NY, USA; and Sigma, St. Louis, MO, USA.

2.4.3. Indirect Method

1. Anti-hapten antibody (such as rabbit anti-biotin from Enzo Biochemicals, New York, NY, USA, or sheep anti-DIG from Boehringer, Mannheim, Germany).

2. Secondary antibody gold conjugate (such as goat anti-rabbit gold conjugate, or rabbit anti-goat gold conjugate, both available from British Biocell International, Cardiff, UK).

3. Methods

3.1. Specimen Preparation

There is great diversity of methods for preparation of biological specimens for standard electron microscopy. Similarly, for *in situ* hybridization many different approaches to specimen preparation can be taken. There are four main requirements for successful labeling: (1) that the specimen is sufficiently robust to withstand the relatively harsh conditions of hybridization, (2) that the target molecules are not obscured or destroyed, (3) that the specimen is made sufficiently hydrophilic for good interaction of specimen and probe solution (important in the selection of a resin when used), and (4) that the labeled specimen is in a state that enables the visualization of both the label and the structures of interest. Providing that these criteria are satisfied, then the approach to specimen preparation should be focused on the best way to visualize the cell component being studied.

Broadly, we can consider three main approaches to preparing the specimen.

1. Thin section postembedding labeling, in which the specimen is either cryosectioned or fixed and embedded in resin prior to *in situ* hybridization.
2. Thin section preembedding labeling, in which the specimen is lightly fixed, permeabilized, and then *in situ* hybridization labeled prior to embedding in resin.
3. Wholemounts, in which the specimen is mounted on a grid, *in situ* hybridization labeled, and then viewed in the microscope in its entirety. For most studies the first method is the most appropriate and reliable, the other techniques being useful for more specialized applications.

Although our experience is exclusively with approach 1, interested readers are directed to other published protocols *(9,14,16,17,19)* for details on approaches 2 and 3.

While it is impossible to prescribe a fixation protocol that will work for all specimen types, the following method is one that we have used with success on a wide range of cell types. An alternate starting point would be to mimic a protocol used for immunocytochemistry on the same specimen type.

1. Prepare a fresh fixative solution of either 4% paraformaldehyde and 0.5% glutaraldehyde in 50 m*M* PIPES buffer, pH 7.0, or 1% glutaraldahyde in 50 m*M* PIPES, pH 7.0 (*see* **Note 1**).
2. Adjust the osmolarity with sucrose or NaCl if necessary (i.e., 250 m*M* sucrose for marine specimens or NaCl for animal tissue).
3. Immerse the specimen in the fixative solution, or add the fixative to cell suspension (suspended in a minimal volume of growth medium). Best results are obtained with very small pieces of tissue (<10 mm^3).

4. Allow to fix for 1–4 h at 4°C.
5. Rinse specimen with buffer 3×.
6. Partially dehydrate the specimen in a graded ethanol series (10%, 30%, 50%, 70% for 30 min in each) (*see* **Note 2**).
7. Now commence the resin embedding by transferring the specimen to a 3:1 mixture of ethanol and LR Gold resin, and rotate overnight at room temperature.
8. Remove this mixture, replace it with 1:1 ethanol and LR Gold resin, and allow to rotate for 4 h.
9. Repeat this procedure but now in a 1:3 ethanol/LR Gold mix.
10. After 4 h replace the mix with 100% resin. Rotate overnight at room temperature.
11. LR Gold is polymerized with benzoyl peroxide paste added at approx 1% (w/v). The hardness of the resin is determined by the concentration of benzoyl peroxide and it is advisable to test the polymerization of the resin before use, as there will be some batch to batch variation.
12. For polymerization transfer the specimen into resin to which the benzoyl peroxide has just been added. Oxygen must be excluded during polymerization, and this is most conveniently achieved by enclosing the specimens in gelatin capsules filled with resin plus accelerator. The polymerization reaction is exothermic so it is advisable to polymerize on ice to minimize excessive specimen heating (*see* **Note 3**).
13. Cut ultrathin sections (silver to gold) with a microtome and diamond knife and pick sections up on pioloform-coated nickel or gold mesh grids (pioloform is stable at hybridization temperatures and therefore superior to formvar). Copper grids are unsuitable as formamide corrodes copper.

3.2. Preparing Hapten-Labeled RNA Probes

To make a single-stranded RNA probe the template DNA must be cloned (or subcloned) into an in vitro transcription vector. There are several commercially available vectors with bacteriophage RNA polymerase promoters (T7, T3, and SP6) engineered into either side of a polylinker region. This enables both strands of the template to be separately transcribed to produce either an antisense or sense (for a negative control) RNA copy. By linearizing the vector immediately downstream of the template with a restriction endonuclease, the transcripts will be limited to "runoffs" of the region of the template and incorporate minimal vector sequence (*see* **Notes 4** and **5**).

3.2.1 In vitro Transcription

1. Linearize the template using a restriction endonuclease that does not generate a 3' overhang (these allow the polymerase to turn back and continue transcribing off the other strand producing a hairpin probe). If a 3' overhang cannot be avoided, blunt the ends using T4 polymerase *(20)*.
2. Determine the DNA concentration of the template *(20)*. Prior to in vitro transcription DNA should be dissolved in DEPC-treated distilled water or TE.

3. At room temperature add the following reagents in this order:
   a. 4 µL of 5× transcription buffer.
   b. 2 µL of 100 m*M* DTT.
   c. 1 µL of RNasin (20 U).
   d. 1 µL of 10 m*M* ATP.
   e. 1 µL of 10 m*M* GTP.
   f. 1 µL of 10 m*M* CTP.
   g. 0.5 µL of 10 m*M* UTP.
   h. 2.5 µL of 2.5 m*M* biotin-11-UTP (or 2.5 m*M* DIG-11-UTP).
   i. Template DNA (0.5-2.0 µg).
   j. 1 µL of RNA polymerase (20 U).
   The total reaction volume should be adjusted to 20 µL by addition of DEPC-water.
4. Mix the components and then incubate at 30°C for 3–4 h.
5. Stop the reaction by adding 1 µL of 20% SDS.
6. Set aside 3 µL for agarose gel electrophoresis (*see* **Subheading 3.2.2.**).

3.2.2. Gel Electrophoresis of Transcription Products

It is advisable to check the success of the transcription reaction by gel electrophoresis. A simple nondenaturing agarose gel stained with ethidium bromide is sufficient *(20)*.

Electrophorese the saved 3 µL against RNA markers (e.g., 3 µg of RNA ladder, Bethesda Research Labs/Gibco-BRL, Bethesda, MD, USA). The apparent molecular weight of the labeled probe will not be highly accurate owing to the nondenaturing conditions; however, it will be a useful approximation and allow an estimate of the quantity of transcription product by comparison to the markers. A yield of several micrograms is typical. The linear template should also be visible in this gel.

3.2.3. Removal of Unincorporated Nucleotides from Labeled Probe

1. Add DEPC-treated water to the reaction products to bring the volume up to 50 µL.
2. Pass this solution though a BioSpin 30 column (or equivalent column) according to the manufacturer's instructions to remove unincorporated nucleotides by exclusion chromatography.

3.2.4. Probe Precipitation

The probe is now ready for precipitation and resuspension in hybridization buffer. We do not recommend hydrolysis of RNA probes to create short fragments. Probes up to 2.5 kb work well with our labeling protocols.

1. Estimate the volume of probe solution and add 1/10 vol of 4 *M* LiCl, and then 2.5 vol of ethanol.
2. Allow to stand in a –20°C freezer for at least 1 h.
3. Centrifuge in a microcentrifuge at 4°C for 15 min at 12,000*g*.

4. Pour off the supernatant and invert the tubes to let drain for 15 min at room temperature.
5. Resuspend the pellet in 10–1000 μL of hybridization buffer. The amount of hybridization buffer used will depend on the efficiency of the transcription reaction. The final probe concentration should be approx 2–10 μg/mL. It is advisable to resuspend the probe in a small volume of buffer and label several grids with a dilution series to determine the optimal probe concentration for labeling, that is, the concentration that gives the maximum signal-to-noise ratio. Excessive probe leads to a nonspecific signal.

3.3. Generating Hapten-Labeled DNA Probes

These probes can be made from genomic clones, cDNAs, or even genomic DNA (suitable for PCR approach only). In all cases the resultant probe is double stranded, with both complementary strands hapten-labeled (*see* **Note 6**). Three different methods of probe preparation can be employed: random priming, nick-translation, or PCR labeling. Both random priming and nick-translation (*20*) produce a product of variable length, whereas PCR labeling creates a probe of fixed length defined by two site-specific primers (labeled nucleotides are incorporated into the PCR product by using a 2:1 ratio of dTTP to DIG-dUTP *[21]*).

The double-stranded nature of these probes results in some inefficiencies in the hybridization process. One reason for this is that the probe is a mixture of complementary molecules that will hybridize with themselves and thus reduce the proportion of hybridization with the *in situ* molecules. Furthermore, if the target molecules are *in situ* RNAs, then only half of the labeled probe molecules will be complements (the anti-sense molecules), so the efficiency of the hybridization is again reduced. Hence for labeling of *in situ* RNA, RNA probes are recommended over DNA probes. For *in situ* DNA labeling, however, double-stranded DNA probes do give satisfactory results.

The random priming method is a popular choice for making DNA probes as it reliably yields an abundant, highly labeled product. In this method the linear template duplex is heat denatured, and then random primers (hexanucleotides) are annealed to the template at sites of complementation. The primers are then extended by DNA polymerase incorporating labeled nucleotides into the probe molecule.

1. Isolate the cloned insert from the vector and determine its concentration (*20*).
2. Combine 100 ng of template with 100 ng of random primers, and bring the volume up to 10 μL with water.
3. Boil this mixture for 3 min, and then chill on ice for 5 min.
4. Add the following:
   a. 5 μL of 10× Messing buffer.
   b. 5 μL of each of dATP, dGTP, dCTP, and the labeled dUTP.
   c. 2 μL of DNA polymerase 1 (Klenow subunit) (20 U).

    d. 13 µL of distilled water.

The final reaction volume will be 50 µL.

5. Incubate this mixture at 37°C either for 1 h, to yield approx 100 ng of probe, or overnight for a yield of up to 10 µg.

    To remove the unincorporated nucleotides follow the procedure described previously for RNA probes (**Subheading 3.2.4.**).

6. To precipitate the probe, estimate the volume of probe solution and add the following:

    a. 1/10 vol of 3 *M* Na Acetate, pH 5.2.

    b. 1 µL of *E. coli* transfer RNA solution (40 µg/mL in water; stored frozen) which serves as a carrier molecule for efficient precipitation of the low levels of DNA.

    c. 2.5 vol of ethanol.

7. Stand the sample in a –20°C freezer for at least 1 h.

8. Centrifuge in a microcentrifuge at 4°C for 15 min at 12,000*g*.

9. Pour off the supernatant and invert the tubes to let drain for 15 min at room temperature.

10. Resuspend pellet in 10–1000 µL of hybridization buffer as per **Subheading 3.2.4**.

3.4. Hybridization

The dissociation and creation of nucleic acid duplexes ("melting" and "annealing") is an equilibrium reaction in any fixed environment and is fundamental to the technique of *in situ* hybridization. Hybridization conditions should aim to maximize the annealing of probe molecules to their true complements while excluding nonhomologous matching. Several parameters influence the hybridization reaction, and the easiest to manipulate to control labeling fidelity is temperature. Other environmental parameters that affect hybridization fidelity (or "stringency") are formamide concentration and monovalent ion concentration. Formamide destabilizes nucleic acid duplexes, reducing the melting temperature. It therefore allows hybridization to be done at lower temperatures than would otherwise be necessary and that would potentially damage the specimen section. Conversely, monovalent ions tend to stabilize nucleic acid hybrids. Hence, low salt concentrations are recommended to maintain stringency. If we define a temperature, T_m, at which only half a population of complementary molecules exist as duplexes at any time, we can then equate the influence of the other parameters that affect hybridization according to the formulae given in **Note 7** *(22,23)*.

3.4.1. RNA Probes

1. Dispense 3-µL aliquots of probe solution onto the bottom of a sterile 4-cm polycarbonate Petri dish.

2. Without pretreatment of the sections, place the grids, section side down, on the probe droplets. Prehybridization was not found to be beneficial.
3. Place the Petri dish into the sealable container along with a small amount of hybridization buffer (0.25 mL) to humidify the container.
4. Seal the container (laboratory film is unsuitable at high temperatures; we find autoclave tape works well).
5. Hybridize at the chosen temperature for at least 1 h.
6. After hybridization, grids are washed by dipping in 4× SSC, then 2× SSC, and then 1× SSC.

   For additional stringency the grids may be incubated on a droplet of 1× SSC in the vapor-tight container at hybridization temperature for an additional hour. We find, however, that this step can frequently be omitted.

3.4.2. DNA Probes

For the use of DNA probes, the double-stranded nature of DNA requires an additional step to the procedure described in **Subheading 3.4.1**. This is the melting of the DNA duplex of both the probe molecules and the *in situ* target molecules, which will then enable the probe to form new duplexes with the target. A convenient way to achieve this is to immerse the grids (with sections on them) in 50 to 100 µL of the probe solution in an Eppendorf tube, and then place this tube into boiling water for 45 s (Roderick Bonfiglioli, personal communication). This affects the dissociation of native duplexes. The tubes are then transferred directly to the hybridization oven. Posthybridization processing is the same for DNA and RNA probes.

3.5. Probe Detection

The researcher is presented with considerable variety when considering the method of probe detection. The principal governing factor in the selection of the system to be used is the desired resolution of labeling. The choices are: the hapten type (DIG [digoxigenin] or biotin) (*see* **Note 6**), the size of colloidal gold (*see* **Note 8**), and whether to use direct immunogold detection (anti-hapten gold conjugate) or indirect immunogold detection (hapten-directed primary antibody, followed by secondary antibody gold conjugate) (*see* **Note 9**).

Following hybridization, grids are transferred to a drop of SC buffer. It is important from this stage onwards to keep the grids wet until the immuogold labeling is complete. The sections are blocked by incubating the grids, section side up, in drops of 1% BSA in SC buffer for 10 min at room temperature.

3.5.1. Direct Method

After blocking, incubate the grids in droplets of the anti-hapten antibody gold-conjugate diluted in SC buffer containing 1% BSA. The dilution ratio (gold-conjugate:buffer) will be indicated by the product manufacturer and may have to be refined empirically. Generally a ratio of 1:20 is appropriate. Incubation should be at room temperature and proceed for at least 60 min. Longer incubations, up to overnight, may be used to saturate probe detection depending on the size of the gold grains. For longer incubations ensure grids are in a vapor-tight container humidified with SC buffer to stop the droplets from evaporating.

3.5.2. Indirect Method

Following blocking, incubate in anti-hapten antibody (5 μg/mL) in SC buffer containing 1% BSA at room temperature for 30 min. Rinse grids under a gentle stream of SC buffer from a wash bottle. Incubate in secondary antibody conjugated to gold as for the direct method.

After incubation with the gold-conjugate, rinse grids with SC buffer, and then with distilled water. Allow grids to air dry. Stain with uranyl acetate and lead citrate as required, remembering that excessive staining may obscure the gold labeling.

4. Notes

1. Aldehyde fixation is commonly used and offers good ultrastructural preservation and nucleic acid retention. Either a combination of paraformaldehyde and glutaraldehyde, or glutaraldehyde alone can be used. Unfortunately the superior preservation quality of osmium tetroxide-fixed material is not viable as this fixative blocks nucleotide basepairing, probably through oxidization of the carbon–carbon bonds and disruption of structure *(24)*. An alternate method of specimen fixation could be through rapid freeze techniques. Limited trials in our laboratory, however, suggest that without some chemical fixation the specimen is not strong enough to withstand the harsh conditions of hybridization.
2. Standard dehydration using an ethanol or acetone series is satisfactory in most cases. However, if the ultrastructural preservation is insufficient it may be improved through the use of freeze-substitution. For this the fixed material is cryoprotected with sucrose, frozen in liquid nitrogen, and then substituted at supercool temperature (approximately –80°C) with acetone, methanol, or ethanol prior to resin embedding.
3. The choice of resin embedding medium will usually parallel the choices made for immunocytochemistry. Most important is the hydrophilicity of the resin to allow good probe–section interaction. For this reason the epoxy resins are not recommended; instead acrylic resins such as LR Gold, LR White, Lowicryl K4M, or Lowicryl HM20 are preferred. It is prudent to avoid UV polymerization of the resins because UV light has the potential to damage target nucleic acids, impairing hybridization with the probe. All the above mentioned resins can be cured using heat or accelerators (*see* manufacturer's recommendations).
4. RNA is easily destroyed by ribonucleases that are extremely resilient enzymes and common contaminants from our skin and untreated solutions. Hence, it is important to wear gloves while preparing probes and to use DEPC-treated water and buffers *(20)*.

5. A pitfall of RNA probes arises from their occasional tendency to hybridize with rRNAs. This is due to the fortuitous identify of certain vector polylinker sequences with rRNA sequences. This phenomenon was first reported by Witkiewicz et al. *(25)* when they noticed that several 9-bp motifs spanning the *Not*I, *Sma*I, and *Sac*I sites of the polylinker of the transcription vector pBluescript™ KS II (Stratagene) had complements in the human large subunit rRNA. These very short complements were demonstrated to give false labeling of the cytoplasm, and Northern blotting confirmed that hybridization was occurring to the large subunit rRNA *(25)*. Further vector sequence identities have been identified in plant rRNAs *(26)* so a general caution in using RNA probes is issued by us. In light of these potential traps, precautions should be built into the research design. Past studies indicate that it would be prudent to exclude these problematic G+C-rich motifs from the transcription template. This could be done either during the cloning process, during the linearization of the vector by cutting upstream of these motifs, or by use of a vector that lacks these motifs (e.g., some of the pGEM™ series: pGEM3z, pGEM3zf, pGEM4z, and pGEM7zf). Negative control experiments should then be used to check for nonspecific binding. Probes made with only the polylinker region and no insert can be used to determine if this region is contributing to any nonspecific binding. In addition, Northern blotting with the full probe can be used to test that hybridization is limited to the correct target molecules.

6. Both DIG and biotin are effective hapten probe labels and should both produce reliable results. The option of haptens is primarily useful when considering double labeling for distinct target molecules where the different hapten types can be used for the different probes.

7. $T_m(DNA) = 81.5°C + 16.6 \log[Na^+] + 0.41(\%G–C) – 0.72(\%F) – 1.4(\%mismatch)$
   $T_m(RNA) = 79.8°C + 18.5 \log[Na^+] + 0.584(\%G–C) – (300 + 2000[Na^+])/L – 0.35(\%F) – 1.4 (\%mismatch)$
   where $[Na^+]$ = sodium ion concentration in moles/L *(M)* and must be between 0.05 and 0.5 *M*; %G–C = percentage of guanine plus cytosine in the duplex; *L* = average probe length; %F = percentage formamide in the hybridization buffer; and %mismatch = percentage of noncomplementary basepairing in the hybridizing strands. For a given probe, 5–10°C below T_m represents a good starting point for the optimization of the hybridization temperature. In 50% formamide and 0.15 *M* salt, this will be approx 42°C for DNA labeling, and between 50°C and 70°C for RNA labeling.

8. Colloidal gold particle size should be selected according to the desired resolution of the labeling, and the magnification of the final images to be presented. In theory, small gold grains should offer less steric hindrance and electrostatic repulsion than larger ones, leading to greater labeling intensities. Comparative studies of the different sized grains (1–10 nm) suggest that in practice this is not the case, with all sizes giving equivalent labeling *(27)*. We find that the most useful size range is from 5 nm to 20 nm. It should be noted that the larger sizes are slower to reach labeling saturation; hence prolonged incubation times with the

gold-conjugate may be necessary. Smaller gold (5 nm) is visible only at relatively high magnification.

9. Either direct or indirect immunogold detection can be used depending on the required resolution of labeling. The direct method gives a much more compact signal with multiple gold grains clustered tightly around the target molecule. The indirect method can result in a more dispersed signal—by up to 10 nm—compared to the direct method *(27)*. The direct method therefore offers greater labeling detail with respect to cell ultrastructure. However, studies at lower magnification may favor the indirect method that when combined with larger gold grains or silver enhancement will allow better visualization of signal.

Acknowledgments

We thank Roderick Bonfiglioli for useful discussions of DNA labeling. R. W. is the recipient of an Australian Postgraduate Award.

References

1. McFadden, G. I., Gilson, P. R., Hofmann, C. J., Adcock, G. J., and Maier, U.-G. (1994) Evidence that an amoeba acquired a chloroplast by retaining part of an engulfed eukaryotic alga. *Proc. Natl. Acad. Sci. USA* **91**, 3690–3694.
2. McFadden, G. I., Reith, M. E., Munholland, J., and Lang-Unnasch, N. (1996) Plastid in human parasites. *Nature* **381**, 482.
3. Köhler, S., Delwiche, C. F., Denny, P. W., Tilney, L. G., Webster, W., Wilson, R. J. M., Palmer, J. D., and Roos, D. (1997) A plastid of probably green algal origin in apicomplexan parasites. *Science* **275**, 1485–1489.
4. Bonfiglioli, R. G., McFadden, G. I., and Symons, R. H. (1994) *In situ* hybridization localizes avocado sunblotch viroid on chloroplast thylakoid membranes and coconut cadang cadang viroid in the nucleus. *Plant J.* **6**, 99–103.
5. Bonfiglioli, R. G., Webb, D. R., and Symons, R. H. (1996) Tissue and intra-cellular distribution of coconut cadang cadang viroid and citrus exocortis viroid determined by *in situ* hybridization and confocal laser scanning and transmission electron microscopy. *Plant J.* **9**, 457–465.
6. Somasundaran, M., Zapp, M. L., Beattie, L. K., Pang, L., Byron, K. S., Bassell, G. J., Sullivan, J. L., and Singer, R. H. (1994) Localization of HIV RNA in mitochondria of infected cells: potential role in cytopathogenicity. *J. Cell Biol.* **126**, 1353–1360.
7. Baccetti, B., Collodel, G., and Piomboni, P. (1996) The debate on the presence of HIV-1 virus in human spermatozoa. *Persp. Drug Discov. Des.* **5**, 129–142.
8. Baccetti, B., Benedetto, A., Burrini, A. G., Collodel, G., Ceccarini, E. C., Crisa, N., Dicaro, A., Estenoz, M., Garbuglia, A. R., Massacesi, A., Piomboni, P., Renieri, T., and Solazzo, D. (1994) HIV-particles in spermatozoa of patients with AIDS and their transfer into the oocyte. *J. Cell Biol.* **127**, 903–914.
9. Huang, S., Deerinck, T. J., Ellisman, M. H., and Spector, D. L. (1994) In vivo analysis of the stability and of nuclear poly(A)$^+$ RNA. *J. Cell Biol.* **126**, 877–899.
10. Huang, S. and Spector, D. L. (1992) U1 and U2 small nuclear RNAs are present in nuclear speckles. *Proc. Natl. Acad. Sci. USA* **89**, 305–308.

11. Gilson, P. R. and McFadden, G. I. (1996) The miniaturized nuclear genome of a eukaryotic endosymbiont contains genes that overlap, genes that are cotranscribed, and the smallest known spliceosomal introns. *Proc. Natl. Acad. Sci. USA* **93**, 7737–7742.

12. Cohen, N. S. (1996) Intracellular localization of the mRNAs of argininosuccinate synthetase and argininosuccinate lyase around liver mitochondria, visualized by high-resolution *in situ* reverse transcription-polymerase chain reaction. *J. Cell. Biochem.* **61**, 81–96.

13. Shaw, M. K., Thompson, J., and Sinden, R. E. (1996) Localization of ribosomal RNA and Pbs21-mRNA in the sexual stages of *Plasmodium berghei* using electron microscope *in situ* hybridization. *Eur. J. Cell Biol.* **71**, 270–276.

14. Singer, R. H., Lawrence, J. B., Silva, F., Langevin, G. L., Pomeroy, M., and Billings-Gagliardi, S. (1989) Strategies for ultrastructural visualization of biotinylated probes hybridized to messenger RNA *in situ*. *Curr. Top. Microb. Immunol.* **143**, 55–69.

15. Bassell, G. J. (1993) High resolution distribution of mRNA within the cytoskeleton. *J. Cell. Biochem.* **52**, 127–133.

16. Bassell, G. J., Powers, C. M., Taneja, K. L., and Singer, R. H. (1994a) Single mRNAs visualized by ultrastructural *in situ* hybridization are principally localized at actin filament intersections in fibroblasts. *J. Cell Biol.* **126**, 863–876.

17. Bassell, G. J., Singer, R. H., and Kosik, K. S. (1994b) Association of poly(A) mRNA with microtubules in cultured neurons. *Neuron* **12**, 571–582.

18. Martinez, A., Miller, M. J., Quinn, K., Unsworth, E. J., Ebina, M., and Cuttitta, F. (1995) Non-radioactive localization of nucleic acids by direct *in situ* PCR and *in situ* RT-PCR in paraffin-embedded sections. *J. Histochem. Cytochem.* **43(8)**, 739–747.

19. Deerinck, T. F., Martone, V., Ram, L., Greene, D., Tsien, R. Y., Spector, D. L., Huang, S., and Ellisman, M. H. (1994) Fluorescence photooxidation with eosin-5-isothiocyanate: a method for high resolution immunolocalization and *in situ* hybridization for light and electron microscopy. *J. Cell Biol.* **126**, 901–910.

20. Sambrook, J., Fritsch, E. F., and Maniatis, T. (1989) *Molecular Cloning: A Laboratory Manual*. Cold Spring Harbor Laboratory Press, Cold Spring Harbor, New York.

21. Grunewald-Janho, S., Keesey, J., Leous, M., van Miltenburg, R., and Schroeder, C. (1996) Nonradioactive *in situ* hybridization application manual. Boehringer Mannheim, Germany.

22. Bodkin, D. K. and Knudson, D. L. (1985) Sequence relatedness of palyam virus genes to cognates of the palyam serogroup viruses by RNA–RNA hybridization. *Virology* **143**, 55–62.

23. Angerer, L. M. and Angerer, R. C. (1989) *In situ* hybridization with ^{35}S-labelled RNA probes. *DuPont Biotechnology Update* **4**, 1–6.

24. Horobin, R. W. (1974) *An Exploratory Outline of Histochemistry and Biophysical Staining*. Gustav Fisher/Butterworths, London.

25. Witkiewicz, H., Bolander, M. E., and Edward, D. R. (1993) Improved design of riboprobes from pBluescript and related vectors for *in situ* hybridization. *BioTechniques* **14**, 408–412.

26. McFadden, G. I. (1995) *In situ* hybridization, in *Methods in Cell Biology*, vol. 49 (Galbraith, D. W., Bohnert, H. J., and Bourque, D. P., eds.), Academic Press, San Diego, pp. 165–183.

27. Egger, D., Troxler, M., and Bienz, K. (1994) Light and electron microscopic *in situ* hybridization — non-radioactive labelling and detection, double hybridization, and combined hybridization-immunocytochemistry. *J. Histochem. Cytochem.* **42**, 815–822.
28. Fakan, S. and Fakan, J. (1987) Autoradiography of spread molecular complexes, in *Electron Microscopy in Molecular Biology—A Practical Approach* (Sommerville, J. and Scheen, U., eds.), IRL Press, Oxford, pp. 201–214.

18

In Situ Hybridization of Whole-Mount Embryos

Murray Hargrave and Peter Koopman

1. Introduction

Since the early analyses of gene expression in the *Drosophila* embryo
(1), whole-mount *in situ* hybridization has become one of the most powerful and versatile tools in developmental biology. The ability to visualize a
gene's expression both in time and space is a necessary first step in investigating the roles of that gene in cell differentiation and morphogenesis in
the developing embryo. Unlike conventional *in situ* hybridization to tissue
sections, the whole-mount procedure provides a three-dimensional readout
of the sites of gene expression. This, combined with sequence analysis,
allows an initial prediction of gene function and provides the basis for further investigation. Unique patterns of gene expression have also been used
to define regions of developing tissue within areas that otherwise appear
anatomically uniform.

Increasing sophistication of the technique, by the use of dual hybridizations
visualized with different colors or by combining hybridization with immunohistochemistry, has now made it possible to observe two or more gene products
simultaneously in the one tissue.

The protocol described in this chapter is similar to that described by
Christiansen et al. *(2)* and Wilkinson and Nieto *(3)*, and has been optimized for
analysis of vertebrate embryos, particularly those of mice. Embryos are particularly well suited to this technique owing to their small size, permeability,
and translucency compared with adult tissues. The whole-mount procedure is
robust and reliable and once some experience has been gained is easy to modify
for use with different developmental systems.

From: *Methods in Molecular Biology*, Vol. 123: In Situ *Hybridization Protocols*
Edited by: I. A. Darby © Humana Press Inc., Totowa, NJ

2. Materials

2.1. Transcription of Labeled Probe

1. A high-purity preparation of a plasmid DNA vector containing the gene fragment of interest and RNA polymerase priming sites (T3, T7, or SP6).
2. Appropriate restriction enzymes and their buffers.
3. Sterile, RNase-free, high-purity water.
4. Diethylpyrocarbonate (DEPC).
5. Phenol buffered to pH 7.5 with Tris-HCl.
6. Chloroform.
7. RNase-free 3 M NaOAc, pH 4.8.
8. RNase-free absolute ethanol.
9. Sp6, T7, or T3 RNA polymerases.
10. 5× RNA polymerase transcription buffer: 200 mM Tris-HCl, pH 7.9, 30 mM MgCl$_2$, 10 mM spermidine, 50 mM NaCl.
11. 0.1 M Dithiothreitol (DTT).
12. Digoxigenin (DIG) RNA labeling nucleotide mix: 10 mM ATP, 10 mM CTP, 10 mM GTP, 6.5 mM UTP, 3.5 mM DIG-11-UTP (Boehringer, Mannheim, Germany) (*see* **Note 1**).
13. Placental ribonuclease inhibitor.
14. Agarose.
15. TAE agarose gel running buffer: 0.04 M Tris-acetate, 0.001 M EDTA.
16. Ethidium bromide (*see* **Note 2**).
17. RNase-free DNase I.

2.2. Production of Embryo Powder

1. Acetone.
2. Filter paper (Whatman).
3. Mortar and pestle.

2.3. Preparation and Hybridization of Embryos

1. Sterile, RNase-free phosphate-buffered saline (PBS), pH 7.4.
2. Triton X-100 detergent.
3. Paraformaldehyde powder.
4. 0.1% Triton X-100 solution in RNase-free PBS (PBTX)
5. RNase-free methanol.
6. 20 mg/mL Solution of proteinase K in RNase-free water.
7. 25% Glutaraldehyde solution.
8. RNase-free formamide.
9. RNase-free 20× saline sodium citrate (SSC) solution, pH 7.0: 3 M NaCl, 0.3 M sodium citrate.
10. Blocking reagent: for nucleic acid hybridization and detection (Boehringer Mannheim, Germany).
11. [(3-chloramidopropyl)-dimethylammonio]-1-propanesulphonate (3-CHAPS) detergent (Sigma).

12. Torula yeast RNA (Sigma).
13. EDTA.
14. Heparin sodium salt.
15. DIG-labeled RNA probes.

2.4. Posthybridization Washes

1. Posthybridization wash solution 1: 50% formamide, 5× SSC, 0.1% Triton X-100, 0.5% CHAPS.
2. 2× SSC solution.
3. 2× SSC, 0.1% CHAPS solution.
4. 0.2× SSC, 0.1% CHAPS solution.
5. 50 mM Tris-HCl, pH 7.5, 150 mM NaCl, 0.1% Triton X-100 solution (TBTX).
6. Sheep serum.
7. Bovine serum albumin (BSA) fraction V.

2.5. Preabsorption of Antibody

1. Sheep anti-DIG Fab fragments conjugated to alkaline phosphatase (Boehringer).

2.6. Post-Antibody Washes and Staining

1. 100 mM NaCl, 100 mM Tris-HCl, pH 9.5, 50 mM MgCl$_2$, 0.1% Tween-20 solution (NTMT).
2. 75 mg/mL Nitroblue tetrazolium (NBT) in dimethylformamide.
3. 50 mg/mL 5-Bromo-4-chloro-3-indolyl-phosphate (BCIP) in dimethylformamide.
4. Sodium azide, 10% stock solution.
5. Glycerol.

3. Methods

All steps up to and including hybridization are carried out in RNase-free conditions, using RNase-free solutions and wearing gloves. Glassware can be rendered RNase free by baking at 180°C and solutions should be made with DEPC-treated water and chemical stocks that are kept separate from the those for general use. Disposable plastic ware is usually RNase free and is ideal for many steps in the protocol.

3.1. Transcription of Labeled Probe

1. Obtain a clone of the gene of interest (*see* **Note 3**).
2. Select two restriction enzymes, each with a unique site at one end of the cloned fragment and linearize 10–20 μg of plasmid with each (*see* **Fig. 1**). Enzymes that produce 3' overhangs (such as *Apa*I, *Bgl*I, *Kpn*I, *Pst*I, *Pvu*I, *Sac*I, *Sac*II, and *Sph*I) should not be used as the RNA polymerases can false prime at these sites.
3. Check that the plasmid is completely linearized by running an aliquot on an agarose gel.

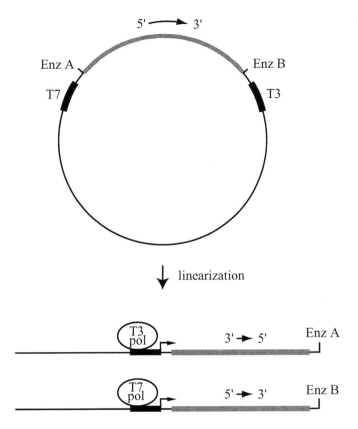

Fig. 1. Production of RNA probes from a plasmid clone. Linearization with restriction enzyme A and transcription with T3 RNA polymerase would give a single-stranded antisense probe that will hybridize to mRNA. Similarly linearization with restriction enzyme B and transcription with T7 RNA polymerase would give a single stranded sense probe that should be used as a control.

4. Phenol/chloroform extract the DNA, add 1/10 vol of RNase-free 3 M NaOAc and 2 vol of RNase-free absolute ethanol, and precipitate at –20°C for 30 min.
5. Pellet the DNA in a refrigerated microfuge by centrifuging at 12,000g for 15 min.
6. Wash the pellet in ice-cold RNase-free 70% ethanol and air dry.
7. Resuspend the DNA in RNase-free water at a final concentration of 1 µg/µL. The linearized plasmid can be stored at –20°C and used for multiple probe synthesis reactions.
8. Select an appropriate RNA polymerase for each of the linearized constructs.
9. Mix the following in a 1.5mL RNase-free tube:
   a. Sterile RNase-free water   8.5 µL
   b. 5× Transcription buffer   4 µL

    c. Linearized plasmid (1 µg/µL)   1 µL

    d. 0.1 *M* DTT   2 µL

    e. 10× DIG RNA labeling mix   2 µL

    f. Placental ribonuclease inhibitor (40 U/µL)   1.5 µL

    g. SP6, T7, or T3 RNA polymerase (20 U/µL)   1 µL

10. Incubate at 37°C for 1 h, then add another 20 U of RNA polymerase.
11. Incubate for a further hour at 37°C.
12. Remove a 1-µL aliquot and run on a 1% agarose/TAE gel to estimate the amount synthesized (*see* **Note 4**). An ethidium bromide-stained RNA band of many fold greater intensity than the plasmid band should be seen. Estimate the amount of probe produced.
13. Dilute the probe to 50 µL with DEPC Milli-Q H_2O, add 5 µL of RNase-free 3 *M* NaOAc, mix, and add 2.5 vol of RNase-free absolute ethanol.
14. Incubate at –20°C for 30 min to precipitate the RNA and centrifuge in a refrigerated microfuge for 15 min at 12,000*g*.
15. Wash the pellet well (twice) with RNase-free 70% ethanol to remove of any unincorporated nucleotides.
16. Redissolve pellet in RNase-free water to a final concentration of 1.0–0.1 µg/µL and store at –20°C.

3.2. Production of Embryo Powder

1. Homogenize advanced-stage embryos (e.g., 12.5–14.5 d *post coitum* mouse embryos, or Hamburger-Hamilton stage 30–32 chick embryos) in a small volume of chilled phosphate-buffered saline (PBS).
2. Add 4 vol of ice-cold acetone, mix well, and place on ice for 30 min.
3. Spin at 10,000*g* for 10 min and discard supernatant.
4. Wash the pellet with ice-cold acetone and centrifuge again.
5. Air dry the pellet on a piece of filter paper and grind into a fine powder with a mortar and pestle.
6. Air dry the powder and store in an air-tight tube at 4°C.

3.3. Embryo Preparation and Hybridization

All steps are carried out on a rocking platform in filled, 2-mL round-bottom tubes to ensure thorough mixing without damage to the embryos, and unless otherwise stated, at room temperature.

1. Dissect embryos in ice-cold PBS (*see* **Note 5**).
2. Fix the embryos in 4% w/v paraformaldehyde (PFA) in PBS at 4°C from 3 h to overnight. PFA is dissolved in PBS by heating at 65°C in a closed container and then cooled before use.
3. Wash twice with PBTX for 10 min each at 4°C.
4. Wash with 25%, 50%, 75% methanol/PBTX, then twice with 100% methanol for 20 min each. The embryos can be stored at 4°C or –20°C at this point, for up to a few months, or preferably in prehybridization solution (*see* **step 12**).

5. Rehydrate by taking the embryos back through the methanol/PBTX series in reverse.
6. Wash 3× with PBTX for 10 min each.
7. Incubate with 10 μg/mL proteinase K in PBTX at room temperature. The length of this treatment depends on the size of the sample and the batch of proteinase K. Ideally, each batch should be tested. As a rough guide, use 5 min for 7.5d*pc* mouse embryo, 10 min for 8.5d*pc*, 15 min for 9.5d*pc*, 20 min for 10.5d*pc*, 25 min for 11.5d*pc*, 30 min for 12.5d*pc*.
8. Wash twice with PBTX for 5 min each. Wash carefully as the embryos can be fragile.
9. Refix the embryos in 0.2% glutaraldehyde/4% PFA in PBTX for 20 min.
10. Wash twice with PBTX for 10 min.
11. Place the embryos in prehybridization mix and allow to sink.
12. Incubate at 65°C for at least 2 h, although it is often convenient to perform this step overnight. For this, and the subsequent washes at 65°C, it is suitable to use a heater block placed on its side on a rocking platform or a hybridization oven with rotating cylinders. The embryos can be stored in this solution at –20°C.
13. Remove prehybridization mix and add hybridization mix including 1.0 μg/mL of DIG-labeled RNA probe. If high background is seen, probe concentration can be decreased to 0.5 μg/mL. The tube needs to be full of hybridization mix or background problems may occur.
14. Incubate at 65°C overnight. If the probe is short or heterologous, 55°C can be used for prehybridization, hybridization, and stringency washes (*see* **Notes 6** and **7**).

3.4. Posthybridization Washes

1. Wash with 100% solution 1 for 5 min at 65°C. From this point on, RNase-free conditions are no longer necessary.
2. Wash with 75% solution 1/ 25% 2× SSC for 5 min at 65°C.
3. Wash with 50% solution 1/ 50% 2× SSC for 5 min at 65°C.
4. Wash with 25% solution 1/ 75% 2× SSC for 5 min at 65°C.
5. Wash with 0.1% CHAPS in 2× SSC twice for 30 min at 65°C. During these washes, start preabsorbing the antibody as described in **Subheading 3.5**.
6. Wash with 0.1% CHAPS in 0.2× SSC twice for 30 min at 65°C.
7. Wash with TBTX twice for 10 min at room temperature.
8. Preblock the embryos with 10% sheep serum and 2% BSA in TBTX for 2–3 h at room temperature.
9. Remove the 10% sheep serum, 2% BSA from the embryos and replace with the preabsorbed antibody (*see* **Subheading 3.5.**) and incubate on a rocker overnight at 4°C.

3.5. Preabsorption of Antibody

1. During the washing of the embryos (**Subheading 3.4., step 5**), weigh out 3 mg of embryo powder into a 1.5-mL tube, add 0.5 mL of 10% sheep serum, 2% BSA in TBTX, and 1 μL of anti-DIG-AP Fab fragment (Boehringer Mannheim, Germany). Embryo powder should match the species being studied.

2. Rock gently at 4°C for 3 h or longer.
3. Spin in a microfuge for 10 min at 4°C.
4. Remove the supernatant without disturbing the pellet and dilute to 2 mL using 10% sheep serum, 2% BSA in TBTX.
5. Store at 4°C until use.

3.6. Post-Antibody Washes and Staining

1. Remove the antibody solution and wash the embryos at least 5× with TBTX containing 0.1% BSA for 1 h at room temperature. The antibody solution can be kept at 4°C and reused up to four times.
2. Wash overnight at 4°C with TBTX containing 0.1% BSA. This wash is optional but usually convenient.
3. Wash twice with TBTX for 15 min.
4. Wash 3× with NTMT for 10 min.
5. Incubate with NTMT including 4.5 µL of NBT and 3.5 µL of BCIP/mL. Rock for the first 20 min, then transfer the embryos to a glass embryo dish or scintillation vial (*see* **Note 8**).
6. When the color has developed to the desired extent, wash with NTMT for 10 min, then with PBTX for 15 min (*see* **Notes 9** and **10**).
7. Wash several times in PBS with 1% Triton X-100. This will blue the stain and decrease background and signal (*see* **Note 11**).
8. Fix the stain by incubating the embryos in 4% PFA in PBS overnight at 4°C.
9. Photograph embryos as soon as possible as the signal can fade or the entire embryo can turn blue upon storage (*see* **Note 12**).
10. If the embryos are to be stored for extended periods, use PBS containing 0.05% w/v sodium azide, or take them through a PBTX/glycerol series into 100% glycerol (*see* **Note 13**).

3.7. Summary

Although the whole-mount protocol may at first seem complex, it can be broken down into four general parts: preparation of the samples, hybridization, application of antibody, and visualization of the staining. All of the steps in between are aimed at producing the greatest signal to background ratio and although the technique is fairly robust, some adjustment for the particular system of interest may be required. Embryo treatments and washing may need to be changed to allow the processes of hybridization and antibody binding to occur with greatest efficiency. Likewise, changes to the hybridization conditions may be necessary for atypical experiments or when ideal probes are not available. Such changes should be made with the general principles of hybridization science in mind.

Most of the common considerations are outlined in **Subheading 4.**; however, some more detailed discussions of certain aspects can be found in the references at the end of the chapter (*4,5*).

4. Notes

1. A number of options exist for both probe labeling and color development. UTP labeled with biotin or fluorescein and the corresponding antibodies, conjugated to either alkaline phosphatase, peroxidase or fluorescent markers, are commercially available. Additionally, there are a number of different substrate systems which allow staining with different colors. Ultilizing these options it is possible to stain for two different transcripts (*6*). Furthermore, *in situ* hybridization can be used in conjunction with immunohistochemistry to analyze the distribution of multiple gene products within the one sample.

2. A number of the reagents in this protocol are known to be irritants or toxins while the safety status of others is unclear. The use of a fume hood and protective clothing is necessary in a number of cases.

3. When designing a probe, select a portion of the gene that lacks highly conserved motifs to minimize cross-hybridization to related gene transcripts. Similarly a probe that contains poly-A tracts, repetitive sequences, or large A–T rich stretches may bind nonspecifically. The size of the probe is ideally about 1 kilobase pairs (kb), but can be anywhere between a few hundred basepairs (bp) to 2 kb in length. Larger probes may need limited alkali hydrolysis to an average size of 500–700 bp for optimal results. Probes that are smaller than 200 bp may require changes to the hybridization and stringency washes.

4. When assessing the success of the probe transcription reaction by running an aliquot on an agarose gel, it is important that the gel be free of RNases as the probe may degrade while running, leading to the conclusion that the labeling reaction was unsuccessful. Since a standard agarose gel is nondenaturing, the RNA probe may not run at the at the size predicted from the insert and may be present as multiple bands. If the majority of product runs at 0–50 bp then the probe has degraded and will have to be remade.

5. A number of species, including mice, have thick extraembryonic membranes that should be torn or removed. Be sure to break the amnion. A number of structures within the embryo can trap reagents which in turn can lead to false staining. These structures, such as the brain ventricles, heart, and otic vesicles, can be punctured with a syringe needle to prevent this.

6. Hybridization of a sense strand probe should be included as a negative control when investigating the expression pattern of a novel transcript.

7. At later developmental stages, such as 11.5 d *post coitum* and older mouse embryos, the density of tissue and the size of the embryo will prevent penetration of the probe to the deeper structures including most of the internal organs. Care should be taken when interpreting the apparent lack of staining in such structures. This problem can be overcome by dissecting out the tissues of interest and performing *in situ* analysis on them alone. The tissues should be dissected from the embryos before the initial paraformaldehyde fixation to prevent excessive background. A number of structures can be held under a large coverslip on a microscope slide and viewed at higher magnification than can be a achieved on a conventional dissecting microscope.

8. During the staining reaction avoid using a plastic Petri dish as crystals tend to form. Staining should be monitored frequently but otherwise kept in the dark as much as possible. Allow the color reaction to proceed until signal is strongest without producing background staining (*see* **Fig. 2**). The staining time may vary from a few hours to 12 h, depending on the expression level of the gene, the specific characteristics of the probe, and the optimization of the protocol. If samples are to be sectioned, overstaining is recommended. The staining reaction should not be left to continue overnight as the samples may overstain and the experiment fail. The staining reaction can be stopped by washing the embryos in NTMT, then TBTX for 15 min each and storing the embryos in fresh TBTX at 4°C in the dark and started again at **Subheading 3.6., step 4**. After staining has been stopped, embryos should not be left in NTMT solution. Even though color reagents may have been substantially diluted the alkaline phosphatase is still active and overstaining will occur. The condition of the NBT and BCIP should be checked before use as using old stocks may increase background. NBT should be bright yellow and BCIP clear. If either have become brown they should be discarded.

9. If the species of embryo or the tissue of interest has pigmentation that may interfere with the staining, the embryos can be bleached in 6% hydrogen peroxide in PBTX for an hour after the methanol rehydration steps.

10. Endogenous phosphatase enzyme may lead to nonspecific staining when using color reaction substrates that react with alkaline phosphatase. Most of the endogenous phosphatases will be rendered inactive by the paraformaldehyde fix and the high temperatures of hybridization; however, if a problem is encountered, bleach the embryos as described previously or add 2 mM levamisole to the staining solution. The phosphatase enzyme conjugated to the anti-DIG antibody is not affected by levamisole.

11. Some observation and judgment is required for destaining. For weak signal, this step can be shortened or omitted. If signal is strong and background is weak, then a total of a few hours is recommended. Overstained or high background samples can be washed for up to several days and the destaining can be stopped and started again with the embryos stored at 4°C in PBTX between destaining treatments.

12. When photographing embryos, position them, immersed in PBS, in grooves cut in a layer of agarose in a Petri dish. The lighting during photography should be adjusted to optimize the translucency of the sample. Transferring embryos into 50% glycerol may help to clear tissue.

13. Sectioning of stained embryos may allow better resolution of positive tissues and make hybridization of tissue sections unnecessary. Sectioning can be performed on a vibratome with agarose embedded samples or on a microtome with paraffin embedding. The latter should only be performed on strongly stained samples as some stains, including that from BCIP and NBT, are soluble in ethanol and xylene. Tissue morphology is likely to be poorer in sectioned whole-mounts than on tissue sections, owing to the proteinase K treatment.

Fig. 2. Sox9 whole-mount. Strong staining can be noted in elements of the developing skeleton, such as the digits, limb bones, scapula, pelvis, vertebrae, and ribs, reflecting a role for this gene in skeletal development (*see* **ref. 8**).

Acknowledgments

This work was supported by Australian Research Council grants to P. K. and an Australian Postgraduate Award to M. H. The CMCB is a Special Research Centre of the Australian Research Council. The advice of Jeff Christiansen is gratefully acknowledged.

References

1. Tuatz, D. and Pfiefle, C. (1989) A non-radioactive *in situ* hybridization method for the localization of specific RNAs in *Drosophila* embryos reveals translational control of the segmentation gene hunchback. *Chromosoma* **98**, 81–85.
2. Christiansen, J. H., Dennis, C. L., Wicking, C. A., Monkley, S. J., Wilkinson, D. G., and Wainwright, B. J. (1995) Murine *Wnt-11* and *Wnt-12* have temporally and spatially restricted expression patterns during embryonic development. *Mech. Dev.* **51**, 341–350.
3. Wilkinson, D. G. and Nieto, M. A. (1993) Detection of messenger RNA by *in situ* hybridization to tissue sections and whole mounts. *Methods Enzymol.* **225**, 361–373.
4. Wilkinson, D. G. (1992) Whole mount *in situ* hybridization of vertebrate embryos, in In Situ *Hybridization: A Practical Approach* (Wilkinson, D. G., ed.), IRL Press, Oxford, UK, pp. 75–83
5. Jowett, T., Mancera, M., Amores, A., and Yan, Y. (1996) *In situ* hybridization to embryo whole mounts and tissue sections: mRNA detection and application to developmental studies, in In Situ *Hybridization* (Clark, M., ed.), Chapman and Hall, London, UK, pp. 91–121.
6. Jowett, T. and Lettice, L. (1994) Whole-mount *in situ* hybridizations on zebrafish embryos using a mixture of digoxigenin- and fluorescein-labelled probes. *Trends Genet.* **10**, 73,74.

7. Hauptmann, G. and Gerster, T. (1994) Two-colour whole-mount *in situ* hybridization to vertebrate and *Drosophila* embryos. *Trends Genet.* **10**, 266.
8. Wright, E., Hargrave, M. R., Christiansen, J., Cooper, L., Kun, J., Evans, T., Gangadharan, U., Greenfield, A., and Koopman, P. (1995) The *Sry*-related gene *Sox-9* is expressed during chondrogenesis in mouse embryos. *Nature Genet.* **9**, 15–20.

19

cRNA Probes: *Comparison of Isotopic and Nonisotopic Detection Methods*

Teresa Bisucci, Tim D. Hewitson and Ian A. Darby

1. Introduction

Since the first description of the technique of *in situ* hybridization to detect specific mRNA species on tissue sections, a variety of methods have been employed. Preparation of tissue and cells ranges from the use of frozen sections, which show good preservation of mRNA but poor morphology, to standard histological methods for tissue preservation, such as paraformaldehyde fixation followed by paraffin embedding. This latter method may show reduced preservation of tissue mRNA levels, but gives more useful structural information as the tissue morphology is better preserved. Similarly, a variety of probe types have been used, starting with double-stranded cDNA probes and moving onto single-stranded RNA probes (riboprobes). In addition, short, single-stranded oligonucleotide probes have also been used, allowing the user to synthesize specific probes from published sequences. Lastly, various methods of probe labeling and detection have been used, including different radioactive isotopes and, more recently, nonradioactive methods such as biotin, fluorescein, and digoxigenin (DIG). For examples of applications of these methods in various tissues *see* **refs.** *1–10*.

The probe labeling techniques used have their own advantages and disadvantages. Radioactive probes require greater levels of laboratory safety measures, such as perspex shielding and a designated preparation room, they also present a greater hazard to the experimenter. Depending on the isotope used, radioactive probes may require long exposure times to visualize results and because of radioactive decay, the probes may undergo radiolysis during storage. The half-life of the isotope used will, in any case, cause the isotopically labeled probe to have a defined shelf life. Nonradioactive techniques have the

From: *Methods in Molecular Biology*, Vol. 123: In Situ *Hybridization Protocols*
Edited by: I. A. Darby © Humana Press Inc., Totowa, NJ

advantage of being safer in the laboratory, probes generally have a longer shelf life and results can be visualized immediately. Until recently, the major disadvantage of nonisotopic labeling techniques for *in situ* hybridization has been their relative lack of sensitivity when compared to radioactivity.

In this chapter we describe our routine methods for detecting target mRNA in tissue sections and in cultured cells. In general, we use single-stranded RNA probes on formalin- or paraformaldehyde-fixed paraffin sections. There are a number of advantages of RNA probes over cDNA or oligonucleotide probes, which make them the standard for *in situ* hybridization. These include the formation of tighter (RNA–RNA) hybrids, the possibility of higher stringency posthybridization washing (using RNase to remove unbound probe), and the lack of a competing reaction that occurs in the case of double-stranded DNA probes, which reanneal as well as binding to target mRNA. The choice of label and detection system depends to some extent on the abundance of the target mRNA; in the case of low-abundance mRNA species, we have continued to use radioactively labeled probes, which give better sensitivity in our hands, with ^{33}P now being the isotope of choice, owing to its relative safety in the laboratory, and its compromise between reasonably short exposure times and good resolution. However, when mRNA species of higher abundance are being detected, probe labeling with nonradioactive methods becomes more feasible. Signal amplification methods for nonisotopic *in situ* hybridization *(6)* may further improve the sensitivity of the technique as discussed in this chapter and in the chapter by Speel et al., this volume.

2. Materials

2.1. Tissue Preparation

1. Paraformaldehyde (Merck, Darmstadt, Germany).
2. Phosphate-buffered saline (PBS), pH 7.4: 0.14 M NaCl, 0.003 M KCl, 0.008 M Na_2HPO_4, 0.0015 M KH_2PO_4.
3. Ethanol, laboratory grade.
4. Chloroform (BDH, Poole, UK).
5. Paraplast (or similar) embedding wax (melting point 56°C).
6. Stainless steel embedding moulds (Tissue Tek).
7. Glass vials for tissue processing.

2.2. Slide Preparation

1. Glass slides.
2. 3-Aminopropyltriethoxy-silane (APES) (Sigma, St. Louis, MO, USA).
3. Acetone (BDH, Poole, UK).

2.3. Pretreatment of Paraffin-Embedded Tissue

All buffers are treated with 0.05% diethylpyrocarbonate (DEPC) (Sigma, St. Louis, MO, USA), with the exception of Tris-based buffers.

1. Histolene (Histolabs/Fronine, NSW, Australia) or xylene.
2. Ethanol.
3. PBS.
4. Pronase buffer (P buffer): 50 mM Tris-HCl, pH 7.5, 5 mM EDTA, pH 8.0.
5. Pronase E (protease from *Streptomyces griseus*) (Sigma, St. Louis, MO, USA).
6. 0.1 M Sodium phosphate buffer, pH 7.2.
7. 4% Paraformaldehyde in PBS.
8. Double-distilled water, DEPC treated.
9. 70% Ethanol.

2.4. Labeling of the Probe

2.4.1. Isotopic Labeling

1. cDNA in appropriate in vitro transcription vector providing polymerase sites for cRNA production (T7, T3, SP6).
2. 5× Transcription buffer (200 mM Tris-HCl, pH 7.5, 30 mM MgCl$_2$, 10 mM spermidine [Sigma, St. Louis, MO, USA], 50 mM NaCl).
3. 100 mM Dithiothreitol (DTT) (Boehringer-Mannheim, Mannheim, Germany).
4. RNasin, ribonuclease inhibitor (Promega, Madison, WI, USA).
5. 10 mM ATP, 10 mM CTP, 10 mM GTP, 12 µM UTP (Promega, Madison, WI, USA).
6. RNA polymerases T7, T3, and SP6 (Promega, Madison, WI, USA).
7. Radionucleotide; 5'-[α-^{33}P]UTP.
8. DNase I (Promega, Madison, WI, USA).
9. Transfer RNA (tRNA) 20 mg/mL of stock (Boehringer-Mannheim, Mannheim, Germany).
10. 7.5 M NH$_4$C$_2$H$_3$O$_2$.
11. 3 M NaC$_2$H$_3$O$_2$·3H$_2$O, pH 5.2.
12. Ethanol.
13. Hydrolysis buffer (80 mM NaHCO$_3$, 120 mM Na$_2$CO$_3$, 20 mM β-mercaptoethanol).
14. Stop buffer (200 mM NaC$_2$H$_3$O$_2$, pH 6.0, 1% glacial acetic acid, 10 mM DTT).
15. DEPC-treated double-distilled H$_2$O.
16. Dry heat block or water bath accurately set at 37°C.
17. Microcentrifuge.

2.4.2. Nonradioactive Probe Labeling Using DIG

All tissue is prepared in the same manner as for radioactive probes, i.e., **Subheadings 2.1.–2.3.**

1. cDNA in appropriate in vitro transcription vector, with polymerase sites for cRNA production (T7, T3, SP6).
2. 5× Transcription buffer (*see* **Subheading 2.4.1., step 2**).
3. DIG 10× labeling mix (Boehringer Mannheim, Mannheim, Germany).
4. RNasin (Promega, Madison, WI, USA).
5. RNA polymerase (T7, T3, or SP6).

6. 0.2 *M* EDTA.
7. 4 *M* LiCl.
8. Ethanol.
9. Hydrolysis buffer: 0.06 *M* Na$_2$CO$_3$, 0.04 *M* NaHCO$_3$.
10. Neutralization buffer: 0.2 *M* NaC$_2$H$_3$O$_2$, 1% acetic acid.
11. DEPC-treated double-distilled H$_2$O.

2.4.3. Nonradioactive Probe Labeling with Biotin

Tissue and slides are prepared the same as for DIG labeling, except that biotin RNA labeling mix (Boehringer Mannheim) is substituted at **Subheading 2.4.2., step 3**.

2.5. Dot Blot Analysis of Nonradioactive Probes

2.5.1. Dot Blot Analysis of DIG-Labeled Probe

To determine the concentration of a DIG-labeled riboprobe a dot blot is necessary.

1. Nylon membrane (Amersham).
2. DIG-labeled RNA control (Boehringer Mannheim, Mannheim, Germany).
3. Buffer 1: 0.1 *M* (maleic acid) C$_4$H$_4$O$_4$, 0.15 *M* NaCl, pH 7.5.
4. Buffer 2: 1% skim milk powder in Buffer 1.
5. Anti-DIG peroxidase, Fab fragments (Boehringer Mannheim, Mannheim, Germany).
6. Buffer 3 (0.1 *M* Tris-HCl, 0.1 *M* NaCl, 0.05 *M* MgCl$_2$).
7. Diaminobenzidine-4HCl (DAB) (Sigma, St. Louis, MO, USA).

2.5.2. Dot Blot Analysis of Biotin-Labeled RNA Probes

1. Same procedure as for DIG-labeled probes, except at **step 5**, substitute avidin–biotin–peroxidase complex (e.g., ABC from Vectastain kit, Vector, Burlingame, CA, USA).

2.6. Hybridization for Radioactive Riboprobe

1. 10× Salts: 3 *M* NaCl, 100 m*M* Na$_2$HPO$_4$, 100 m*M* Tris-HCl, pH 7.5, 50 m*M* EDTA, 0.2% BSA, 0.2% Ficoll, 0.2% polyvinylpyrrolidone.
2. Formamide (BDH, Poole, UK).
3. Dextran sulfate (Amrad Pharmacia Biotech, Uppsala, Sweden).
4. tRNA.
5. DEPC-treated distilled H$_2$O.

2.7. Posthybridization Washes for Radioactive Riboprobes

1. 20× Standard saline citrate (SSC) solution: 3 *M* NaCl, 0.3 *M* sodium citrate.
2. Wash buffer: 2× SSC, 50% formamide.
3. RNase A (Sigma, St. Louis, MO, USA).
4. RNase buffer: 10 m*M* Tris-HCl, pH 7.5, 1 m*M* EDTA, 0.5 *M* NaCl.

2.8. Autoradiography and Emulsion for Radioactive Riboprobe

1. X-ray film cassette.
2. Film (XAR-5 or Hyperfilm).
3. Liquid nuclear research emulsion (gel form) (Ilford, Cheshire, UK).
4. Developer, Phenisol (Ilford), diluted 1:4 with distilled H_2O.
5. Hypam fixer (Ilford), diluted 1:4 with distilled H_2O.
6. Harris' hematoxylin stain.
7. Eosin stain.
8. Scott's tap water, 82 mM $MgSO_4$, 42 mM $NaHCO_3$.
9. Mounting medium, nonaqueous.

2.9. DIG-Labeled Probe Detection

1. PBS.
2. Quench solution: 0.3% H_2O_2 in 100% methanol.
3. Antibody diluent: 1× PBS, 0.1% gelatin, 0.2% BSA.
4. Anti-DIG-peroxidase, Fab fragments (Boehringer Mannheim).
5. DAB.
6. Harris' hematoxylin stain.
7. Nonaqueous mountant.

2.10. Biotin-Labeled Probe Detection

1. Tris-buffered saline (1× TBS): 0.15 M NaCl, 5 mM KCl, 0.25 M Tris-HCl.
2. TBS-Tween: TBS with 5% Tween-20 (BDH Chemicals, Poole, UK).
3. 0.3% H_2O_2 in 100% methanol.
4. Block solution: 1.5% serum in TBS, 0.1% gelatin, 0.2% BSA.
5. DAKO GenPoint kit for catalyzed signal amplification. Primary streptavidin–horseradish peroxidase, biotinyl-tyramide, secondary streptavidin–horseradish peroxidase (DAKO, Carpinteria, CA).
6. DAB.

3. Methods

3.1. Tissue Fixation, Processing, and Embedding

1. Place tissue biopsy in 4% paraformaldehyde/PBS, overnight at room temperature.
2. Wash the tissue in 7% sucrose/0.1 M sodium phosphate buffer overnight at 4°C.
3. Dehydrate tissue through graded alcohols (50%, 70%, 90%, 100%) and then two changes of 100% chloroform.
4. Place tissue in molten paraffin wax (approx 58°C) and leave tissue in wax for a minimum of 4 h.
5. Discard primary wax and replace with fresh wax and leave for 4 h.
6. Ensuring correct orientation of the tissue, embed the tissue in wax using stainless steel molds. Place the molds at –20°C for 1 h and then remove the wax block from the mold.

3.2. Cell Preparation

The most convenient way to perform *in situ* hybridization on cultured cells, particularly on monolayer cultures, is to grow the cells directly on glass slides, coverslips, or plastic. Slides with sterile plastic culture chambers are available commercially (Lab Tek, Nunc, Roskilde, Denmark), and these are ideal for hybridization of monolayer cultures. For cells growing in suspension or cells that are harvested, cytospin preparations may be used. Examples of *in situ* hybridization results using monolayer cultures are shown in **Fig. 1E** and **1F**.

1. Aspirate cell culture medium.
2. Wash cells twice with PBS.
3. If using slide chambers, remove the plastic chamber and silicone gasket.
4. Fix the cells by immersion in 4% paraformaldehyde, 20 min at room temperature.
5. Dehydrate through graded ethanols: 50%, 70%, 95%, 100%.
6. Store in an airtight container at 4°C until use. Pretreat cells according to the labeling technique to be employed, *see* **Subheading 3.5.** or **3.5.1**.

3.3. Coating Slides with APES

1. Place glass slides in racks and wash in an alkaline detergent overnight.
2. Rinse the slides thoroughly with running water and then allow them to dry.
3. Wrap the slides in aluminum foil and sterilize by baking at 180°C for 3 h.
4. Place slides in racks and immerse in a 2% solution of APES in acetone for 20 s.
5. Rinse slides in acetone for 20 s, and then in distilled H_2O, twice.
6. Dry the slides at 37°C overnight and store in an airtight container.

3.4. Tissue Sectioning

1. Fill a small container with distilled H_2O and prepare a water bath at 42°C.
2. Cut 5-μm sections of the paraffin-embedded tissue on a microtome.
3. Place the sections into the H_2O and then with an uncoated glass slide transfer the section into the water bath. The section should flatten.
4. Mount the section with a coated slide and allow the section to dry overnight at 42°C.

3.5. Pretreatment of Tissues and Cells

1. Dewax the sections in histolene or xylene and rehydrate through graded alcohols and finally DEPC-treated distilled H_2O (not required for cell culture).
2. Rinse the sections in prewarmed (37°C) P buffer.
3. Digest tissue with Pronase E in P buffer (125 μg/mL) at 37°C for 10 min.
4. Rinse twice in 0.1 *M* sodium phosphate buffer.
5. Post-fix the sections in 4% paraformaldehyde/PBS at room temperature for 10 min.
6. Rinse twice in 0.1 *M* sodium phosphate buffer.

7. Wash the sections in distilled H_2O and dehydrate in 70% ethanol twice.
8. Air dry sections and store at room temperature in a closed container until required.

3.5.1. Pretreatment of Tissue for DIG- and Biotin-Labeled Probes

1. The steps for pretreatment of tissue are identical to above except at **step 3**, Pronase E is performed for 20 min.

3.6. Labeling the Probe

3.6.1. Radioactive Probe Preparation

Template concentration is important in the labeling procedure and for riboprobe synthesis 500–1000 ng of template is recommended.

1. For one transcription reaction the following final concentrations of reagents are required: 1× transcription buffer, 16 mM DTT, 20 U RNasin, 400 µM ATP, 400 µM CTP, 400 µM GTP, 12 µM UTP, template (500–1000 ng), 20 U of appropriate RNA polymerase, 50 µCi of 5'[α-^{33}P]UTP, and distilled H_2O to a final volume of 20 µL.
2. Incubate the reaction mixture at 37°C for 1 h in a dry heat block or water bath.
3. Digest the template DNA with 1 U of DNase I and incubate the reaction at 37°C for a further 15 min.
4. Add 40 µg of tRNA and adjust the reaction volume to 100 µL with DEPC-treated distilled H_2O.
5. Set aside 1 µL for scintillation counting.
6. Precipitate the riboprobe by adding 50 µL of 7.5 M ammonium acetate and 300 µL of 100% ethanol and place at –70°C for 20 min.
7. Pellet the riboprobe by centrifugation at 10,000g for 20 min at room temperature.
8. Remove the supernatant and wash the pellet with 70% ethanol.
9. Resuspend the riboprobe in 100 µL of DEPC-treated distilled H_2O and remove 1 µL for scintillation counting.
10. In the case of long probes, access to the target mRNA in the tissue may be limited. To improve penetration of the probe hydrolysis may be necessary. We have chosen a probe length of approx 0.15 kb. For hydrolysis, add 100 µL of hydrolysis buffer to the riboprobe and incubate at 65°C for the appropriate length of time (*see* formula in **Note 1**).
11. Terminate the hydrolysis reaction by adding stop buffer and then precipitate the hydrolyzed probe by adding: 40 µL of 3 M sodium acetate, 40 µg of tRNA, and 800 µL of 100% ethanol.
12. Precipitate as described in **steps 7** and **8**.
13. Resuspend in 100 µL of DEPC-treated distilled H_2O, and take 1 µL for scintillation counting.

3.6.2. Labeling the Probe — DIG

1. To a microfuge tube add the following final concentration of reagents: 2 µg of template cDNA, 1× transcription buffer, 1× DIG labeling mix, 20 U of RNasin,

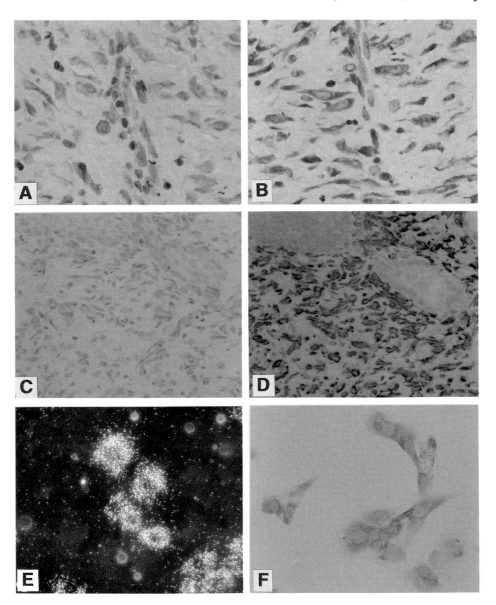

Fig. 1. **(A)** Paraffin section of wound tissue hybridized with a α1 procollagen I cRNA probe labeled with biotin shows weak labeling of fibroblasts and no labeling of blood vessels. **(B)** Adjacent section of wound tissue hybridized with α1 procollagen I cRNA probe labeled with DIG shows markedly stronger labeling of the same tissue with no increase in background. **(C)** Wound tissue section hybridized with a biotinylated probe for α1 procollagen I shows weak labeling of positive cells. **(D)**

(continued)

20 U of RNA polymerase, and DEPC-treated double-distilled H_2O to a final volume of 20 mL.

2. Incubate the reaction mixture at 37°C for 2 h in a dry heat block or water bath.
3. Remove template DNA by digestion with DNase I, *see* **Subheading 3.6.1., step 3**.
4. Precipitate DIG-labeled riboprobe by stopping the reaction with addition of 2 µL of 0.2 *M* EDTA, then add 2.5 µL of LiCl, 75 µL of 100% ethanol, and place at –70°C for 2 h.
5. Pellet DIG-riboprobe by centrifugation at 10,000*g* for 20 min at room temperature.
6. Remove supernatant and wash pellet with 70% ethanol.
7. Resuspend pellet in 50 µL of DEPC-treated double-distilled H_2O.
8. The riboprobe may require hydrolysis to ensure correct size (*see* **Subheading 3.6.1., step 10**). Add 100 µL of hydrolysis buffer and incubate at 60°C for the appropriate length of time; *see also* **Note 1** for calculation of hydrolysis times.
9. Stop hydrolysis by adding 150 µL of neutralization buffer and 900 µL of 100% ethanol.
10. Place at –70°C for 2 h and pellet as discussed earlier.

3.6.3. Labeling the Probe — Biotin

1. As for labeling with DIG except substitute 1× biotin labeling mix for 1× DIG labeling mix.

3.7. Dot Blot of Nonradioative Probes

3.7. Dot Blot of DIG-Labeled Probe

Prepare serial dilutions of a DIG-labeled RNA control, available from Boehringer Mannheim.

1. Spot 1 µL of the DIG-labeled controls and experimental probe onto a nylon membrane.
2. Fix the RNA onto the membrane by baking at 120°C for 30 min.
3. Rinse in buffer 1.
4. Incubate membrane in buffer 2 for 30 min at room temperature.
5. Incubate membrane in anti-DIG-peroxidase (POD), diluted 1:1000 in buffer 2, for 45 min.

Adjacent section of wound tissue hybridized with a biotinylated probe for α1 procollagen I detected using tyramide amplification system. The tyramide amplification system results in labeling stronger than that seen with either biotin or DIG. (**E**) Cultured fibroblasts hybridized with ^{33}P-labeled cRNA probe for α1 procollagen I, detected using liquid emulsion autoradiography and darkfield microscopy. The radioactively labeled probe gives strong labeling of positive cells, some background, and some scatter of silver grains beyond the cytoplasm of the positive cells. (**F**) Cultured fibroblasts hybridized with biotin-labeled α1 procollagen I probe, detected using tyramide amplification system. Specific labeling is weak, but shows good morphology of the cells and no background labeling.

6. Wash membrane in buffer 1, twice.
7. Equilibrate membrane for 2 min in buffer 3.
8. Wash membrane in 1× PBS, 3×.
9. Make DAB according to manufacturer's instructions. Filter through a 0.45-μm syringe filter, add 12 μL of 30% H_2O_2, add to membrane, and watch for brown color to develop.

3.7.2. Dot Blot of Biotin-Labeled Probe

1. As above, i.e., from **steps 1–4** and **steps 6–9**; only **step 5** differs. Instead of **step 5**, incubate membrane with avidin–biotin complex (ABC) made up using Vectastain kit.

3.8. Hybridization

3.8.1. Radioactive Probes

1. 500 μL Hybridization buffer consists of: 1× salts, 50% formamide, 10% dextran sulfate, 360 μg of tRNA. Five hundred microliters is sufficient for approx 10 sections.
2. Add labeled riboprobe to the hybridization buffer at a concentration of 20×10^6 per 500 μL of hybridization buffer.
3. Heat the probe/hybridization buffer mix to 85°C for 5 min before placing on the sections.
4. Coverslip the sections and place in a humidified airtight chamber.
5. Hybridize overnight at 60°C.

3.8.2. DIG-Labeled Probes

Follow **Subheading 3.8.1.**, **steps 1–4**.

1. Hybridization of DIG-labeled cRNA is performed at 42°C, overnight.

3.8.3. Biotin-Labeled Probes

Follow **Subheading 3.8.1.**, **steps 1–4**.

1. Hybridization of biotin-labeled riboprobe is performed at 37°C, overnight.

3.9. Posthybridization Washes

3.9.1. Radioactive Probes

1. Heat wash buffer to 55°C and soak slides to remove coverslips.
2. Wash slides at 55°C for 30 min. Replace wash buffer and wash slides for a further 30 min.
3. Wash slides in three changes of RNase buffer and then incubate the sections with RNase A (150 μg/mL) in RNase buffer at 37°C for 1 h, with agitation (shaking water bath).
4. Wash the sections in 2× SSC for 45 min at 55°C and then dehydrate through graded alcohols and air dry.

3.9.2. Posthybridization Washes — DIG

1. Heat wash buffer to 42°C and soak slides to remove coverslips.
2. Wash slides for a total for 60 min in wash solution, replacing solution after 30 min,

3. Rinse slides in three changes of RNase buffer and then incubate slides with RNase A (150 µg/mL) at 37°C for 1 h.
4. Wash slides in 2× SSC, 45min, 37°C.

3.9.3. Posthybridization Washes – Biotin

1. As previously described, except the washes are performed at 37°C, instead of 42°C.

3.10. Autoradiography

1. Sections that have been hybridized with ^{33}P-labeled probes can be placed on X-ray film, to provide an idea of the success or otherwise of the hybridization reaction. This preliminary autoradiography can also serve as a guide for exposure times required in the liquid emulsion autoradiography step. However, for small pieces of tissue or where few cells are labeled in the tissue section, this step may be omitted.
2. In a darkroom under safelight illumination (Ilford safelight filter number 904 or Kodak safelight filter number 2), weigh out 10 g of emulsion, add 10 mL of distilled H_2O, and incubate at 42°C for 2 h to allow the emulsion to melt.
3. Pour the liquefied emulsion into a dipping chamber (available from Amersham) and "dip" the experimental slides, ensuring all slides are coated evenly and that there are no air bubbles.
4. Remove excess emulsion by allowing slides to drain vertically on absorbent paper in the dark.
5. Place slides into a plastic slide rack and store in a lightproof box containing desiccant.
6. Expose in the lightproof container for 10–20 d, depending on the strength of the hybridization signal.

3.11. Signal Development of Radioactive Riboprobe

1. In a darkroom under safelight illumination, place slides in the diluted developer for 2 min with mild agitation.
2. Stop development by immersion in 0.5% acetic acid for 30 s.
3. Immerse the slides for 2 min in rapid fixer.
4. Rinse slides in running tap water for 5 min.
5. Stain slides with Harris' hematoxylin, rinse in tap water, and place in Scott's tap water for 30 s or until hematoxylin appears blue, rinse in water, and then stain with eosin.
6. Dehydrate sections through graded alcohols, rinse in two changes of histolene and mount using a nonaqueous mountant.

3.12. Detection of DIG-Labeled cRNA

1. Rinse slides in 1× PBS, 3×.
2. Quench endogenous peroxidase activity by incubating with 0.3% H_2O_2/methanol for 20 min at room temperature.

3. Wash the slides in 1× PBS, 3×.
4. Incubate the sections with 1.5% fetal calf serum made in diluent, for 30 min.
5. Incubate sections in anti-DIG-POD in dilute serum 1:30, for 45 min at room temperature.
6. Wash the slides 3× in 1× PBS.
7. Apply activated DAB and watch for color to develop.
8. Counterstain with hematoxylin.

3.13. Detection of Biotin-Labeled Probe

1. Rinse slides in 1× TBS
2. Quench slides in 0.3% H_2O_2 in methanol, 30 min at room temperature
3. Wash slides in 1× TBS, 3×, 5 min each.
4. Block sections using 1.5% serum in 1× TBS, 30 min at room temperature.

The following steps are performed using reagents from the Dako GenPoint kit:

5. Add primary streptavidin (1:100–1:1000) and incubate sections for 15 min at room temperature.
6. Wash slides in 1× TBS + 0.5% Tween-20, for 5 min, 3×.
7. Add biotinyl tyramide (undiluted) to sections and incubate 15 min.
8. Repeat **step 6**.
9. Add secondary streptavidin (undiluted) and incubate for 15 min.
10. Repeat **step 6** (*see* **Note 2**).
11. Apply activated DAB and monitor color development.
12. Counterstain sections using hematoxylin.

4. Notes

1. Hydrolysis formula;

   Hydrolysis time

   $$(\text{min}) = \frac{\text{length of probe (kb)} - \text{length of desired end product (kb)}}{[0.11 \times \text{length of probe (kb)} \times \text{length of desired end product (kb)}]}$$

   For example, starting with a probe that is 1.5 kb and requiring an end product of 0.15 kb, the hydrolysis time is

   $$t = 1.5 - 0.15 / 0.11 \times 1.5 \times 0.15 = 54.5 \text{ min}$$

2. The tyramide amplification system can be used to increase the sensitivity of *in situ* hybridization, when the target mRNA is in low abundance. However, the conditions, including dilutions of the various components, need to be optimized for each probe. Increasing the number of cycles of amplification also leads to increased background. Even a single amplification step shows much stronger labeling than is seen with the same probe using biotin and a single streptavidin detection step (for examples, *see* **Fig. 1**).

Acknowledgment

We would like to thank DAKO Corporation (USA) for their generous sponsorship of color reproduction costs of this chapter.

References

1. Komminoth, P. (1992) Digoxigenin as an alternative probe labeling for *in situ* hybridization. *Diagn. Mol. Pathol.* **1,** 142–150.
2. Komminoth, P., Merk, F. B., Leav, I., Wolfe, H. J., and Roth, J. (1992) Comparison of ^{35}S- and digoxigenin-labeled RNA and oligonucleotide probes for *in situ* hybridization. *Histochemistry* **98,** 217–228.
3. Crabb, I. D., Hughes, S. S., Hicks, D. G., Puzas, J. E., Tsao, G. J. Y., and Rosier, R. N. (1992) Nonradioactive *in situ* hybridization using digoxigenin-labeled oligonucleotides. Applications to muskuloskeletal tissues. *Am. J. Pathol.* **141,** 579–589.
4. Kerstens, H. M., Poddighe, P. J., and Hanselaar, A. G. (1995) A novel *in situ* hybridization signal amplification method based on the deposition of biotinylated tyramine. *J. Histochem. Cytochem.* **43,** 347–352.
5. Darby, I. A., Evans, B. E., Fu, P., Lim, G. B., Moritz, K. M., and Wintour, E. M. (1995) Erythropoietin gene expression in fetal and adult sheep kidney. *Br. J. Haematol.* **89,** 266–270.
6. Herbst, H., Wege, T., Milani, S., Pellegrini, G., Orzechowski, H. D., Bechstein, W. O., Neuhaus, P., Gressner, A. M., and Schuppan, D. (1997) Tissue inhibitor of metalloproteinase-1 and -2 RNA expression in rat and human liver fibrosis. *Am. J. Pathol.* **150,** 1647–1659.
7. Darby, I. A, Bisucci, T., Hewitson, T. D., and MacLellan, D. G. (1997) Apoptosis is increased in a model of diabetes-impaired wound healing in genetically diabetic mice. *Int. J. Biochem. Cell Biol.* **29,** 191–200.
8. Hewitson, T. D., Darby, I. A., Bisucci, T., Jones, C., and Becker, G. (1998) Evolution of tubulointerstitial fibrosis in experimental renal infection and scarring. *J. Am. Soc. Nephrol.* **9,** 632–642.
9. Wookey, P. J., Tikellis, C., Nobes, M., Casley, D., Cooper, M. E., and Darby, I. A. (1998) Amylin as a growth factor during fetal and postnatal development of the rat kidney. *Kidney Int.* **53,** 25–30.
10. Lindahl, P., Hellstrom, M., Kalen, M., Karlsson, L., Pekny, M., Soriano, P., and Betsholtz, C. (1998) Paracrine PDGF-B/PDGF-Rbeta signaling controls mesangial cell development in kidney glomeruli. *Development* **125,** 3313–3322.

PART 3

APPLICATIONS

20

Molecular Cytogenetic Analysis of Sperm from Infertile Males Undergoing Intracytoplasmic Sperm Injection

William G. Kearns, Myung-Geol Pang, Darren Griffin, Lesil Brihn, Michael Stacey, Gustavo F. Doncel, Sergio Oehninger, Anabil A. Acosta, Stanton F. Hoegerman

1. Introduction

1.1. Background

The successful implementation of intracytoplasmic sperm injection (ICSI) has provided a unique means to allow couples suffering from severe male infertility to achieve their reproductive goals. If the infertile man has a germ cell in the ejaculate or retrievable from the reproductive tract, ICSI can be carried out with remarkable fertilization results. However, despite the great therapeutic advantages of the technique, ICSI provides solutions to clinicians often in the absence of an etiologic or pathophysiologic diagnosis.

ICSI has become the final route to treat a variety of male infertility disorders. Cases that have failed reconstructive urological surgery, varicocele ligation, medical treatment of defined endocrinopathies or genital tract infections, ejaculatory disorders, autoimmunity, and men with "idiopathic" oligoasthenoteratozoospermia (OAT) can be directed to ICSI. Consequently, ICSI can be indicated after the unsuccessful use of specific therapies or as the initial treatment of choice depending on the degree of sperm dysfunction and numbers. Typically, patients are selected for ICSI according to the following indications: (1) poor semen parameters, which is predictive of in vitro fertilization failure; (2) previous failed fertilization; and (3) presence of obstructive or nonobstructive azoospermia, where ICSI is combined with sperm extraction from the testes or epididymis. Although the potential use of ICSI is promising,

From: *Methods in Molecular Biology*, Vol. 123: In Situ *Hybridization Protocols*
Edited by: I. A. Darby © Humana Press Inc., Totowa, NJ

there are concerns about its widespread use owing to the potential transmission of chromosomal or genetic disease and DNA integrity. Therefore, it is imperative that we continue our efforts toward the identification of the specific sperm defects involved as well as their origin (genetic, developmental, environmental, or others).

Unfortunately, more than 40% of male infertility cases are still considered "idiopathic," i.e., the causes of infertility leading to sperm deficiencies are unknown. Patient and familial observations and new molecular biology techniques developed to investigate chromosomal aberrations and molecular abnormalities have dramatically increased our understanding of infertility in general, and of male reproductive dysfunction in particular.

A personal and family history followed by a thorough physical examination is indicated in all infertile couples. The semen analysis still remains the cornerstone of the evaluation of the male potential for fertility. In addition, specialized genetic counseling is indicated if: (1) the female partner is > 35 years old, (2) there are known genetic problems in the family, (3) there are possible genetic problems because of mental retardation in family members, or (4) there are several infertile family members. The necessity for systematic genetic screening for all infertile couples has yet to be established.

There is still some controversy as to when chromosomal/genetic screening in infertile men is appropriate. In those male cases with severe OAT and in those with nonobstructive azoospermia, a karyotype is indicated. This should be extended by a search for Yq deletions, testicular biopsy, and chromosomal/ genetic analysis of spermatozoa if feasible. In cases of congenital bilateral absence of vas deferens (CBAVD) and normal renal ultrasound, a standardized test set of mutational analysis of the cystic fibrosis gene should be performed.

Only the identification of specific sperm defects will allow the development of directed therapies. Andrology testing extended by genetic screening and counseling remains an ever-growing component in the workup of the infertile couple. We enter the next millennium with many questions that remain to be answered by the hand of efficacious (but definitely improvable) screening techniques and a new formidable therapy using ICSI.

1.2. Male Contribution to Chromosome Abnormalities In Pregnancies

Aneuploidy is a common occurrence in liveborn, term pregnancies and fetal wastage. Approximately 1 in 300 newborns and more than one third of all spontaneous abortuses have an extra or missing chromosome *(1–3)*. Aneuploidy arises as a result of a meiotic nondisjunctional (NDJ) error during gametogenesis or, less commonly, by mitotic NDJ in the developing embryo. Molecular studies indicate that approx 10% of autosomal trisomy and at least 50% of sex chromosome aneuploidy arise as a result of nondisjunction during spermatogenesis *(4)*.

1.3. Occurrences of Chromosome Abnormalities in Somatic Cells of Infertile Men

Cytogenetic studies of somatic cells from infertile men have shown an elevated frequency of constitutional chromosome anomalies as compared to controls *(5–11)*. These include numerical abnormalities (most commonly mosaic and nonmosaic Kleinfelter syndrome) and structural abnormalities (especially Robertsonian translocations and deletions of the Y chromosome). De Braekeleer and Dao *(8)* reviewed mitotic studies on 2105 infertile males with either azoospermia or severe oligozoospermia (count $< 10 \times 10^6$/mL) and showed that 12% had cytogenetic abnormalities compared to less than 2% for normal controls. Aberrant semen parameters may be associated with an increased frequency of somatic chromosome abnormalities *(9)*.

1.4. Chromosome Deletions and Their Association with Aberrant Spermatogenesis

Chromosome abnormalities (numerical and structural) in somatic cells are not the only genetic aberrations found in infertile men. In 1976, Tiepolo and Zuffaridi reported cytogenetic detection of *de novo* deletions in band q11 of the Y chromosome in six azoospermic males *(10)*. They correctly hypothesized that a gene or genes critical for spermatogenesis exist within Yq11. Using polymerase chain reaction (PCR) and Southern blot strategies, Rejjo et al. *(11)* showed that approx 13% of males with nonobstructive azoospermia have deletions of the azoospermia factor gene (AZF) within interval 6 of Yq11. Testicular biopsies showed histological variation that suggested that Sertoli-cell-only syndrome and testicular maturation arrest are associated with AZF deletions. Rejjo et al. *(12)* also showed *de novo* Yq11 chromosome deletions in two men with severe oligozoospermia. One male showed the same deletion in his sperm. It appeared that the absence of AZF did not preclude spermatogenesis but reduced the rate. They hypothesized that less severe spermatogenic arrests could be caused by variations in the size and position of deletions on the Y chromosome and that another gene or genes may also play a role in spermatogenesis.

Some infertile men have Y chromosome deletions in somatic cells. A small percentage of these deletions are detected cytogenetically; however, most are found only using sequence tag site (STS)-PCR or Southern blot strategies. The frequency and location of such microdeletions in somatic cells and gametes among infertile men is unknown.

In one study, Pryor et al. *(13)* determined that 7% (14 of 200) of infertile men had deletions on the Y chromosome of somatic cells. Nine of the 14 males were azoospermic or severe oligospermic, four had oligospermia, and one was normospermic. The size and location of the deletions varied and did not correlate with the severity of spermatogenic arrest. The fathers of six infertile men were studied and 33% (2/6) had the same deletion as their sons. The authors concluded that the size and location of the deletions did not correlate with the severity of spermatogenic failure. However, of the infertile males with more severe spermatogenic arrest, all had a deletion within Yq11. In addition, of the fathers studied, 33% transmitted their genetic mutation to their sons.

The only transcription unit identified in this commonly deleted region of Yq11 is DAZ (deleted in azoospermia). DAZ encodes a putative RNA-binding protein that is expressed only in the testes *(11,12)*. The DAZ gene is transcribed in spermatogonia and perhaps in early spermatocytes. DAZ is transcribed in premeiotic cells and may function in maintaining the germ stem cell population. Mutations within DAZ appear to inhibit spermatogenesis. The size and position of the deletions within Yq11 correlate poorly with the severity of spermatogenic failure and a deletion does not appear to preclude the presence of viable sperm and possible fertilization. Deletions within Yq11, specifically DAZ, may not be the only cause for a decreased sperm count.

1.5. Use of ICSI in IVF Clinics

In vitro fertilization protocols incorporating ICSI now are performed in many centers to treat severe male infertility. In the majority of cases, the primary diagnosis is "idiopathic" infertility. This term indicates our lack of understanding of the causes of infertility in most infertile males. Most infertile men present with a variety of sperm disorders including oligo-, astheno-, and teratozoospermia, alone or in combination. One of the most severe forms is oligoasthenoteratozoospermia. From 1990 to 1996, the Andrology Laboratory at the Jones Institute performed 10,039 semen analyzes. Approximately 80% of male patients presented with abnormal semen parameters (low count, poor morphology, and abnormal motility, alone, or in combination). The only known therapy for infertile males with severe OAT to successfully fertilize oocytes is ICSI. ICSI is the direct introduction of a spermatozoan into an oocyte to achieve fertilization and through embryo transfer, pregnancy when the number of spermatozoa in the ejaculate is very low or absent. Many IVF centers isolate the motile sperm fraction to provide an enriched selection of "normal" sperm for use by ICSI. However, ICSI can be performed using spermatozoa isolated in the semen, obtained directly from the epididymis *(14,15)* or from testicular tissue *(16,17)*. Recently, pregnancies were reported following round spermatid injection by ICSI into an oocyte *(18)*. Using ICSI, fertilization and pregnancy rates approach those of natural conception, even in the presence of compromised semen parameters *(19,20)*. However, the miscarriage rate and the potential for genetic abnormalities in the offspring remain unclear. While ICSI is in clinical use, its efficacy and safety need to be clarified.

1.6. Is ICSI Associated with an Increased Miscarriage Rate and an Increased Incidence of Aneuploid Offspring Using Sperm from OAT Males?

Recently, Mercan et al. *(21)* performed a preliminary retrospective study to determine whether pregnancy outcome following ICSI is affected by the severity of sperm abnormalities. They analyzed ICSI results in patients with severe OAT and compared them to patients undergoing ICSI for other indications. Group A included ICSI patients with severe OAT (< 20 million/mL, motility < 20% and normal morphology < 4% by strict criteria) and group B included patients undergoing ICSI for other indications (i.e., failure to fertilize by standard in vitro fertilization [IVF] protocols). The mean age of the female partner was 34 for group A and 35 for group B. The mean number of embryos transferred was 4 for group A and 3.9 for group B. The mean semen parameters for group A were 5.7 million/mL sperm count, 15% motility, and 1.3% normal sperm morphology. In contrast, the mean semen parameters for group B were 48.2 million/mL count, 40% motility, and 4.5% normal sperm morphology. The miscarriage frequency observed in group A was 39% and in group B it was 17%. The observed miscarriage rate in couples undergoing standard IVF protocols for factors other than male infertility was 26%. These data suggest that clinical pregnancies following ICSI using sperm from males with severe OAT may carry an increased risk of miscarriage. Owing to the low number of patients in the investigation, further studies are required to achieve statistical meaningful results.

The possibility that aneuploid offspring may be relatively common following ICSI is suggested by the results of Tournaye et al. *(22)* and cumulative data reported by Van Steirteghem (IX World Congress on In Vitro Fertilization and Assisted Reproduction, April 3–7, 1995, Vienna, Austria and Andrology in the Nineties, March 13–14, 1997, Jones Institute for Women's Health, Norfolk, VA, USA). They reported on 1275 consecutive treatment cycles using ICSI performed in 919 couples with male factor infertility. All couples had at least one failed conventional IVF treatment cycle. Some males had semen parameters incompatible with standard IVF or were afflicted with excretory azoospermia, which required microsurgical epididymal sperm aspiration or testicular sperm retrieval. Not all males were severe OAT. The mean maternal age was 32 years. One percent of 491 fetuses diagnosed prenatally had sex chromosome aneuploidy (47,XXX; 47,XYY; two 47,XXY; 46,XX/47,XYY) and an additional fetus had trisomy 20. An additional pregnancy yielded a Down syndrome newborn. The expected collective frequency of 47,XXX, 47,XYY, and 47,XXY is approx 3 per 1000 (0.3%) *(1–4)*. The frequency of sex chromosome aneuploidy observed in this study is therefore three to four times the frequency seen in newborns from the general population. In't Veld et al. *(23)* performed prenatal diagnoses for advanced maternal age on 12 patients whose pregnancies were established by ICSI in IVF clinics in Belgium and the Netherlands. Four of fifteen (27%) were diagnosed with chromosomal abnormalities. Three patients had a twin pregnancy. Cytogenetic results identified two individuals with XXY, one with 45,X/46,X,dic(Y)(q11)/46,X,del(Y)(q11), and two with Turner syndrome. Subsequent to this report, In't Veld suggested that this preliminary study might be an atypical sample. Further studies are necessary to clarify whether ICSI is associated with an increased miscarriage rate and an increased incidence of aneuploid offspring using sperm from males with severe OAT.

1.7. Aneuploidy in Sperm from OAT Males

Although ICSI has revolutionized assisted reproduction technologies; the genetic risks associated with ICSI remain unclear. Pang et al. *(24)* performed a preliminary study on 15 patients using fluorescent *in situ* hybridization (FISH) to determine aneuploidy for chromosomes 1, X, and Y. OAT was defined by the criteria of both Kruger et al. *(25)* for morphology and the WHO *(26)* for concentration and motility. There were significant increases in aneuploidy for all analyzed chromosomes in patients vs controls. Moosani et al. *(27)* reported both sperm karyotypes and FISH analyzes of sperm from five infertile men with normal somatic karyotypes. Two were teratozoospermic, two were oligospermic, and one was asthenozoospermic. FISH analyzes showed a significant increase in the frequency of disomy for chromosome 1 in three teratozoospermic or oligozoospermic patients. One oligozoospermic patient showed a significant increase in the frequency of disomy for chromosome 12 vs controls. The frequency of XY disomy was significantly increased in four of five patients studied. Sperm karyotypes showed a significant increase in the frequency of numerical abnormalities relative to controls. Newberg et al. *(28)* showed significantly higher frequencies of disomy for chromosomes 18, X, and Y in sperm from a male with moderate OAT and somatic cell mosaicism; 46,XY (90%)/45,X (10%). Miharu et al. *(29)* analyzed sperm from 9 fertile and 21 infertile males for aneuploidy for chromosomes 1, 16, X, and Y using one-color FISH. They found no significant differences between fertile and infertile males in aneuploidy frequency. However, the type of male infertility was not reported. Francisco et al. *(30)* analyzed whole, pellet, and swim-up semen fractions for aneuploidy from four infertile males: two with teratozoospermia, one with mild teratozoospermia, and one with normal semen parameters. All fractions showed a significantly higher ($p < 0.05$) frequency of chromosome 18 disomy but not of sex chromosome aneuploidy in pooled data. Diploid sperm remained in the pellet fraction. In't Veld et al. *(31)* performed molecular cytogenetic analysis on sperm from a male with OAT. FISH analysis showed *de novo* chromosome abnormalities in nearly all sperm cells analyzed. Fewer than 2% of cells were haploid, 40% were diploid, 24% were triploid, and 22% were aneuploid for the sex chromosomes. Lahdetie et al. *(32)* analyzed sperm by FISH for chromosomes 1 and 7 from four males with severe OAT. Pooled data showed a significant increase ($p < 0.05$) in aneuploid and diploid sperm from patients vs controls. They noted that this was due primarily to one patient with a high frequency of hyperhaploid sperm. These data suggest there may be a correlation between increased frequencies of nondisjunction and impaired semen parameters. Furthermore, preliminary data suggest that techniques that isolate the more motile sperm for use in ICSI, may not select for haploid sperm.

1.8. Experimental Design

1.8.1. Optimization of Experimental Protocols

Incomplete or overdecondensing conditions for sperm head decondensation will contribute to an overestimation of aneuploid sperm. Conditions were optimized for sperm decondensation, head swelling, and FISH analysis. We compared the physical parameters of sperm (length, area, perimeter, and degree of roundness) and the copy number determined by FISH of an autosome and the gonosomes in sperm decondensed using different protocols. Optimal decondensation and FISH conditions were determined based upon a lower frequency of cells with spurious "nullisomy" due to overdecondensation or hybridization failure without inducing spurious "disomy" resulting from increased distances between split signals. Based upon these measurements, we are able to exclude all overdecondensed or underdecondensed sperm from our FISH analysis.

1.8.2. Determination of Aneuploidy Prevalence Among Sperm from Fertile Males **(33)**

Semen was obtained by masturbation from four proven fertile controls and the sperm was immediately decondensed using dithiothreitol (DTT) and a 37°C incubation. Over- and underdecondensed sperm were excluded from our analysis. Two and three-probe FISH was performed for chromosomes 4, 6, 7, 8, 9, 10, 11, 12, 13, 17, 18, 21, X, and Y on decondensed sperm. One probe-set hybridization mix included chromosomes 18, X, and Y so as to differentiate meiosis I from meiosis II errors for the gonosomes. More than 30,000 sperm were scored. The frequency of total aneuploidy was estimated by calculating, using formulae developed by Hoegerman et al. *(34)*, per chromosome aneuploidy frequencies as twice the disomy frequency. These calculations assume independence of nondisjunctional events within the same cell. The total aneuploidy for these controls ranged between 3.9 and 7.7%. The mean frequency of diploid sperm was 0.04%. These data suggest that between 3.9 and 7.7% of sperm from controls is aneuploid.

1.8.3 Determination of Aneuploidy Prevalence Among Sperm from Males With Severe OAT (35,36)

This study was designed to detect aneuploidy in sperm from males with severe OAT. FISH determined aneuploidy frequencies of 12 autosomes and the gonosomes on sperm from fresh ejaculate of OAT patients. The 19 patients ranged in age from 25 to 47 years and were classified as severe OAT. Sperm counts were between 2 and 15×10^6/mL. Motility ranged between 17.9% and 41.1% and between 1% and 4.4% sperm showed normal forms. FISH, using direct-labeled (fluorochrome-dUTP) satellite or cosmid contig DNA probes specific for chromosomes 4, 6, 7, 8, 9, 10, 11, 12, 13, 17, 18, 21, X, and Y, was performed on decondensed sperm. More than 23,000 sperm were scored (**Fig. 1**). According to formulae of Hoegerman et al. *(34)* discussed previously, the total aneuploidy in sperm from these 19 OAT patients ranged from 33% to 74%, of which diploid sperm ranged between 0.4% and 9.6%. For controls, total aneuploidy ranged between 4.1% and 7.9%. Chi-square or Fisher's exact tests were performed. There was a significant increase ($p < 0.01$) in total aneuploidy in sperm from males with severe OAT vs normal fertile controls. This preliminary study shows that between 33% and 74% of sperm from the males with severe OAT are aneuploid.

1.8.4 Determination of Pregnancy Results from 15 Males With Severe OAT Donating Sperm for ICSI (35,36)

Using sperm from 15 of the OAT males described previously, we determined the outcome of oocyte fertilization and embryo transfer following ICSI/IVF. The mean maternal age of female partners was 32 years. All couples' infertility was classified as male factor. Sixteen series of ICSI were performed and 68% of oocytes fertilized. Following fresh embryo transfer, 13 of 16 (81%) failed to progress to a clinical pregnancy. One preclinical spontaneous abortion, one first trimester loss, or no resultant pregnancy occurred for 14 couples. One couple has an ongoing singleton pregnancy subsequent to transfer of five embryos. In this preliminary study, the ongoing clinical pregnancy rate was 6% (1/16). In contrast, the ongoing clinical pregnancy rate following ICSI using sperm from nonOAT males and in couples undergoing standard IVF is approx 28%. Our preliminary data show a significant decrease ($p < 0.05$) in the establishment of an ongoing clinical pregnancy using sperm from patients with severe OAT undergoing ICSI compared to couples undergoing standard IVF protocols or ICSI/IVF for reasons other than OAT *(21)*. The outcome of these IVF transfers suggest that the frequency of implantation with subsequent on-

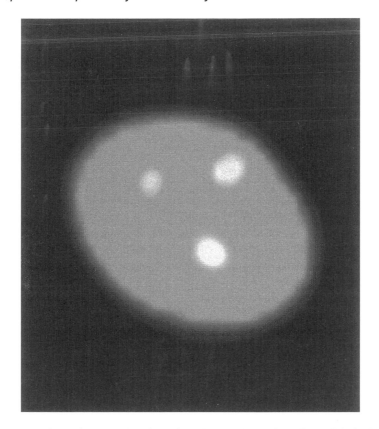

Fig. 1. Decondensed sperm head analyzed by FISH using direct labeled satellite DNA probes for chromsomes 7 and 18. Chromosome 7 is in medium gray tint and chromosome 18 is in light gray tint Sperm head is counterstained with DAPI (dark gray tint).

going pregnancy is reduced using sperm from males with severe OAT. Genetic abnormalities may be associated with these failed IVF cycles.

1.8.5. Determining Whether Swim-Up Separation Isolates Haploid from Aneuploid Sperm (36)

Most IVF centers isolate more motile sperm using methods such as swim-up or Percoll separation to select for the "best" sperm for ICSI. Sperm from the motile fraction aliquot is then used for ICSI. We isolated the motile sperm fraction from 10 of the previously described OAT patients. Aneuploidy was compared between whole semen, pellet, and swim-up fractions. The data showed significantly elevated ($p < 0.05$) frequencies of autosomal disomy and sex chromosome aneuploidy in all fractions for all patients vs controls. There was a significantly higher frequency ($p < 0.05$) of diploid sperm in the pellet fraction vs the swim-up supernatant in all patients. This suggests that swim-up separation does not separate aneuploid from haploid sperm.

1.8.6. Cytogenetics of Somatic Cells and Sperm from a 46,XY/45,X Mosaic Male with Moderate OAT (28)

Cytogenetic analysis of somatic cells and determination by FISH of aneuploidy frequencies for the sex chromosomes and chromosome 18 in sperm was performed on a 30-year-old male with primary infertility and moderate OAT. Mean semen analysis showed a sperm count of 35.8×10^6/mL, motility of 17.4%, and 7% normal morphology. Compared to controls, significantly higher ($p < 0.05$) frequencies of gonosomal and chromosome 18 aneuploidy were detected in the sperm of the patient compared to those observed in normal controls. This case report showed that a male with moderate OAT has a high frequency of aneuploid sperm and gonadal mosaicism.

1.8.7. Determination of Aneuploidy Frequencies in Sperm from Four Infertile, Non-OA, Males by FISH (30)

This study was designed to determine whether infertile males with less severe semen parameters also exhibit an increased frequency of aneuploidy in sperm. We determined aneuploidy for chromosomes 18, X, and Y by FISH in sperm from four infertile males: two with teratozoospermia, one with mild teratozoospermia, and one normospermic. Analysis showed a higher frequency of chromosome 18 disomy but not of sex chromosome aneuploidy vs controls. Elevated frequencies of chromosome 18 disomy but not sex chromosome aneuploidy suggest there may be an association between nondisjunction and impaired semen parameters. Infertile males with less severe semen parameters may have more haploid sperm compared with males with severe OAT.

1.9. Conclusions

Collectively, our data show significant increases ($p < 0.05$) in the frequencies of diploidy and both autosomal and gonosomal aneuploidy in sperm from OAT patients compared with controls. Only one successful pregnancy was achieved using these patients' sperm. The elevated levels of aneuploidy in sperm from these males with severe OAT may be causally related to this lack of success.

The elevated frequency of autosomal nondisjunction, which we observed, may not markedly impact the frequencies of autosomal abnormalities in newborns because all resulting autosomal monosomies and most trisomies are lethal. The elevated frequencies of aneuploidy observed in this population of OAT males suggest they may be at increased risk for producing aneuploid offspring.

2. Materials

1. Fresh semen.
2. Pipets.
3. Centrifuge tubes.
4. Phosphate-buffered saline (PBS), pH 7.0.
5. Centrifuge.
6. Vortex mixer.
7. Fixative: 3 parts methanol:1 part glacial acetic acid.
8. Glass slides.
9. Phase contrast microscope.
10. Denaturing solution: 1.5 M NaCl, 0.5 M NaOH.
11. Neutralizing solution: 0.5 M Tris, 3 M NaCl.
12. 2× SSC: 0.3 M NaCl, 0.03 M Na citrate.
13. 70%, 85%, and 100% Ethanol.
14. Hot block (dry heat block).
15. DNA probes (direct labeled with fluorescein isothiocyanate [FITC], rhodamine, texas red, cascade blue, Cy3 or Cy5 dNTPs).
16. Incubator.
17. FISH stringency washes: 50% formamide/2× SSC, 2× SSC, and 4× SSC, 0.05% Tween-20.
18. Water bath.
19. Coverslips.
20. Rubber cement.
21. DAPI (4',6-diamino-2-phenylindole) with antifade (p-phenylenediamine dihydrochloride).
22. Microscope: epifluorescence microscope with multiple excitation, barrier, and dichromic filters.

3. Methods

3.1. Pretreatment of Samples

1. For fresh semen: 1 mL aliquots are mixed with 4 mL of PBS, pH 7.0, to create a 1:4 ratio of semen to PBS. The sample is then centrifuged at 600g for 5 min.

    For a swim-up supernatant, swim-up pellet, or Percoll separated fractions; the sample is mixed with 4 mL of PBS, pH 7.0, and centrifuged at 600g for 5 min.
2. The supernatant from either method is then discarded. If no pellet is evident, keep the bottom 250 µL.
3. Resuspend the pellet in 1 mL of PBS containing 6 mM EDTA, pH 7.0, at room temperature.
4. Mix thoroughly or vortex and then centrifuge at 2000 rpm for 5 min.
5. Discard the supernatant.
6. Add 1 mL of PBS containing 2 mM DTT, pH 7.0, at room temperature and mix or vortex thoroughly and incubate at 37°C for 45 min.
7. Resuspend pellet with 2 mL of PBS, pH 7.0.

8. Mix or vortex thoroughly and centrifuge at 2000 rpm for 5 min.
9. Discard supernatant and mix remaining fluid by tapping with your fingers.
10. Resuspend thoroughly by vortexing.
11. Gently add 5 mL of fixative (3:1 methanol:glacial acetic acid) in a slow, dropwise manner.
12. Leave the sample at room temperature for 30 min.
13. Centrifuge the sample at 2000 rpm for 5 min and then discard supernatant.
14. Gently add fixative to make a total volume of 5 mL.
15. Fixed samples are stored at –20°C.

3.2. Making Slides

1. Clean the glass slides with ethanol.
2. Drop from a small distance 10 µL of sample onto a slide.
3. Tilt the slide at a 45° angle, allowing the sample to run down the slide for the best distribution.
4. Immediately heat slides over a small beaker of boiling water for 10 s to 2 min.
5. Place in 60% glacial acetic acid for 5–15 s and let the slides air dry.
6. Observe the slides under a phase microscope to confirm even distribution of sample and the absence of clumping.
7. More fixative and vortexing can be used to try to decrease semen cohesion and/or concentration of semen in sample.

3.3. Probe Labeling

1. DNA of the desired chromosomes α-satellite sequence is labeled with a "reporter" by nick-translation, random priming, or PCR incorporation. Labeled DNA is stored at –20°C.

3.4. Denaturation

Target DNA denaturation is accomplished by alkaline denaturation. The probe is denatured separately, on a 75°C heat block for 3–5 min. To accomplish this, a series of washes for the fixed slides must be prepared.

1. Denature using 1.5 *M* NaCl and 0.5 *M* NaOH (*see* **Note 1**).
2. Neutralize with 0.5 *M* Tris and 3 *M* NaCl.
3. Wash in 2× SSC, pH 7.0. (Slides should be washed in each solution for 3 min.)
4. Wash in ice cold ethanol series of 70%, 85%, and 100% for 2 min each.
5. For the last 5 min of the ethanol washes heat the probe made earlier on a 65–70°C hot block for 5 min and then immediately put on ice.
6. Add 10 µL of denatured probe to each air dried slide.
7. Cover with coverslip and apply rubber cement with a pipet to the rims of the coverslip.
8. Place in a moist chamber and incubate at 37°C for at least 2 h.

3.5. Washes

Heat water bath to 37°C.

1. All washes were added to Coplin jars and allowed to reach water bath temperature.
2. Remove slides from incubator and remove rubber cement. Let the coverslip wash off during the first wash.
3. Wash 3× for 5 min each in 50% formamide/2× SSC, pH 7.0.
4. Wash for 5 min in 2× SSC, pH 7.0.
5. Wash for 5 min in 4× SSC Tween-20, pH 7.0.
6. Wash for 5 min in 4× SSC Tween-20, pH 7.0, at room temperature.
7. Apply 10 μL of DAPI with antifade to each slide and cover with coverslip.
8. View under an epifluorescent microscope using appropriate filters.

4. Notes

1. The denaturation step may require some experimentation to find the optimum time. Some samples have required up to 10 min in denaturation solution to yield a scorable signal.

References

1. Hassold, T. J., Matsuyama, A., Newlands, I. M., Matsuura, J. S., Jacobs, P. A., Manuel, B., and Tseui, J. (1978) A cytogenetic survey of spontaneous abortions in Hawaii. *Ann. Hum. Genet.* **41,** 443–454.
2. Thompson, M. W., McInnes, R. R., and Willard, H. F. (1991) *General Principles and Autosomal Abnormalities Genetics in Medicine*, 5th ed., W. B. Saunders, Philadelphia, pp. 201–202.
3. Jacobs, P. A. (1992) The chromosome complement of human gametes. *Oxford Rev. Reprod. Biol.* **14,** 47–72.
4. Koehler, K. E., Hawley, R. S., Sherman, S., and Hassold, T. (1996) Recombination and nondisjunction in humans and flies. *Hum. Mol. Genet.* **5,** 1495–1504.
5. Baschat, A. A., Kuper, W., Al Hasan, S., Diedrich, K., and Schwinger, E. (1996) Results of cytogenetic analysis in men with severe subfertility prior to intracytoplasmic sperm injection. *Hum. Reprod.* **11,** 330–333.
6. Chandley, A. C. (1979) The chromosomal basis of human infertility. *Br. Med. Bull.* **25,** 181–186.
7. Bourrouillou, G., Dastugue, N., and Colombie, P. (1985) Chromosome studies in 952 infertile males with a sperm count below 10 million/mL. *Hum. Genet.* **71,** 366–370.
8. De Braekeleer, M. and Dao, T.-N. (1991) Cytogenetic studies in male infertility: a review. *Hum. Reprod.* **6,** 245–250.
9. Kjessler, B. (1974) Chromosomal constitution and male reproductive failure, in *Male Fertility and Sterility* (Mancini, R. E. and Martini, L., eds.), Academic Press, New York, pp. 231–247.
10. Tiepolo, L. and Zuffardi, O. (1976) Localization of factors controlling spermatogenesis in the nonfluorescent portion of the Y chromosome long arm. *Hum. Genet.* **34,** 119–124.
11. Rejjo, R., Lee, T.-Y., Salo, P., Alagappan, R., Brown, L. G., Rosenberg, M., Rozen, S., Jaffe, T., Straus, D., Hovatta, O., et al. (1995) Diverse spermatogenic defects in humans caused by Y chromosome deletions encompassing a novel RNA-binding protein gene. *Nat. Genet.* **10,** 383–393.

12. Rejjo, R., Alagappan, R., Patrizio, P., and Page, D. C. (1996) Severe oligozoospermia resulting from deletions of azoospermia factor gene on Y chromosome. *Lancet* **347,** 1290–1293.

13. Pryor, J. L., Kent-First, M., Muallem, A., Van Bergen, A., Nolten, W. E., Meisner, L., and Roberts, K. P. (1997) Microdeletions in the Y chromosome of infertile men. *N. Engl. J. Med.* **336,** 534–539.

14. Liu, J., Lissens, W., Silber, S. J., Devroey, P., Liebaers, I., and Van Steirteghem, A. (1994) Birth after preimplantation diagnosis of the cystic fibrosis delta F508 mutation by polymerase chain reaction in human embryos resulting from intracytoplasmic sperm injection with epididymal sperm. *JAMA* **272,** 1858–1860.

15. Tournaye, H., Devroey, P., Liu, J., Nagy, Z., Lissens, W., and Van Steirteghem, A. (1994) Microsurgical epididymal sperm aspiration and intracytoplasmic sperm injection: a new effective approach to infertility as a result of congenital bilateral absence of vas deferens. *Fertil. Steril.* **61,** 1045–1051.

16. Devroey, P., Liu, J., Nagy, Z., Tournaye, H., Silber, S. J., and Van Steirteghem, A. (1994) Normal fertilization of human oocytes after testicular sperm extraction and intracytoplasmic sperm injection. *Fertil. Steril.* **62,** 639–641.

17. Silber, S. J., Van Steirteghem, A. C., Liu, J., Nagy, Z., Tournaye, H., and Devroey, P. (1995) High fertilization and pregnancy rate after sperm injection with spermatozoa obtained from testicle biopsy. *Hum. Reprod.* **10,** 752–762.

18. Tesarik, J., Mendoza, C., and Testard, J. (1995) Viable embryos from injection of round spermatids into oocytes. *N. Engl. J. Med.* **333,** 525.

19. Van Steirteghem, A. C., Liu, J., Joris, H., Nagy, Z., Janssenswillen, C., Tournaye, H., Derde, M. P., Van Assche, E., and Devroey, P. (1993) Higher success rate by intracytoplasmic sperm injection than by subzonal insemination. A report of a second series of 300 consecutive treatment cycles. *Hum. Reprod.* **8,** 1055–1060.

20. Nagy, Z. P., Liu, J., Joris, H., Brocken, G., Tournaye, H., Devroey, P., and Van Steirteghem, A. C. (1994) Extremely impaired semen parameters and the outcome of intracytoplasmic injection. *Hum. Reprod.* **9(Suppl 4),** 27–29.

21. Mercan, R., Lanzedorf, S., Nassar, A., Mayer, J., Muasher, S. J., and Oehninger, S. (1997) The outcome of clinical pregnancies following ICSI is not affected by semen quality. Abstract, American Society for Reproductive Medicine, 53rd Annual Meeting, October 18–22, Cincinnati, OH.

22. Tournaye, H., Liu, J., Nagy, J., Joris, H., Wisanto, A., Bondelle, M., Van der Elst, J., Staessen, C., Smitz, J., Silber, S., Devroey, P., Liebaers, I., and Van Steirteghem, A. (1995) Intracytoplasmic sperm injection (ICSI): the Brussels experience. *Reprod. Fertil. Dev.* **7,** 269–279.

23. In't Veld, P., Brandenburg, H., and Verhoeff, A. (1995) Sex chromosome abnormalities and intracytoplasmic sperm injection. *Lancet* **346,** 773.

24. Pang, M. G., Zackowski, J. L., Ryu, B. Y., Moon, S. Y., Acosta, A. A., and Kearns, W. G. (1994) Aneuploidy detection for chromosomes 1, X and Y by fluorescence *in situ* hybridization in human sperm from oligoasthenoteratozoospermic patients. *Am. J. Hum. Genet.* **pA113** (abstract 644).

25. Kruger, T. F., Menkveld, R., Stander, F. S., Lombard, C. J., Van der Merwe, J. P., Van Zyl, J. A., and Smith, K. (1986) Sperm morphology features as a prognostic factor in in vitro fertilization. *Fertil. Steril.* **46,** 1118–1123.

26. World Health Organization (1992) *WHO Laboratory Manual for the Examination of Human Semen and Sperm–Cervical Mucus Interaction*, 3rd ed. Cambridge University Press, Cambridge, UK.
27. Moosani, N., Pattinson, H. A., Carter, M. D., Cox, D., Rademaker, A. W., and Martin, R. H. (1996) Chromosomal analysis of sperm from men with idiopathic infertility using sperm karyotyping and fluorescence *in situ* hybridization. *Fertil Steril.* **64,** 811–817.
28. Newberg, M. T., Francisco, R. G., Pang, M. G., Brugo, S., Doncel, G. F., Acosta, A. A., Hoegerman, S. F., and Kearns, W. G. (1998) Cytogenetics of somatic cells and sperm from a 46,XY/45,X mosaic male with moderate oligoasthenoteratoospermia. *Fertil. Steril.* **69,** 146–148.
29. Miharu, N., Best, R. G., and Young, S. R. (1994) Numerical chromosome abnormalities in spermatozoa of fertile and infertile men detected by fluorescence *in situ* hybridization. *Hum. Genet.* **93,** 502–506.
30. Francisco, R. G., Newberg, M. T., Pang, M. G., Doncel, G. F., Acosta, A. A., Stacey, M., Osgood, C., Hoegerman, S. F., and Kearns, W. G. (1999) Determining aneuploidy and Y chromosome deletions in four infertile, non-oligoasthenozoospermic (OA) males. *Fertil. Steril.*, under review.
31. In't Veld, P., Broekmans, F. J. M., de France, H. F., Pearson, P. L., Pieters, M. H. E. C., and van Kooij, R. J. (1997) Intracytoplasmic sperm injection (ICSI) and chromosomally abnormal spermatozoa. *Hum. Reprod.* **12,** 752–754.
32. Lahdetie, J., Saari, N., Ajosenpaa-Saari, M., and Mykkanen, J. (1997) Incidence of aneuploid spermatozoa among infertile men studied by multicolor fluorescence *in situ* hybridization. *Am. J. Med. Genet.* **71,** 115–121.
33. Pang, M. G., Hoegerman, S. F., Cuticchia, A. J., Friedman, E., Moon, S. Y., Acosta, A. A., and Kearns, W. G. (1998) Detection of cytogenetic abnormalities in human sperm by fluorescence *in situ* hybridization (FISH) II: Detection of aneuploidy for chromosomes 4, 6, 8, 9, 10, 11, 12, 13, 17, 18, 21, X and Y and diploidy in sperm from fertile males. *Cyto. Cell Genet.*, submitted.
34. Hoegerman, S. F., Jenkins, C. D., Pang, M. G. and Kearns, W. G. (1999) Detection of cytogenetic abnormalities in human sperm by fluorescence *in situ* hybridization (FISH): theoretical considerations and a comparison of FISH data with results of studies of the chromosome of male pronuclei. *Human Reprod.*, in press.
35. Pang, M. G., Hoegerman, S. F., Cuticchia, A. J., Moon, S. Y., Doncel, G., Acosta, A. A. and Kearns, W. G. (1998) Detection of aneuploidy for chromosomes 4, 6, 8, 10, 11, 12, 13, 17, 18, 21, X and Y by fluorescence *in situ* hybridization (FISH) in sperm from nine oligoasthenoteratozoospermic (OAT) patients undergoing ICSI. *Hum. Reprod.*, in press.
36. Pfeffer, J., Pang, M. G., Hoegerman, S. F., Cuticchia, A. J., Oehninger, S. and Kearns, W. G. (1998) Detection of aneuploidy using fluorescence *in situ* hybridization (FISH) in sperm from patients with oligoasenoteratozoospermia after isolation of the motile sperm fraction. *Hum. Reprod.*, submitted.

21

Localization of HIV-1 DNA and Tumor Necrosis Factor-α mRNA in Human Brain Using Polymerase Chain Reaction *In Situ* Hybridization and Immunocytochemistry

Steven L. Wesselingh and Jonathan D. Glass

1. Introduction

We have had a long term interest in the neuropathogenesis of human immunodeficiency virus (HIV)-associated neurological diseases. Neurological disease is a prominent feature of HIV-1 infection, usually occurring during the late stages of acquired immunodeficiency syndrome (AIDS). Opportunistic infections and neoplasms in the brain arise as a consequence of severe immunosuppression, and HIV infection of the brain has been implicated in the pathogenesis of a progressive cognitive decline termed HIV-associated dementia. In light of the clinical evidence for neuronal dysfunction in HIV-related neurological disease and the pathological evidence for loss of neurons, a puzzling feature of brain infection with HIV is that the overwhelming productively infected cells are brain macrophages and microglia, and not cells of neuroectodermal origin such as neurons, astrocytes, and oligodendrocytes. This has been demonstrated repeatedly in many laboratories by the methods of immunocytochemistry for detection of HIV proteins and by *in situ* hybridization (ISH) for HIV RNA.

There are several possible explanations for the observed discrepancy between the neuronal dysfunction clinically evident in some AIDS patients and an inability to detect infection of neurons, astrocytes, or oligodendrocytes. One is that neither standard immunocytochemistry nor ISH techniques are sensitive enough to show HIV infection of low abundance. A second possibility is that "nonproductive" infection, i.e., the presence of HIV DNA or HIV transcripts

From: *Methods in Molecular Biology*, Vol. 123: In Situ *Hybridization Protocols*
Edited by: I. A. Darby © Humana Press Inc., Totowa, NJ

without HIV particle production, might alter cellular function. This scenario could arise by latent incorporation of virus into the host genome, or the presence of replication-defective virus. Indeed, recent studies have demonstrated the presence of HIV nucleic acids in the absence of HIV proteins in astrocytes from pediatric AIDS brains. A third possibility is that direct infection of neuroectodermal cells is not an important factor in the pathogenesis of HIV-induced neurological disease, and that infection of microglia and macrophages somehow leads to neuronal dysfunction via indirect neurotoxic mechanisms such as the production of neurotoxic cytokines by macrophages or microglia.

To understand more completely the mechanism by which neuronal dysfunction or death occurs in HIV-infected patients it has been necessary to perform detailed examinations of brain tissue to identify which cells are HIV infected and which cells are producing cytokines. *In situ* hybridization has had limited usefulness in identifying central nervous system (CNS) cells producing cytokine mRNAs because of the low abundance of the mRNA in an individual cell. This problem is exacerbated in human autopsy tissue because of variable postmortem times, and suboptimal fixation.

To detect all cells containing HIV DNA it was necesary to detect one or two copies of HIV DNA per cell. We have therefore developed a sensitive and specific assay to detect mRNA and DNA in paraffin-embedded CNS tissue *(1–3)*. Cytokine mRNA was detected using reverse transcriptase synthesis of cDNA and primer-specific polymerase chain reaction amplification of the cDNA to amplify the presence of cytokine mRNA and detection by ISH. HIV DNA was detected by primer-specific polymerase chain reaction amplification of HIV DNA and detecetion by ISH. These assays were combined with immunocytochemistry using antibody to glial fibrillary acidic protein (GFAP) to identify astrocytes and ricinus communis agglutinin (RCA) to identify macrophages and microglia. Further reading is available in the form of articles which are related to the development of this technique *(4–26)*.

Development of Polymerase Chain Reaction (PCR) techniques for amplification of mRNA and DNA has enabled very sensitive analysis of changes mRNA and DNA levels in tissue, but does not provide information on the cell type, number or anatomical localization of the cells expressing cytokine mRNAs of interest. *In situ* detection of mRNAs has proven to be useful for cellular localization of some cytokine messages and viral DNA or RNA, but current techniques do not detect RNA or DNA at low levels. The development of IS/PCR combined with immunocytochemistry has enabled sensitive and specific detection and cellular localization of mRNA in paraffin-embedded CNS tissue.

2. Materials

2.1. Pretreatment

1. Super frost Plus glass slides, Menzel-Glaser, Germany.
2. Proteinase K, Boehringer Mannheim, Mannheim, Germany.
3. Proteinase K buffer: 100 mM Tris-HCl, pH 8.0, 50 mM EDTA.
4. 0.1 M Triethanolamine, pH 8.0.
5. Acetic anhydride.
6. 20× SSC: 3 M NaCl, 0.3 M sodium citrate.
7. DNase (10 U/μL), Boehringer Mannheim, Mannheim, Germany.

2.2. cDNA Probes

1. For heating steps, cDNA, PCR, and hybridization use the HYBAID Omnislide (Hybaid, Teddington, UK).
2. 5×AMV Reverse Transcriptase buffer, Boehringer Mannheim, Mannheim, Germany.
3. 10 mM dNTP, Pharmacia Biotech.
4. 3' Primer, Amrad, Australia.
5. Oligo DT, Boehringer Mannheim, Germany.
6. RNase inhibitor, Boehringer Mannheim, Germany.
7. AMV reverse transcriptase, Boehringer Mannheim, Germany.

2.3. PCR

1. 25 mM MgCl$_2$, Advanced Biotechnologies.
2. 10× PCR buffer, Advanced Biotechnologies.
3. 10 mM dNTPs, Pharmacia.
4. *Taq* polymerase, Advanced Biotechnologies.
5. 20 μM Primers, Amrad.
6. Bovine serum albumin (BSA) (50 μg/μL), Sigma.

2.4. Prehybridization and Hybridization

1. Hybridization solution: 6× SSC, 1× Denhardt's solution, 400 μg/mL of sperm DNA, 2 mM Na$_4$ pyrophosphate, 0.2% sodium dodecyl sulfate (SDS), H$_2$O to final volume.
2. Probe labeling: DIG-11-dUTP, from terminal transferase kit, Boehringer Mannheim, Germany.

2.5. Immunohistochemical Detection

1. Buffer 1: 100 mM Tris-HCl, 150 mM NaCl, pH 7.5.
2. Buffer 1 containing 2% normal goat serum and 0.3% Triton X-100.
3. 200 μL Nitroblue tetrazolium (NBT) + x-phosphate, from nucleic acid detection kit, Boehringer Mannheim, Germany.
4. Levamisole.
5. Buffer 2: 100 mM Tris-HCl, 100 mM NaCl, 50 mM MgCl$_2$, pH 9.5 (*see* **step 6**).
6. Buffer 3: 10 mM Tris-HCl, 1 mM EDTA, pH 8.0.

Buffer 2 is very hard to dissolve if made up from powders; therefore make up from solutions.

Stock solution	Volume to make 4 L	Final concentration
1 M Tris-HCl	400 mL	100 mM
5 M NaCl	80 mL	100 mM
1 M MgCl$_2$	200 mL	50 mM

pH 9.5, add distilled H$_2$O to 4 L

3. Methods

3.1. In Situ *RT/PCR*

3.1.1. Pretreatment of Slides

1. Heat to 60°C for 15 min.
2. Two changes of xylene washes of 2 min each.
3. 100% (two changes), 95%, 70%, 50% then distilled H$_2$O for 2 min each.
4. 0.1% Triton X-100 for 3 min.
5. Wash in phosphate-buffered saline (PBS).

3.1.2. Permeabilization (see **Note 1**)

1. Incubate in proteinase K (10 µg/mL) in proteinase K buffer (100 mM Tris pH 8.0/ 50 mM EDTA) for 10–15 min at 37 C.
2. Wash in PBS for 5 min.
3. Post-fix in 4% paraformaldehyde for 5 min.
4. Dehydrate.

3.1.3. Acetylation

1. Equilibrate specimens in 0.1 M triethanolamine, pH 8.0, for 2 min.
2. Add 0.25% acetic anhydride (625 µL in 250 mL) for 5 min with agitation, add additional acetic anhydride to reach 0.5% (625 µL) for 5 min with agitation.
3. Block acetylation in 2× SSC for 5 min.

3.1.4. DNase Treatment (see **Note 2**)

1. Incubate in DNase (10 U/µL) at 750 U/mL in PBS, overnight at room temperature.
2. Wash in PBS.
3. Dehydrate.
4. Heat at 92°C for 5 min.

3.1.5. Reverse Transcription

1. Make up mix:

5× Avian myeloblastosis virus (AMV) RT buffer	4 µL
10 mM dNTP	2 µL
3' Primer	2 µL
Oligo DT	0.5 µL

RNase inhibitor	0.5 μL
AMV RT	1 μL
H₂O	10 μL

2. Add 20 μL/slide at 42°C for 60 min; cover with parafilm coverslip.
3. Wash in PBS.
4. Dehydrate.
5. Heat to 92°C for 5 min.

3.1.6. PCR Reaction (see **Note 3**)

1. Heat slide to 55°C.
2. Make up PCR mix:

25 m*M* MgCl₂	2 μL
10× PCR buffer	3 μL
10 m*M* dNTPs	2 μL
Taq polymerase	0.5 μL
20 μ*M* primers	2.5 μL (each)
BSA (50 μg/μL)	2 μL
H₂O	15.5 μL
Total	30 μL

3. Heat solution to 75°C, add 30 μL/slide, then coverslip and seal with nail polish. Start cycles: 94°C, 3 min; 55°C, 2 min; 94°C, 1 min; 15–20 cycles. Remove coverslip.
4. Wash in PBS for 5 min.
5. Dehydrate through graded ethanol solutions.

3.1.7. Prehybridization and Hybridization

1. Prehybridize in hybridization solution, 200 μL/slide, at 37°C for 30 min.
2. Rinse in PBS. Add DIG-labeled probe (15% formamide + hybridization solution); 60 μL of probe + 225 μL of formamide + 1.215 mL of hybridization solution. Incubate for 90°C for 2–3 min, then overnight at 37°C in a humid chamber.
3. Immerse slides in 6× SSC for 10 min at room temperature.
4. Wash slides in 2× SSC/0.1% SDS at 55°C, 2×, 5 min each.
5. Wash slides in 0.5× SSC/0.1% SDS at 55°C, 2×, 5 min each.
6. Immerse slides in 2× SSC, 1 min, at room temperature.

3.1.8. Immunological Detection (see **Note 4**)

Use Boehringer Mannheim nucleic acid detection kit.

1. Wash slides for 1 min in buffer 1 (100 m*M* Tris-HCl/150 m*M* NaCl, pH 7.5) at room temperature.
2. Incubate slides in buffer 1 containing 2% normal goat serum (5 mL in 250 mL) and 0.3% Triton X-100 (0.75 mL in 250 mL) for 30 min at room temperature.
3. Dilute anti-DIG antibody 1:500 with buffer 1 containing 1% normal goat serum (2.5 mL) and 0.3% Triton X-100.

4. Apply 100 µL of diluted antibody to the slides and incubate in a humid chamber at room temperature for 3 h.
5. Wash slides for 10 min with shaking in buffer 1 at room temperature.
6. Wash slides for 10 min with shaking in buffer 2 (100 mM Tris-HCl, 100 mM NaCl, 50 mM MgCl$_2$, pH 9.5) at room temperature.
7. Apply 500 µL of color solution to each slide.
   Color solution reagents:
       200 µL of NBT + x-phosphate
       Two drops (20 µL) of levamisole solution (2.4 mg/10 mL) to 10 mL of buffer 2
8. Incubate slides in a humid chamber in absence of light for minimum of 2 h or overnight.
9. Stop reaction by incubating slides for 5 min in buffer 3 (10 mM Tris-HCl, 1 mM EDTA, pH 8.0) at room temperature.
10. Dehydrate slides in graded ethanol washes (50%, 70%, 95%, 100%) 2 min each, then 2 min in xylene.
11. Mount coverslips with nonaqueous mountant.

3.2. Quantitation of HIV DNA-Positive Cells

For each tissue block, the PCR/ISH reaction for HIV DNA was done on serial tissue sections that were also stained for one of the immunohistochemical cellular markers for macrophages/microglia, astrocytes, or neurons. At 200× magnification, all of the HIV DNA-positive cells (between 10 and 500) were identified on each of these serial sections, and the number double stained with the cellular marker was counted. The ratio of HIV DNA-positive macrophages/microglia, astrocytes, and neurons was calculated independently for each tissue block, and the ratio of HIV DNA-positive cells also stained with the cellular marker and the total HIV DNA-positive cells was expressed as a percentage.

3.3. Results of IS/PCR

Twenty-nine tissue blocks from 21 cases were studied: 18 tissue blocks from group 1 (10 AIDS cases), 5 from group 2 (5 AIDS cases), and 6 from HIV-seronegative controls (*see* **Notes 5** and **6**).

Double-label immunohistochemistry showed that all gp41 cells were macrophages/microglia; there were no examples of gp41-positive astrocytes, endothelial cells, or neurons. Gp41-positive cells were present in all of the group 1 cases, and were absent in group 2 and in seronegative controls.

3.4. Detection of HIV-1 DNA by PCR/ISH and Identification of Infected Cells

Primers and probes to detect HIV group-specific antigen (gag) were used. Positive cells showed a dark blue signal in the nuclei. There was variability from case to case in the ability to detect gag, which correlated with the presence of HIV as determined by immunocytochemistry for gp41. PCR/ISH detected the presence of HIV in all 18 tissue blocks showing immunocytochemical and/or morphological evidence of productive HIV infection (group 1), and in 1/5 tissue blocks not showing evidence of productive HIV infection. The PCR/ISH protocol without amplification (no *Taq* polymerase) reduced both the number and intensity of stained nuclei (**Fig. 1E**). The use of irrelevant oliogonucleotide probes gave no signal (**Fig. 1F**). There were no positive signals in the HIV-seronegative control tissue.

In every case positive for HIV DNA, the majority of cells containing the virus were identified immunohistochemically as macrophages/microglia. Both cells with round macrophage morphology and stellate microglia morphology were found to be positive. They were located around blood vessels, within inflammatory foci, and singly within the parenchyma. In serial sections from any individual tissue block more cells were positive for HIV DNA by PCR/ISH than were positive for HIV gp41 by immunocytochemistry. The gp41 immunohistochemistry was incompatible with the PCR/ISH process, so double staining with this antibody could not be performed, but in general the types of cells and distribution in serial sections were similar for the two techniques.

The quantitative method showed that virtually all of the HIV DNA-positive cells were accounted for by immunoidentification with RCA-1 and GFAP. HIV DNA-positive astrocytes were identified in 18/19 positive tissue blocks, accounting for 3–30% of positive cells. Seven of these blocks showed > 20% of HIV DNA-positive cells were astrocytes. gp41 immunohistochemistry did not reveal any examples of productively HIV-infected astrocytes or neurons. There were also no convincing examples of HIV infection of endothelial cells or oligodendroglia either by PCR/ISH or gp41 immunohistochemistry, although specific antibodies for identification of these cells were not used.

The sensitivity of our methodology was determined by detection of host gene DNA. We chose TFIID (TATA-binding protein), a host gene with only one genomic copy, by using primers designed to amplify a target comparable in size to the gag target. We were able to detect TFIID consistently in the majority of nuclei on a section of brain or spinal cord (not shown).

3.5. Detection of Tumor Necrosis Factor-α (TNFα) mRNA by IS RT/ PCR (see Note 7)

We had previously demonstrated increased levels of TNFα mRNA in the brains of patients with HIV-associated dementia by RT/PCR of total RNA extracted from homogenized brain tissue. We therefore examined brain sections from patients with HIV dementia for TNFα mRNA using IS RT/PCR. We examined sections from the brains and spinal cords of five demented and five nondemented AIDS patients and four HIV-seronegative control patients.

Numerous cells positive for TNFα mRNA were present both in the gray and white matter but they were most abundant in white matter.

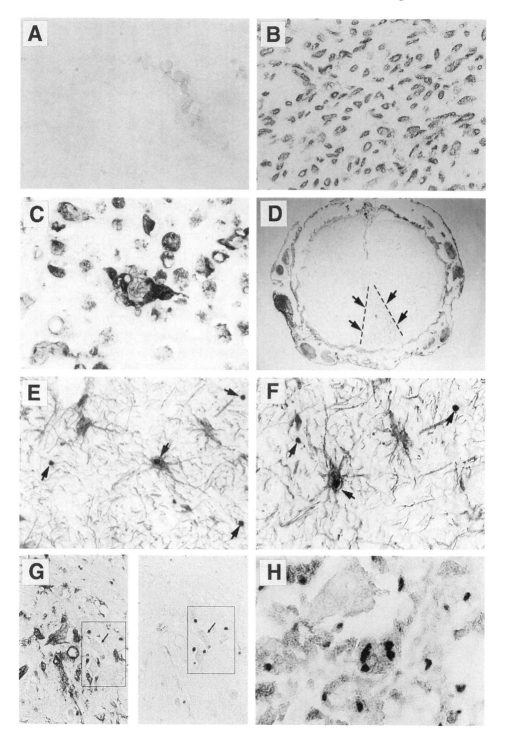

Combined RT/PCR/ISH and immunohistochemistry revealed that the cells positive for TNFα mRNA were most abundant in the white matter and the majority were RCA positive and therefore of macrophage lineage. Only very occasional TNFα mRNA-positive cells were GFAP-positive and therefore astrocytes.

Spinal cords from six AIDS patients with vacuolar myelopathy, six AIDS patients without vacuolar myelopathy, and three seronegative controls were examined. Combined RT/PCR/ISH and immunohistochemistry revealed that in cases the cells positive for TNFα mRNA were RCA-positive and therefore of macrophage lineage.

Utilizing tumors from *scid* mice inoculated with CHO cells transfected with the *HuMIG* gene in both orientations in respect of the promotor we were able to test the specificity of the technique. The tumors with the promotor and gene in the correct orientation had detectable *HuMIG* mRNA as determined by IS RT/PCR whereas the tumors with the gene in the reverse orientation were negative.

Fig. 1. (**A,B**) Human MIG transcripts in CHO cells injected into scid mice were used as negative and positive controls. (**A**) Absence of signal when CHO cells transfected with the gene in reverse orientation. (**B**) CHO cells transfected with HuMIG in the correct orientation, signal is well demonstrated and discretely localized by RT/PCR/ISH. (**C,D**) High- and low-power photomicrographs of RT/PCR/ISH coupled with immunocytochemistry performed on a section of spinal cord from a patient with severe vacuolar myelopathy. Localization of TNFα transcripts in macrophages by combined RT/PCR/ISH for TNFα and immunocytochemistry with HAM56. Note the affected region demonstrated by the lines and arrows at low power. The macrophages, including the perivascular macrophages, have a dark blue-black reaction, reflecting the brown immunocytochemistry and the blue RT/PCR/ISH reaction products. The endothelial cells have only the brown immunocytochemistry Most or all of the TNFα positive cells are macrophages. (**E,F,G,H**) Representative examples of PCR/ISH for HIV DNA, and immunoidentification of infected cell types. (**E,F**) PCR/ISH with immunoidentification of astrocytes. Many glial fibrillary acidic protein (GFAP)-positive astrocytes are seen in (**E**), only one of which has a nuclear PCR/ISH signal (*arrow*) and two positive cells that are not astrocytes (*arrows*). Another region from the same section (**F**) shows three PCR/DNA-positive nuclei (blue) (*arrow*), only one of which is GFAP positive (*arrow*). (**G**) Relationship of HIV DNA-positive cells to neurons, pontine region. PCR/ISH-positive cells are demonstrated adjacent to large neurons (*inset*). Staining with RCA-1 demonstrates the positive cells to be macrophages/microglia. Myelin-associated protein type 2 (MAP-2)-positive cells are show next to MAP-2-negative, DNA-positive cells (**H**) Region of microglial/macrophage infiltration, typical of a large microglial nodule, in the region of the caudate. PCR/ISH doubles-labeled with RCA-1 for macrophages/microglia PCR/ISH shows many DNA-positive cells, the positive cells are all identified as microglial/macrophages by RCA-1.

3.6. Discussion

The PCR/ISH technique is an extraordinarily sensitive method for detecting the presence of DNA in specific cell types. Because of the ability to use formalin-fixed, paraffin-embedded archival tissue blocks, PCR/ISH is useful for answering questions about the presence of even low abundance infectious agents in human autopsy tissue. For detection of HIV infection in the brain, we found that PCR/ISH was at least as sensitive as immunocytochemistry for showing infection of macrophages and microglia. All of the tissue sections identified to be HIV-infected by the presence of HIV-multinucleated giant cells and/or gp41 immunoreactivity were positive by PCR/ISH. In most cases PCR/ISH revealed a higher abundance of macrophage/microglia infection than did gp41 immunohistochemistry in individual tissue sections. It is not clear whether these DNA-positive but gp41-negative cells had low levels of HIV proteins (undetectable by immunocytochemistry) or were latently infected with the virus, or contained defective virus.

As has been demonstrated by several other laboratories, the majority of cells containing HIV DNA were macrophages/microglia. In several cases, however, we found that a significant minority of cells harboring HIV DNA were astrocytes, although we found no examples of gp41-positive astrocytes by immunocytochemistry. This finding suggests that PCR/ISH is useful for the identification of latently infected cells, or cells producing levels of HIV proteins too low for immunocytochemical detection. The presence of HIV nucleic acid without HIV proteins has been reported previously in AIDS brains.

These data leave open the possibility that nonproductive infection of astrocytes may contribute to HIV neuropathogenesis. A role for astrocytes in HIV-related neurotoxicity has been suggested from in vitro culture systems.

We did not specifically identify oligodendrocytes and endothelial cells by immunocytochemical markers; however, quantitative analysis showed that virtually all of the DNA-positive cells were identified as macrophages/microglia or astrocytes. The summed percentages of DNA-positive cells was remarkably close to 100% for each case, suggesting that another infected population of cells was not overlooked. These quantitative results also speak to the reproducibility of the PCR/ISH technique on serial tissue sections, as the summed percentages were never significantly greater than 100%.

Our data are consistent with the majority of previous data showing that productive HIV infection in the brain is restricted to macrophages and microglia. The importance of this conclusion is that it reiterates the need to focus on the role of infected macrophages, and possibly astrocytes, in the pathogenesis of HIV-related neurological disorders. It is clear from prior work that there is an association between neurological disease during AIDS and excessive macrophage activation with tissue elevations of macrophage-derived products, such as TNFα.

IS/RT-PCR enabled us to detect TNFα mRNA in archival paraffin-embedded tissue and is clearly more sensitive than conventional ISH (*see* **Note 8** on general considerations).

Numerous TNFα positive cells were detected in the spinal cords, brains, and nerves of patients with HIV associated neurological disease. The majority of the TNFα positive cells were identified as macrophages by combined immunohistochemistry and *in situ* RT/PCR. These data suggest that the increased levels of TNFα expression previously noted in neurological diseases associated with HIV infection are predominantly due to macrophage or microglial production of TNFα. It was of interest to find a small number of GFAP-positive cells also positive for TNFα mRNA, but these were clearly in the minority. There has been an ongoing debate concerning the production of TNFα by astrocytes, as there is considerable evidence that astrocytes are capable of producing TNFα in vitro but less in vivo evidence. The brains of patients with multiple sclerosis, subacute sclerosing panencephalitis, and adrenoleukodystrophy (ALD) have, however, by immunohistochemical techniques numerous TNFα-positive cells that appeared to be astrocytes. It is of course not necessary that the same cell types make TNFα in all pathological conditions.

4. Notes

1. The permeabilization step is critical and needs to be optimized for each mRNA and each tissue. The concentration of proteinase K, the temperature of incubation, and the length of time of incubation should all be varied in a checker-board fashion.
2. To amplify and detect TNFα mRNA, DNase and cDNA steps were added to the above technique.

   Following washes in PBS, the slides were incubated in DNase (BM) overnight, washed in PBS, and 20 µL of DNA synthesis mix applied containing 50 mM Tris-HCL (pH 8.3 at 42°C), 25 U of AMV reverse transcriptase (BM), and the 3' primer for 60 min at 42°C. Each slide was amplified for 27 cycles.
3. Following washes and dehydration, 30 µL of a PCR reaction mix containing 330 µM of each dNTP, 30 pmol of each specific primer, PCR buffer (3.5 mM $MgCl_2$, 10 mM Tris-HCl, 50 mM KCl, gelatin 0.1 mg/mL, pH 8.3 at 20°C), 16 µg bovine serum albumin (BM), and 5 U *Taq* polymerase (BM) were applied at 63°C (on a heating block). A silanized coverslip was applied and sealed with nail polish. The PCR was performed for 35 cycles (94°C, 30 s; 55°C, 30 s; and 72°C, 30 s) in a Biosycler oven (Integrated Separation Systems). The coverslips were removed and the slides were washed in PBS.

   Before hybridization, the amplified DNA was denatured with prehybridization solution (3× SSC, 1× Denhardt's solution [Sigma], 40% formamide, 50 mM Na_4 pyrophosphate, 200 µg/mL of each poly[A] and denatured salmon sperm DNA [BM], 0.1 M ATP) at 95°C for 5 min, and then chilled in ice-cold 0.5× SSC. The sections were then hybridized in the same buffer at 45°C for 18 h to a digoxigenin end-labeled oligonucleotide probe internal to the PCR primer pair. Following stringent washes in 2× SSC and 0.5× SSC at 45°C, and 0.1× SSC at 40°C, the sections were incubated for 3 h at room temperature with an anti-digoxigenin antibody coupled to alkaline phosphatase. The blue reaction product was revealed with nitroblue tetrazolium/bromochloro-indolyl phosphate.

Controls were included in each run of IS/PCR. For the PCR step, sections from the same tissue blocks were treated without *Taq* DNA polymerase and/or with irrelevant primer pairs. For the ISH step, sections without preceding PCR and/or sections treated with irrelevant oligonucleotide probes were used. Tissue from HIV-negative patients was also included in each run.

4. Immunocytochemistry was performed for identification of macrophages/microglia (anti-ferritin, 1:8000, Sigma; biotinylated RCA-1, 1:5000, Vector), astrocytes (anti-GFAP, 1:2000, Dako), neurons (anti-MAP-2, 1:2000, BM), and for HIV gp41 (anti-gp41, 1:1000, Genetic Systems). The avidin-biotin complex method was used with diaminobenzidine as the colorimetric substrate. Staining was done either singly or after the PCR/ISH.

5. Human brain tissue was obtained from the HIV-neuropathology brain bank at the Johns Hopkins Medical Institutions. Brains were fixed in 20% buffered formalin for at least 2 wk and tissue sections were embedded in paraffin.

6. Tissue from 15 cases with AIDS were studied. Eighteen tissue blocks from 10 of these cases (group 1) were chosen because they showed pathological evidence of HIV infection in the brain by the presence of HIV-multinucleated giant cells and/or immunoreactivity for HIV antigen using a monoclonal antibody to gp41. Five tissue blocks from the remaining five AIDS cases (group 2) showed no morphologic or immunocytochemical evidence of HIV infection. Four cases in group 1 and one case in group 2 were prospectively characterized as having HIV-associated dementia as previously described *(2)*. Tissue sections from six HIV-seronegative people were studied as negative controls, including brains with subacute sclerosing panencephalitis (SSPE) and herpes simplex type-1 encephalitis (HSE).

7. Sections from the brains of five demented patients and five nondemented AIDS patients and four HIV-negative controls were examined. Sections from the spinal cords of six AIDS patients with vacuolar myelopathy (VM), six AIDS patients without VM, and three seronegative controls were examined.

8a. Start with conventional RT/PCR or PCR, utilizing homogenized tissue. Use the primers and probe you intend to use for the IS/PCR. These steps ensures the mRNA/DNA is there and the primers and probes work well.

8b. Start with conventional ISH as this may give you some idea about abundance of the mRNA and ensures that everything is ready for the hybridization steps.

8c. Hot start is essential for optimal specificity and sensitivity. It is dependent on the *Taq* polymerase becoming active at a temperature above the annealing temperature so as to avoid nonspecific anealing and amplification.

It can be achieved in a number of ways:

i. By preheating the reaction mix and slide (above the anealing temperature).

ii. By adding the reaction minus the *Taq* to the slide, heating the slide above the anealing temperature, and then adding the *Taq*.

iii. Utilizing *Taq* start antibody (Clontech) that binds the *Taq*, rendering it inactive until it reaches a certain temperature (approx 70°C).

8d. Multiple controls are essential for this technique, as artifacts are common. Biological positive controls and multiple negative controls are necessary.

8e. Depending on the abundance of the amplified product either a single labeled oligonucleotide probe or a cocktail of three or more probes can be used. The label can be either radioactive or nonradioactive. We have had good success with digoxigenin, which is clearly more sensitive than biotin.

8f. Using immunohistochemistry, anti-GFAP antibody binding could be detected both before and after the RT/PCR reaction whereas RCA binding was difficult to detect after RT/PCR and this staining needed to be performed first. However, when the RT/PCR hybridization was performed after immunohistochemistry the signal was decreased, suggesting that some mRNA had been lost. Therefore, optimum results for each technique were obtained when performed alone. When the immunohistochemistry was performed prior to the RT/PCR/ISH all solutions and glassware were treated to ensure the absence of RNase.

References

1. Wesselingh, S. L, Takahashi, K., Glass, J. D., McArthur, J. C., Griffin, J. W., and Griffin, D. E. (1997) Cellular localization of TNFα mRNA in neurological tissue from HIV-infected patients by combined RT/PCR *in situ* and immunohistochemistry. *Neuroimmunology* **74**, 1–8.

2. Takahashi, K., and Wesselingh, S. L., Griffin, D. E., McArthur, J. C., Johnson, R. T., and Glass J. D. (1996) Localization of HIV-1 in human brain using PCR *in situ* hybridization and immunocytochemistry. *Ann. Neurol.* **39**, 705–711.

3. Esolen, L. M., Takahashi, K., Johnson, R. T., Vaisberg, A., Moench, T. R., Wesselingh, S. L., and Griffin, D. E. (1995) Brain endothelium cell infection in children with acute fatal measles. *J. Clin. Invest.* **96**, 2478–2481.

4. Bagasra, O., Seshamma, T., and Pomerantz, R. J. (1993) Polymerase chain reaction *in situ:* intracellular amplification and detection of HIV-1 proviral DNA and specific genes. *J. Immun. Methods* **158**, 131–145.

5. Chen, R. H. and Fuggle, S. V. (1993) *In situ* cDNA polymerase chain reaction. A novel technique for detecting mRNA expression. *Am. J. Pathol.* **143**, 1527–1534.

6. Chiu, L. P., Cohen, S. H., Morris, D. W., and Jordan, G. W. (1992) Intracellular amplification of proviral DNA in tissue sections using the polymerase chain reaction. *J. Histochem. Cytochem.* **40**, 333–341.

7. Embleton, M. J., Gorochov, G., Jones, P. T., and Winter, G. (1992) In-cell PCR from mRNA: amplifying and linking the rearranged immunoglobulin heavy and light chain V-genes within single cells. *Nucleic Acids Res.* **20**, 3831–3837.

8. Embretson, J., Zupancic, M., Ribas, J. L., Burke, A., Racz, P., Tenner-Racz, K., and Haase, A. T. (1993) Massive convert infection of helper T lymphocytes and macrophages by HIV during the incubation period of AIDS. *Nature* **362**, 359–362.

9. Embretson, J., Zupanic, M., Beneke, J., Till, M., Wolinsky, S., Ribas, J. L., Burke, A., and Haase, A. T. (1993) Analysis of human immunodeficiency virus-infected tissues by amplification and *in-situ* hybridization reveals latent and permissive infections at single-cell resolution. *Proc. Natl. Acad. Sci. USA* **90,** 357–361.

10. Haase, A. T., Retzel, E. F., and Staskus, K. A. (1990) Amplification and detection of lentiviral DNA inside cells. *Proc. Natl. Acad. Sci. USA* **87,** 4971–4975.

11. Komminoth, P., Long, A. A., and Wolfe, H. J. (1993) Indirect and direct in-situ polymerase chain reaction (*in-situ* PCR) for viral DNA detection in tissue sections. *Lab. Invest.* **68,** 800/137A

12. Komminoth, P., Long, A. A., Ray, R., and Wolfe, H. J. (1992) *In situ* polymerase chain reaction detection of viral DNA, single copy genes, and gene rearrangements in cell suspensions and cytospins. *Diagn. Mol. Pathol.* **1,** 85–87.

13. Komminoth, P. and Long, A. A. (1993) *In situ* polymerase chain reaction. An overview of methods, applications and limitations of a new molecular technique. *Virchows Arch. B Cell Pathol.* **64,** 67–73.

14. Long, A. A., Komminoth, P., and Wolfe, H. J. (1992) Detection of HIV provirus by *in situ* polymerase chain reaction: correspondence to the editor. *N. Engl. J. Med.* **327,** 1529.

15. Long, A. A., Komminoth, P., Lee, E., and Wolfe, H. J. (1993) Comparison of indirect and direct *in-situ* polymerase chain reaction in cell preparations and tissue sections. *Histochemistry* **99,** 151–162.

16. Nuovo, G. F., Margiotta, M., MacConnel, P., and Becker, J. (1992) Rapid *in situ* detection of PCR-amplified HIV-1 DNA. *Diagn. Mol. Pathol.* **1,** 98–102.

17. Nuovo, G. J., Becker, J., Margiotta, M., MacConnell, P., Comite, S., and Hochman, H. (1992) Histological distribution of polymerase chain reaction-amplified human papillomavirus 6 and 11 DNA in penile lesions. *Am. J. Surg. Pathol.* **16,** 269–275.

18. Nuovo, G. J., Forde, A., MacConnell, P., and Fahrenwald, R. (1993) *In situ* detection of PCR amplified HIV-1 nucleic acids and tumor necrosis factor cDNA in cervical tissues. *Am. J. Pathol.* **143,** 40–48.

19. Nuovo, G. J., Gallery, F., Hom, R., MacConnell, P., and Bloch, W. (1993) Importance of different variables for enhancing *in situ* detection of PCR-amplified DNA. *PCR Methods Appl.* **2,** 305–312.

20. Nuovo, G. J., Gallery, F., MacConnell, P., Becker, J., and Bloch, W. (1991) An improved technique for *in situ* detection of DNA after polymerase chain reaction amplification. *Am. J. Pathol.* **139,** 1239–1244.

21. Nuovo, G. J., Gorgone, G. A., MacConnell, P., Margiotta, M., and Gorevic, P. D. (1992) *In situ* localization of PCR-amplified human and viral cDNAs. *PCR Methods Appl.* **1,** 117–123.

22. Nuovo, G. J., MacConnell, P., Forde, A., and Delvenne, P. (1991) Detection of human papillomavirus DNA in formalin-fixed tissues by *in situ* hybridization after amplification by polymerase chain reaction. *Am. J. Pathol.* **139,** 847–854.

23. Patterson, B. K., Till, M., Otto, P., Goolsby, C., Furtado, M. R., McBride, L. J., and Wolinsky, S. M. (1993) Detection of HIV-1 DNA and messenger RNA in individual cells by PCR-driven *in situ* hybridization and Flow Cytometry. *Science* **260,** 976–979.

24. Sällström, J. F., Zehbe, I., Alemi, M., and Wilander, E. (1993) Pitfalls of *in-situ* polymerase chain reaction (PCR) using direct incorporation of labeled nucleotides. *Anticancer Res.* **13,** 1153–1154.

25. Staecker, H., Cammer, M., Rubinstein, R., and Van De Water, T. R. (1994) A procedure for RT-PCR amplification of mRNAs on histological specimens. *BioTechniques* **16,** 76–80.

26. Staskus, K. A., Couch, L., Bitterman, P., Retzel, E. F., Zupancic, M., List, J., and Haase, A. T. (1991) *In-situ* amplification of visna virus DNA in tissue sections reveals a reservoir of latently infected cells. *Microb. Pathogen.* **11,** 67–76.

Index